AF248816

Neural Networks in Chemical Reaction Dynamics

Neural Networks in Chemical Reaction Dynamics

Lionel M. Raff, Ranga Komanduri,
Martin Hagan, and Satish T.S. Bukkapatnam

OXFORD
UNIVERSITY PRESS

OXFORD
UNIVERSITY PRESS

Oxford University Press, Inc., publishes works that further
Oxford University's objective of excellence
in research, scholarship, and education.

Oxford   New York
Auckland   Cape Town   Dar es Salaam   Hong Kong   Karachi
Kuala Lumpur   Madrid   Melbourne   Mexico City   Nairobi
New Delhi   Shanghai   Taipei   Toronto

With offices in
Argentina   Austria   Brazil   Chile   Czech Republic   France   Greece
Guatemala   Hungary   Italy   Japan   Poland   Portugal   Singapore
South Korea   Switzerland   Thailand   Turkey   Ukraine   Vietnam

Copyright © 2012 by Oxford University Press

Published by Oxford University Press, Inc.
198 Madison Avenue, New York, New York 10016
www.oup.com

Library of Congress Cataloging-in-Publication Data

Neural networks in chemical reaction dynamics / Lionel M. Raff ... [et al.].
        p. cm.
Includes bibliographical references and index.
ISBN 978-0-19-976565-2 (hardcover : acid-free paper)
1. Chemical reactions—Data processing.   2. Neural networks (Computer science)
        I. Raff, Lionel M.   II. Title.

QD501.N48 2011
541′.390285632—dc22

2010054098

1 3 5 7 9 8 6 4 2
Printed in the United States of America
on acid-free paper

Dedicated with Admiration and Respect to the Memory of

Professor Ranga Komanduri

Oklahoma State University Regents Professor
A. H. Nelson, Jr. Endowed Chair in Engineering
Scholar, Teacher, Mentor, Scientist, and Engineer

CONTENTS

PREFACE

During the past decade, research efforts by a number of different groups have demonstrated that feedforward neural networks (NN) can play an important role in facilitating the development of potential-energy surfaces (PES) of sufficient accuracy to execute accurate molecular dynamics (MD), Monte Carlo (MC), quantum scattering calculations, determination of rovibrational energy levels within an accuracy of a few cm^{-1}, and perhaps photodissociation, if the excited-state electronic surfaces can be computed. Specifically, it has been found that NNs can greatly facilitate the use of potential energies and gradients obtained from first-principles electronic structure calculations to develop what are generally termed *ab initio* PESs for use in MD, MC, and quantum dynamics investigations. Likewise, NNs have been shown to be highly useful in making older methods, such as many-body expansions and genetic algorithms (GA), far more robust and powerful.

Neural network (NN) methods have been developed to greatly assist in the adjustment of parameters of empirical PESs to fit a database of some type. They have also been found highly useful in reducing the statistical error that is always present in MD studies. This has been accomplished by employing NNs to predict the results of trajectories without actually executing numerical integrations. When this is done, the results of a huge number of molecular trajectories can be obtained with far less computational effort. The result is a significant reduction in the statistical errors of the calculations and a corresponding increase of predictive accuracy. Most recently, it has been shown that NNs can be used to predict the results of high-level electronic structure calculations from lower-level Hartree-Fock calculations. This allows the computational bottleneck of executing the extensive *ab initio* electronic structural calculations to be substantially reduced.

While all these methods have been reported by several research groups in the leading journals of the field, there is no succinct compilation of all these results and methods that describes both the methods and the techniques involved in their application. This monograph provides clear descriptions of how NNs have been used to move MD investigations into the realm of *ab initio* calculations, which as recently as a few years ago were considered beyond the reach of the computational facilities available. The monograph also describes in detail how NNs can be effectively employed to execute all of the studies previously mentioned. In addition to the formal descriptions of the methods, example applications are provided to illustrate the power and limitations of NN methods along with a clear road map describing how to execute each of the studies.

The authors feel confident that this emerging technology will play an increasingly important role in MD, MC, and quantum mechanical studies of chemical reaction dynamics in the years to come.

The authors would like to thank the National Science Foundation (NSF) for continued support over the past 20 years of the work presented here. In particular, the authors thank the Manufacturing Processes and Machines Program of the Civil, Mechanical, and Manufacture Innovation (CMMI) Division of NSF [Drs. B. M. Kramer, J. Cao, J. Larsen-Basse, K. Srinivasan, K. Rajurkar, G. Hazelrigg]. The authors also thank the graduate students, in particular Milind Malshe, Rutuparna Narulkar, A. Pukrittayakamee, and H. Lu who made significant contributions as part of their Ph.D. dissertations. The atomistic simulation group at Oklahoma State University has had many graduate students who worked on various MD/MC projects over time in the past two decades. To all of them, and especially to Dr. Naga Chandrasekaran, the authors are grateful for their valuable contributions. The authors also thank Professor M. Rockley and Dr. P. M. Agrawal for their valuable contributions and Mrs. Sharon Green for her superb job of providing editorial assistance in the preparation of this monograph and several papers the authors have published in this area.

This monograph is based in large part on the recent work by the authors' interdisciplinary group on neural networks in chemical reaction dynamics. Consequently, no effort was made to include a comprehensive review of the literature on chemical reaction dynamics or neural networks. Instead, only references directly related to the subject matter are covered to limit the scope of the monograph. Also, references covering the entire manuscript are given toward the end of the monograph instead of at the end of each chapter to avoid duplication and save space.

Special thanks are extended to Professor Tucker Carrington Jr. and Dr. Sergei Manzhos for reviewing portions of the manuscript and for providing cogent suggestions for improving it. We also thank them for granting permission to use material from their published work. We also wish to thank Professors J. Behler and M. Parrinello and Drs. B. G. Sumpter and D. W. Noid for allowing us to use figures and tables from their papers as well as for providing helpful discussion of various points.

Finally, the authors sincerely thank various publishers for generously granting permission to use material the authors have published in their journals, including the *Journal of Chemical Physics*, *Physical Review*, *Journal of Physical Chemistry*, and *Philosophical Magazine*. All the figures and tables used in the monograph from these journals are duly acknowledged with a reference number giving details of the article in the reference section and a statement stating "Reprinted with the Permission from the publisher (name)" at the end of each figure caption and table caption.

ACRONYMS

BFGS	Broyden-Fletcher-Goldfarb-Shannon Method
BP	Behler - Parrinello Method
CASPT2	Second-Order Complete Active Space Calculations
CFDA	Combined Function Derivative Approximation
CNP	Complete Nuclear Permutation
CPU	Central Processing Unit
CVD	Chemical Vapor Deposition
DAC	Density Adative Cutoff
DD	Direct Dynamics
DD/MD/NS/NN	Direct Dynamics/Molecular Dynamics/Novelty Sampling/Neural Network Method
DFT	Density Functional Theory
EAM	Embedded-Atom Method
EMS	Efficient Microcanonical Sampling
FRC	Fixed Radius Cutoff
GA	Genetic Algorithm
HDMR	High-Dimensional Model Representation
HF	Hartree-Fock
IMLS	Interpolating Moving Least Squares
IP	Invariant Polynomial
IR	Infrared
IRP	Intrinsic Reaction Path
IVR	Intramolecular Vibrational Relaxation
LBS	Large Basis Sets
L-IMLS	Local Interpolative Moving Least Squares
LM	Levenberg-Marquardt
LSTH	Liu, Siegbahn, Truhlar, Horowitz
MAE	Mean Absolute Error
MC Simulation	Monte Carlo Simulation
MD Simulation	Molecular Dynamics Simulation
MD/NS/NN	Molecular Dynamics/Novelty Sampling/Neural Network Method
ME	Moiety Energy
MEAM	Modified Embedded-Atom Method
MEP	Minimum-Energy Path
MLP	Multi-Layer Perception

MNS	Modified Novelty Sampling
MSI	Moving Shepard Interpolation
NN	Neural Network
NS	Novelty Sampling
PES	Potential-Energy Surface
RBFNN	Radial Basis Function Neural Network
RC	Redundant Coordinates
RHF	Restricted Hartree-Fock
RKHS	Reproducing Kernel Hilbert Space
RPE	Relative Percentage Error
SBS	Small Basis Sets
RMSE	Root Mean Square Error
SNN	Symmetric Neural Network
SVD	Singular Value Decomposition
UNN	Unsymmetric Neural Network
VB	Vinyl Bromide

Neural Networks in Chemical Reaction Dynamics

1

FITTING POTENTIAL ENERGY HYPERSURFACES

1.1. INTRODUCTION

Molecular dynamics (MD) and Monte Carlo (MC) simulations are the two most powerful methods for the investigation of dynamic behavior of atomic and molecular motions of complex systems. To date, such studies have been used to investigate chemical reaction mechanisms, energy transfer pathways, reaction rates, and product yields in a wide array of polyatomic systems. In addition, MD/MC methods have been successfully applied for the investigation of gas-surface reactions, diffusion on surfaces and in the bulk, membrane transport, and synthesis of diamond using chemical vapor deposition (CVD) techniques. The structure of vapor deposited rare gas matrices has been studied using trajectories procedures.

If the chemical reaction of interest contains three atoms or fewer, various types of quantum and semiclassical calculations can be brought to bear on the problem. These methods include wave packet studies, close-coupling calculations at various levels of accuracy, and S-matrix theory. Several excellent review articles have been published describing the principal techniques and problems involved in conducting MD studies; the reader may wish to consult these as background material for this discussion.[1-7]

With the advent of relatively inexpensive, powerful personal computers, MD/MC simulations have become routine. Once the potential-energy hypersurface for the system has been obtained, the computations are straightforward, though time-consuming. In the majority of cases, the computational time required is on the order of hours to a few days. However, the accuracy of these simulations depends critically on the accuracy of the potential hypersurface used.

The major problem associated with MD/MC investigations is the development of a potential-energy hypersurface whose topographical features are sufficiently close to those of the true, but unknown, surface that the results of the calculations are experimentally meaningful. Once the potential surface is chosen or computed, all the results from any quantum mechanical, semiclassical, or classical scattering or equilibrium calculation are determined. The only purpose of the MD calculations is to ascertain what these results are. Therefore, the most critical part of any MD/MC study is the development of the potential-energy hypersurface and the associated force field. Surprisingly, this is often the portion of the investigation to which least effort is devoted. This situation arises because of the inherent difficulty associated with this part of the overall problem.

1.2. EMPIRICAL AND SEMI-EMPIRICAL POTENTIAL SURFACES

When four or more many-electron atoms are present, global potentials have usually been obtained using empirical methods that rely heavily upon ad hoc parametrized functional forms suggested by physical and chemical considerations. The parameters contained in these forms are generally fitted to equilibrium thermochemical, spectroscopic, and structural data. If experimental activation energies are known, the barrier heights predicted by the empirical potential are adjusted to these values. Such methods have been employed by numerous investigators since the mid-1970s.

Many engineering applications of MD/MC studies, including machining, indentation, uniaxial tension, and tribological simulations to date have relied on empirical potentials, such as pairwise sums of Morse[8,9] or Lennard-Jones potentials[10] for metals. Baskes and his colleagues have developed an embedded-atom method (EAM)[11] and a modified embedded-atom method (MEAM)[12] for application to fcc and bcc metals, and even to hcp metals. For covalently bonded, semiconductor materials, such as Si, Ge, and diamond many-body potentials, such as Tersoff's potential,[13] the Bolding–Anderson potential,[14] and the Brenner potential[15] have been developed. These potentials provide approximate descriptions of most materials at or near their equilibrium configuration. However, when the lattice atoms are in configurations far removed from equilibrium, the accuracy of any empirical potential can be expected to decrease significantly. Such loss of accuracy also occurs in many chemical reactions that sample regions of configuration space that are far removed from equilibrium.

The limitations and liabilities associated with the use of empirical potential surfaces in MD calculations may be summarized as follows:

1. Since the ad hoc empirical functional form is simply concocted with little or no theoretical foundation, it is highly unlikely that it will accurately predict the correct experimental force field.
2. The parameters contained in an empirical potential are generally adjusted to fit equilibrium and stationary-point data, such as bond lengths, bond angles, vibrational frequencies, reactant and product equilibrium energies, elastic moduli, sublimation energies, and potential barrier heights. As a result, the potential usually fits the stationary points on the surface acceptably well. However, MD and MC simulations are primarily concerned with the energies and forces at nonstationary points. It is very unlikely that these quantities are accurately represented in configurations far removed from equilibrium.

3. If the database contains data from *ab initio* quantum mechanical calculations at nonstationary points as well as experimental data, the problem of fitting the empirical parameters becomes very difficult. Such fitting must be accomplished using iterative adjustment methods that often require months of effort. For example, nine months of laborious iterative fitting was required to obtain an acceptable fit of the 100 parameters contained in the empirical function used to fit the global potential surface for vinyl bromide[16 a,b.]

4. Once an empirical potential is developed for a particular complex system, there is no straightforward means by which it can be improved. Any alteration of the potential requires that the entire fitting process be repeated.

5. Empirical potentials are specific for the system under consideration. The procedures used to obtain such potentials cannot be easily automated and the methods applied to all systems. Every system must be treated individually. Therefore, the research is tedious and labor intensive.

Each of the problems outlined here is essentially intractable. Because of this, the accuracy of MD simulations on complex systems has, for practical purposes, reached its limit. If we wish to advance beyond this natural limit for empirical and hybrid potentials, we must develop ways to effectively utilize quantum mechanical methods that have the power to produce more accurate surfaces and force fields.

1.3. *AB INITIO* POTENTIAL-ENERGY SURFACES (PESs)

In principle, *ab initio* methods can produce much more accurate potential-energy surfaces for molecular dynamics. The fundamental problem is that electronic structure calculations become computationally intractable when the number of atoms and electrons present in the system becomes large. Nevertheless, excellent results have been obtained for numerous three- and four-body systems. In this section, we present a brief review of some of the systems studied and the methods employed to obtain the required *ab initio* potentials.

In many investigations, an empirical or semi-empirical potential-energy surface (PES) is employed to represent the force field of the system under investigation. Such potentials generally yield only qualitative or semiquantitative descriptions of the system dynamics. Empirical potential surfaces can be significantly improved by fitting the chosen functional form for the potential to the force fields obtained from trajectories using *ab initio* Car-Parrinello[17] molecular dynamics simulations. This method has been employed by several research groups.[18–21]

When the system of interest contains four atoms, usually it is no longer possible to conduct electronic structure calculations in all regions of configuration space. Consequently, importance sampling procedures must be employed to identify the critical regions of configuration space. In addition, interpolation between the computed points on the surface becomes increasingly difficult.

If the system under investigation is not too large and complex, *ab initio* electronic structure calculations offer a route to obtaining extremely large databases, which can be employed to obtain the PES for the system. Malshe et al.[22] have recently developed a method that reduces or eliminates many of the problems associated with fitting empirical potentials particularly when some of the parameters are made functions of the system configuration. The method completely obviates the problem of selecting the form of the functional dependence of the parameters upon the system's coordinates. This form is, in effect, determined automatically by a neural network (NN). This use of NNs will be discussed in more detail in Chapter 8.

1.4. OTHER FITTING METHODS FOR POTENTIAL-ENERGY SURFACES

Several research groups have employed methods that obviate the need to select arbitrary functional forms for the potential surface. These methods attempt to accurately sample the configuration space of the system and then fit the resulting database of *ab initio* energies using some generalized numerical procedure. Five such methods that have been frequently employed are (i) moving interpolation techniques,[23–25] (ii) reproducing kernel Hilbert space (RKHS),[26,27] (iii) interpolating moving least squares (IMLS),[28–33] (iv) expansions in terms of invariant polynomials (IP),[34–39] and (v) neural network (NN) methods.[40–50] The following discussion briefly describes these methods. The first four are discussed in more detail in Chapter 2 while the remainder of this monograph is devoted to an extensive discussion of various applications of NN methods.

Ischtwan and Collins[23] have developed a moving interpolation technique in which the potential energy in the neighborhood of any point is approximated by a Taylor series expansion using inverse bond length coordinates as the expansion variables. The overall method is usually denoted as a moving Shepard interpolation (MSI). Initially, a set of system configurations along the reaction path is selected by using chemical intuition. *Ab initio* electronic structure calculations of the system energy and gradients in these configurations are then executed. The data thus produced are employed to obtain the set of Taylor series expansions. A Shepard method[24] is used to effect the required interpolations. This procedure expresses the potential at any configuration as a weighted average of the Taylor series about all N points in the data set, or alternatively, those points within a specified cutoff radius. Subsequently, the results are iteratively improved by computing trajectories on the Taylor series fitted surface and recording the internal coordinates at a series of successive configurations encountered in the trajectories. In part, these new points are added to the data set according to a weight factor that is determined by the relative density of points in the data set in the region of the new points. For systems that are undergoing two-center, bond dissociation or formation reactions, the fitting error with a test set of *ab initio* energies is usually found to be in the range of from 0.0010 eV to 0.031 eV depending upon the complexity of the system. For an overview of this method and results, the review by Collins may be consulted.[25]

The second method for obtaining *ab initio* potential surfaces for three- and four-body systems employs the criterion of reproducing kernel Hilbert space (RKHS) as the means to effect the required fitting of the *ab initio* energies. Ho et al.[26] have employed this procedure to investigate the $O(^1D) + H_2 \rightarrow OH + H$ reaction. When 1280 *ab initio* data points were employed to define the surface, the fitting yielded a surface whose absolute root mean square error compared to *ab initio* energies was 0.0131 eV. Pederson et al.[27] have used the same method to study the dynamics of the $N(^2D) + H_2$ reaction.

Maisuradze et al.[28] have introduced the third method, namely, an interpolating moving least-square (IMLS) method to effect the fitting between the computed *ab initio* points. When the method is unrestricted, the least-square coefficients are obtained from the solution of a large matrix equation that must be solved repeatedly during a trajectory study. In this form, the required computational time increases with NM^2, where N is the number of data points to be fitted and M is the number of basis functions used in the linear combination that provides the fit. Guo et al.[29] have evaluated the IMLS method using the analytic surface of Kuhn et al.[51] for simple O-O bond rupture of HOOH. With $N = 300$, the results showed the statistical fitting errors to lie in the range of 0.085 eV to 0.171 eV. Kawano et al.[30] have compared the accuracy of the IMLS method with that of the MSI procedure for the HOOH bond scission reaction. With $N = 6489$, the root mean square errors for IMLS and MSI were 0.0468 eV and 0.0202 eV, respectively. Guo et al.[31] have recently reported IMLS results for the unimolecular dissociation reaction of $H_2CN \rightarrow H + HCN$. Their results show root mean square fitting errors that vary between 0.0055 eV and 0.0290 eV.

The major problem with the IMLS method is the extremely large computational time required for its execution. When more than four atoms are involved, the computation time increases rapidly due to the M^2 factor. The time also increases linearly with the number of *ab initio* data points employed. Guo et al.[31] have reported a computational time of about 100 minutes for one H_2CN dissociation trajectory lasting 2.5 ps when 830 data points are employed.

The computational difficulties associated with IMLS methods have recently been addressed by Dawes et al.[32a] and by Guo et al.[32b] By converting the IMLS method to what is essentially a modified Shepard method,[25] the large computational bottleneck present in IMLS calculations is avoided. The authors[32] term this hybrid method a "local IMLS" or L-IMLS method. To date, this hybrid procedure has been applied only to molecules containing four or fewer atoms undergoing a single, two-center, bond dissociation reaction. Ishida and Schatz[33a] have also proposed a procedure that combines the IMLS and Shepard methods. This combined method has been applied to the investigation of the three-body reaction dynamics of the $O(^1D) + H_2$ reaction.[33b]

Bowman, Braams, and co-workers[34–39] have developed a different type of least-squares fitting procedure that utilizes least-squares fitting of expansions in terms of polynomials of functions of the interatomic distances to fit *ab initio* databases obtained from electronic structure calculations. The expansion polynomials are totally symmetrized so that they are invariant to the exchange of any two identical atoms. For this reason, the procedure is generally called an "invariant polynomial" (IP) method. The IP method has been applied to an impressive set of relatively complex reactions of molecules that contain between five and seven atoms. In most cases, the database to which the IP expression is fitted contains more than 20,000 points.

The fifth method for obtaining accurate analytic fits to databases obtained from *ab initio* electronic structure calculations involves the use of NNs. Neural networks provide a powerful and robust method for surface fitting and other related tasks associated with reaction dynamics investigations. The remainder of this monograph is devoted to an exposition of these NN methods.

In order to avoid the extremely difficult task of interpolating *ab initio* electronic structure data, some investigators have employed a method known as "direct dynamics" (DD). In this approach, trajectories are computed by direct calculation of the force field at each integration point using some *ab initio* quantum mechanical method. Because of the huge number of computations required for each trajectory, the *ab initio* method chosen must usually be some form of density functional theory (DFT). For example, it may be necessary to follow the dynamics for several picoseconds (ps). With an integration step size on the order of a femtosecond (fs), 1,000 or more integration steps will be required. If the forces are computed at four points in each step, something on the order of 4,000 gradients will have to be computed for each trajectory. As a result, only a very limited number of trajectories can be obtained.

As the system under investigation increases in size, the computational requirements of DD will quickly overwhelm the available computational resources. The basic problem is that all the information obtained about the potential-energy surface and the corresponding force field during the integration of the trajectory is discarded after the completion of the calculation. Therefore, each subsequent trajectory fails to profit from all the computational effort expended in obtaining the previous trajectories. The conventional wisdom has been that *ab initio* MD calculations for complex systems containing five or more atoms with several open reaction channels are presently beyond our computational capabilities. The rationales for this view are (a) the inherent difficulty of high-level *ab initio* quantum calculations on complex systems that may make numerous, large-scale computations impossible, (b) the large dimensionality of the configuration space for such systems that makes it necessary to examine prohibitively large numbers of nuclear conformations, and (c) the extreme difficulty associated with accurate interpolation of numerical data obtained from electronic structure calculations when the dimensionality of the system is nine or greater. In subsequent chapters, we present general NN methods to handle each of these difficulties for many complex systems. Furthermore, as our computational resources increase, the range of systems that can be so treated will expand without the need to significantly modify the methods. Therefore, the theoretical approach is robust and powerful.

1.5. NEURAL NETWORK (NN) APPROACH

In this monograph, we present an integrated approach for obtaining *ab initio* quantum mechanical potential-energy surfaces (PES) and force fields for use in MD simulations for systems of five or more atoms in which several reaction channels may be energetically open. The method involves (a) electronic structure calculations of energies and force fields for an ensemble of atoms whose interatomic distances are such that all atoms fall within a previously specified distance from some central point, generally called the cutoff radius; (b) the implementation of sampling methods that utilize MD calculations and novelty sampling to permit the regions of configuration space that are important in the dynamics to be accurately identified; and (c) the use of NNs, early stopping, and regularization methods to provide rapid and accurate fitting of the *ab initio* database and convenient tests for convergence that do not require MD calculation of dynamical results.

1.6. ESSENTIAL STEPS IN MOLECULAR DYNAMICS SIMULATIONS

For pedagogical purposes, we may divide an MD investigation into several distinct parts. The division used here is not unique, but it serves the purpose of guiding the discussion that follows. The essential steps of any MD study may be considered to be the following:

Step 1: A convenient coordinate system must be chosen that leads to a set of coupled, ordinary differential equations whose solution determines the nature of a molecular trajectory for a given set of initial conditions. This step concludes with the preparation of computer codes that perform numerical integrations to solve the coupled set of differential equations.

Step 2: The molecular motions during a trajectory are determined by the force field in which the atomic particles move. This force field is obtained from the gradients of the potential-energy surface at each integration point. The development of the PES is, therefore, a crucial step in the MD study.

Step 3: Appropriate methods must be devised to determine when to terminate trajectories and how to treat the final results to obtain the experimental quantities of interest in the investigation. Since the calculations are classical rather than quantum mechanical, if the quantities of interest involve quantization of the final states, methods to convert the final classical results to corresponding quantum results with reasonable accuracy must be developed.

Step 4: Since experimental observations are usually ensemble averages of the results of an extremely large number of molecular collisions that is often on the order of the Avogadro number (10^{23}), an MD investigation must incorporate appropriate statistical averages over the same variables averaged in the corresponding experiments. This is generally accomplished by the computation of a number of molecular trajectories in which the initial conditions are chosen randomly from the appropriate distribution functions for those variables over which the investigator seeks to average. Obviously, it is computationally impossible to compute 10^{23} trajectories, so these averages must of necessity be over much smaller sets that usually comprise from 10^2 to 10^4 trajectories. This naturally leads to the problem of estimating the statistical error present in the calculations and the development of methods that reduce this error as much as possible.

Although each of these steps is important in an MD investigation, the most important and the most difficult step in the execution of accurate MD, MC, semi-classical, or quantum scattering calculation is Step 2, the development of a sufficiently accurate PES. This is the case because once the PES is in hand, all of the dynamics for a given set of experimental conditions are determined. The MD, MC, or other scattering calculations are simply computational procedures used to discover the nature of the scattering or reaction dynamics that have already been determined by the formulation of the PES.

During the past decade, research efforts by a number of different groups have demonstrated that feed-forward, neural networks (NN) can play an important role in facilitating the development of PESs of sufficient accuracy to execute accurate MD, MC, or quantum calculations. Specifically, it has been found that NNs can greatly facilitate the use of potential energies and gradients obtained from first-principles electronic structure calculations to develop what are generally termed *ab initio* PESs for use in MD, MC, and quantum dynamics investigations. Likewise, NNs have been shown to be highly useful in making alternative methods to derive empirical PES, such as many-body expansions and genetic algorithms (GA), far more robust and powerful. Methods have been developed to greatly assist in the adjustment of parameters of empirical PESs to fit a database of some type.

In addition to facilitating the development of PESs, NNs have been found to be highly useful in Step 4 where one seeks to reduce the statistical error that is always present in MD studies. This has been accomplished by employing NNs to predict the results of trajectories without actually executing the numerical integrations of Step 1. When this is done, the results of a huge number of molecular trajectories can be obtained with far less computational effort. The result is a significant reduction in the statistical errors of the calculations and a corresponding increase of predictive accuracy. This procedure permits far more detailed investigation of the dependence of various experimentally measured properties on the independent variables of the process under examination.

1.7. ORGANIZATION OF THE MONOGRAPH

The objective of this monograph is to provide the reader with a brief but clear presentation of NNs along with descriptions of how they can be used to move MD investigations into the realm of *ab initio* calculations, which as recently as a few years ago were considered to be beyond the reach of present computational facilities. The discussion also describes in detail how NNs can be effectively employed to study non-adiabatic chemical reactions on *ab initio* potentials, obtain PESs for large ensembles of atoms that are invariant to permutations of identical atoms, enhance GA and energy transfer calculations, render many-body expansions far more robust and powerful, facilitate quantum scattering calculations, predict the results of trajectories without actually integrating the equations of motions, obtain parameters of empirical PESs from *ab initio* data or vibrational spectra, and finally how NNs can play a key role in the determination of electronic structure energies and gradients from high-level computations using primarily calculations conducted at the Hartree-Fock (HF) level of theory.

The monograph is organized into 11 chapters. Chapter 1 begins with a brief introduction to various methods of fitting potential energy surfaces. They include multilayer perceptron neural networks; empirical and semi-empirical potential surfaces; *ab initio* potential-energy surfaces; other fitting methods, namely, (a) moving interpolation techniques, (b) reproducing kernel Hilbert space (RKHS), and (c) interpolating moving least squares (IMLS); (d) invariant polynomicals (IP), and the neural network (NN) approach. The chapter ends with four essential steps in MD simulations and the organization of the monograph.

Chapter 2 provides a reasonably complete description of MSI, IMLS, and IP methods along with the presentation of typical results of applications of these methods. In addition, some hybrid methods that combine these techniques are also presented and typical results given. Results from the application of the RKHS method are presented.

Chapter 3 provides an introduction to the standard NN methods. It concentrates on the most common neural network architecture, namely, the multilayer perceptron (MLP). It describes the basics of this architecture, discusses its capabilities, and shows how it has been used on several different chemical reaction dynamics problems. This is followed by a more specific discussion of how NN methods can be employed in chemical reaction dynamics. For readers already familiar with NNs, parts of this chapter will provide a quick and easy-to-read starting material for their application to important problems in chemical reaction dynamics. Readers unfamiliar with NNs will find Chapter 3 highly useful as a means to bring themselves quickly to the operational level of the use of NNs. Finally, Chapter 3 takes the reader to the edge of the research frontiers in this area by presentation of a powerful new type of NN training algorithm that fits not only a function but also its gradient. This type of NN training is called the "Combined Function Derivative Approximation" (CFDA). Since it is the gradient of the potential energy that is of importance in molecular dynamics (MD) simulations, the CFDA promises to be a major player in future applications of NNs to chemical reaction dynamics. The use of the CFDA leads to a new type of overfitting that is not seen when NNs are employed in their usual form. Chapter 3 concludes with a discussion of this type of overfitting and "pruning" methods to permit its elimination.

In any fitting method, the procedures used to sample the configuration space of the system are of critical importance. In Chapter 4, we describe the theory and operational details of the three most powerful methods associated with configuration space sampling, namely, (i) trajectory and novelty sampling (NS) methods, (ii) initiation using direct dynamics (DD), and (iii) configuration sampling using a gradient fitting method. These descriptions are made clear and instructive by numerous illustrations from the literature of actual applications to real chemical systems.

Chapter 5 is devoted to an extensive discussion of actual examples of NN applications to chemical reaction dynamics. The chapter begins with an overview of the applications reported to date. This overview is then followed by a series of illustrative examples of increasing complexity beginning with an NN method for the *ab initio* computation of the molecular vibrational levels of the H_3^+ molecular ion. The dual NN methods for obtaining highly accurate *ab initio* molecular vibrational levels, recently developed by Manzhos et al.[47] are then described and illustrated by application to H_2O, H_2O_2, and H_2CO.

Chapter 5 then moves on to more complex systems involving chemical reactions. An application of the CFDA method to the gas-phase abstraction of exchange reactions of hydrogen reacting with HBr is described. It is shown that the CFDA method can achieve fitting accuracies on the order of 1 cm^{-1} error over the entire configuration space that is important in the reaction dynamics. Applications to the *cis-trans* isomerization and N–O bond dissociation reactions of HONO are then described. An application of configuration space sampling using gradient fitting is provided by a discussion of the dissociation dynamics of hydrogen peroxide.

Chapter 5 culminates with detailed descriptions of some of the most complex systems yet investigated using purely *ab initio* methods, namely, the unimolecular dissociation of vinyl bromide, $H_2C = CHBr$, into six, simultaneously open reaction channels, and the non-adiabatic dissociation of SiO_2 to either $Si + O_2$ or to $SiO + O$, both of which are non-adiabatic processes involving three different potential surfaces, and the treatment of a large ensemble (64) of silicon atoms using an NN method that automatically incorporates complete permutational symmetry of identical atoms.

Expansion methods to obtain PESs were proposed in the 1980s (see, for example, Murrell et al.).[52,53] However, owing to the bottleneck of the arbitrary and unknown nature of the expansion functions, only very limited use of these methods has been made to date. With the advent of NNs, these expansion methods have now been empowered and promise to be powerful techniques for the future.

Chapter 6 describes in detail two methods, namely, a high-dimensional model representation (HDMR) and a many-body expansion with a moiety-energy (ME) approximation that have been developed to date that combine NNs with expansion methods to produce powerful techniques for PES surface fitting to *ab initio* databases. The first of these is a combination of the high-dimensional model representation (HDMR) method, recently discussed in a series of publications by Rabitz and co-workers,[54–70] and NNs that has been proposed and implemented by Manzhos and Carrington.[48] The second method employs NNs with a many-body expansion that has been developed by Malshe et al.[71] Both methods provide solutions to the combinatorial problem that plagues expansion methods and both methods produce excellent results in terms of fitting accuracy.

Genetic algorithms (GA) have been a standard method for obtaining fits to databases. However, when the database is large, they are plagued with a heavy computational burden due to the computational time required to repeatedly compute the objective function. By utilizing NNs to predict the value of the objective function without actual computation, the entire GA method has been rendered far more powerful. The first portion of Chapter 7 describes this combined GA-NN method with an example of its utilization for obtaining a PES for Si_5 clusters.[72] The second portion of the chapter is devoted to a discussion how the use of NNs can greatly accelerate studies of intramolecular vibrational energy relaxation (IVR).[73]

Often it is useful to employ an analytical functional form for the PES. When this is done, it is always necessary to extensively parameterize the analytic function to allow accurate fitting to the database. The parameter fitting for arbitrary functional forms can be extremely laborious. This is particularly the case whenever it is desired to make one or more of the parameters functions of the instantaneous configuration of the system. Malshe et al.[22] have recently utilized NNs to greatly facilitate parameter fitting and to make it possible to obtain the near optimum functional dependence of the parameter values on the molecular configuration of the system. Chapter 8 describes these methods in detail. This chapter concludes with an application of NNs that permits the determination of empirical parameters from measured vibrational spectra for macromolecular systems.[74]

Since the final result of an MD trajectory is uniquely determined by the starting conditions of the trajectory—namely, orientation variables, impact parameter, relative translational energy, etc.—if a reasonable database exists that maps these initial starting conditions onto the final results of the trajectory, an NN can, in principle, be trained to predict the final results using an input vector containing the starting conditions for the trajectory. In effect, the NN replaces all the laborious numerical integrations required to compute the trajectory. If this can be done, then it becomes possible to compute N trajectories for a system, fit an NN to the results of these N trajectories, and use the NN to predict the results for a huge number of trajectories without having to perform any numerical integrations. These results can then be utilized to greatly enhance the statistical accuracy of the MD simulations with near negligible computational effort. Chapter 9 describes the details of a method that brings this concept to fruition (see Agrawal et al.)[75] An example application of the method to the reactions of a C_2 dimer with an activated diamond (100) surface is described as an illustration of the method.

Chapter 10 presents several applications of NNs to problems arising in quantum mechanics. The first of these describes the manner in which radial basis function neural networks (RBFNN) can be employed to solve the molecular vibrational Schrödinger equation to obtain wavefunctions and associated vibrational energies for multiple levels in a single calculation. In effect, an RBFNN is employed as an approximate wave function in a formalism in which the neurons of the RBFNN play the role of basis functions in the usual expansion methods for solution of the problem. Details of the method developed by Manzhos and Carrington[76,77] are described in the first section of Chapter 10.

The second section of Chapter 10 describes an application of NNs to simplify the electronic structure calculations required to obtain *ab initio* databases. As such, the method has the potential to become one of the most important applications of NNs to chemical reaction dynamics.

The biggest bottleneck to the execution of *ab initio* computation of reaction dynamics for complex systems is the extremely large computation times required to execute electronic structure calculations at a sufficiently high level of accuracy to obtain the extensive databases required to characterize complex chemical processes. This is, without doubt, the limiting factor in the execution of such investigations.

Malshe et al.[78] have recently found that because the Hartree-Fock (HF) energies are highly correlated with the energies obtained from higher-level electronic structure calculations, it is possible to accurately predict the results of higher-level computation from HF results. This finding implies that large databases can be computed at high level by computing only a small subset of the required database at high level, training an NN with the results, and then using the trained NN to predict the higher-level results from the HF energies. If this method proves to be robust, this will enormously expand the complexity of systems than can be investigated using purely *ab initio* methods. Also discussed in this section are analogous studies by Balabin and Lomakina[79] that demonstrate that DFT results with large basis sets can be obtained from DFT calculations that employ much smaller basis set by using a trained NN.

Chapter 10 concludes with a discussion of results obtained by several research groups that demonstrate how the accuracy of electronic structure calculations of equilibrium properties can be significantly improved by employing NNs trained by fitting computational results to experimental data.

Finally, *Neural Networks in Chemical Reaction Dynamics* concludes in Chapter 11 with a summary of the various applications of NNs related to chemical reaction dynamics that are covered in this monograph posing challenges, identifying some opportunities and limitations, and outlining future trends. The authors strongly believe that the approaches presented in this monograph using neural networks to develop potential-energy surfaces will play an increasingly important role in MD, MC, and quantum mechanical studies of chemical reaction dynamics in the years to come.

2

OVERVIEW OF SOME NON-NEURAL NETWORK METHODS FOR FITTING *AB INITIO* POTENTIAL-ENERGY DATABASES

2.1. INTRODUCTION

In this chapter, we describe results obtained by five methods that have been employed to fit *ab initio* potential-energy. These methods are (i) moving or modified Shepard interpolation (MSI),[23–25] (ii) interpolative moving least squares (IMLS),[28–33] (iii) invariant polynomials (IP),[36,37] (iv) reproducing kernel Hilbert space (RKHS),[26,27] and (v) a hybrid method that combines MSI and IMLS methods.[33a,33b] The MSI and IMLS methods are described in some detail in the following. The IP and RKHS procedures are significantly more complex, and the reader is referred to the original papers for a more complete discussion of the details by which these methods are executed.

2.2. MOVING SHEPARD INTERPOLATION (MSI) METHODS

The moving or modified Shepard interpolation (MSI) method was developed primarily by Collins and co-workers.[23-25] The method employs electronic structure calculations to obtain the molecular potential energy at configuration points generated by an automated procedure. These data are then employed in a Shepard interpolation procedure[24] to obtain the potential energies of the system at points other than those in the database. This procedure involves expressing the local potential about each configuration point in a Taylor series expansion. The term "moving" in the title derives from the fact that the set of internal coordinates employed in the interpolation varies from point-to-point in the database.

2.2.1. Required Input Data

Like all fitting methods, the MSI procedure requires the potential energy at a set of configuration points in the $(3N\text{-}6)$ dimensional internal space of the system under

13

investigation. These energies are generally obtained using *ab initio* electronic structure methods at some level of accuracy.

In addition to the potential energies at each configuration point, the method also requires at least the first and second derivatives of the potential with respect to the coordinates being employed at each configuration point. These derivatives are needed to allow the local potential about a given configuration point in the database to be expressed in terms of a Taylor series expansion about that point. In principle, the MSI method may be extended to include third or fourth derivatives,[80] but in most applications, the expansions are truncated after the quadratic terms. In some cases, depending upon the electronic structure method being employed, these derivatives can be computed analytically. In other cases, they must be obtained numerically using finite differences in the neighborhood of each configuration point.

Since the MSI procedure requires at least first and second derivatives, the number of data items needed at each configuration point is significantly greater than just the energy. If there are N atoms in the system, there will be $3N$-6 internal coordinates at each data point. Consequently, there will be $3N$-6 first derivatives of the energy and $(3N$-$6)(3N$-$5)/2$ second derivatives required since there are $(3N$-$6)(3N$-$7)/2$ off-diagonal, mixed second derivatives and $3N$-6 diagonal second derivatives. The total number of derivatives needed is, therefore, $(3N$-$6)(3N$-$3)/2$. If $N = 4$, 27 derivatives will be needed in addition to the energy, so that 28 input data will be required at each configuration point. For a five-atom system, this total is 55.

At first glance, this would seem to impose a severe computational bottleneck on the MSI methods, but this is often not the case. If the electronic structure method being employed permits first and second derivatives to be evaluated analytically, the computation of these quantities increases the required computational effort only slightly. On the other hand, if the electronic structure method does not allow analytic computation of the second derivatives, then there will be a significant increase in the required computational time. However, even in this case, there is a compensating feature of the MSI method. The availability of so much information about the topology of the PES at each data point usually means that the database required by the method to attain convergence will be much smaller than would be the case if only the energy at each point were available.

2.2.2. MSI Method for Molecules with Four or Fewer Atoms

When $N \leq 4$, the number of internal coordinates and the number of interparticle distances are the same. Consequently, the PES may be constructed using either the vector of interparticle distances, $\mathbf{R} = \{R_1, R_2, \ldots, R_{3N-6}\}$ or the vector of reciprocal interparticle distances, $\mathbf{Z}$, where $Z_k = R_k^{-1}$. Since potential energies generally vary with some power of the reciprocal distances, a more physically reasonable asymptotic behavior is obtained if one employs the Z_i as expansion variables rather than the R_i. This is the procedure used in all MSI applications.

At this point, it is assumed that the potential-energy database to which the analytic surface is to be fitted is available. The iterative procedure used to obtain this database in the MSI method is discussed in Section 2.2.4. With the energy and all first and second derivatives of the potential known at point i in the database, the potential in the local environment near point *i* can be expressed as a Taylor series expansion about that point,

$$T_i(\mathbf{Z}) = V[\mathbf{Z}(i)] + \sum_{k=1}^{3N-6} [Z_k - Z_k(i)] \frac{\partial V}{\partial Z_k}\bigg|_{Z=Z(i)}$$

$$+ \frac{1}{2!} \sum_{k=1}^{3N-6} \sum_{j=1}^{3N-6} [Z_k - Z_k(i)]$$

$$\times [Z_j - Z_j(i)] \frac{\partial^2 V}{\partial Z_k \partial Z_j}\bigg|_{Z=Z(i)} + \cdots \tag{2-1}$$

where $V[\mathbf{Z}(i)]$ is the value of the potential at $\mathbf{Z}(i)$ and the derivatives are taken with respect to the inverse distances at $\mathbf{Z}(i)$.

Usually, electronic structure programs provide the derivatives of the potential with respect to the Cartesian coordinates of the atoms, $\mathbf{X} = (X_1, X_2, \ldots, X_{3N})$, but not derivatives with respect to inverse distances. Therefore, the derivatives that appear in Equation (2-1) must be obtained by inverting the chain rule equations

$$\frac{\partial V}{\partial X_n}\bigg|_{X(i)} = \sum_{k=1}^{3N-6} \frac{\partial Z_k}{\partial X_n} \frac{\partial V}{\partial Z_k}\bigg|_{Z(i)}$$

$$\frac{\partial^2 V}{\partial X_n \partial X_m}\bigg|_{X(i)} = \sum_{k=1}^{3N-6} \sum_{j=1}^{3N-6} \frac{\partial Z_k}{\partial X_n} \frac{\partial Z_j}{\partial X_m} \frac{\partial^2 V}{\partial Z_k \partial Z_j}\bigg|_{Z(i)} \tag{2-2}$$

$$+ \sum_{k=1}^{3N-6} \frac{\partial^2 Z_k}{\partial X_n \partial X_m} \frac{\partial V}{\partial Z_k}\bigg|_{Z(i)}$$

where $n, m = 1, 2, \ldots, 3N$. In Equation (2-2), the derivatives needed for the Taylor series expansion in Equation (2-2) are the $\frac{\partial V}{\partial Z_k}$ and the $\frac{\partial^2 V}{\partial Z_k \cdot \partial Z_j}$. The corresponding derivatives with respect to the X_n and X_m on the left-hand side of Equation (2-2) are obtained from the electronic structure program at each point. The derivatives of the Z_k with respect to X_n are easily computed from the known coordinates.

Once the Taylor series expansions for all points in the database are obtained, a modified Shepard interpolation[24,81] is used to provide the potential energy at any arbitrary configuration of the system $\mathbf{Z}$. This procedure expresses $V(\mathbf{Z})$ as a weighted average of the Taylor series about all N points in the database and their symmetry equivalents.

$$V(\mathbf{Z}) = \sum_{g \in G} \sum_{i=1}^{N} W_{g,i}(\mathbf{Z}) T_{g,i}(\mathbf{Z}), \tag{2-3}$$

where each Taylor series expansion T_i has an associated weight, W_i. In Equation (2-3), G represents the symmetry group of the molecular system, which for these purposes generally means the complete nuclear permutation (CNP) group. For example, in practice, this might mean that the ith data point, $\mathbf{Z}(i)$, is replaced with a data point that just permutes the labels of two indistinguishable nuclei. The summation over G in Equation (2-3) means that all permutationally equivalent data points are included in the data set. The required exchange symmetry for identical particles is thereby incorporated into the PES for the system given by Equations (2-1) through (2-3).

The weight function, $W_{g,i}(\mathbf{Z})$, is designed to give more weight to the Taylor series expansions about points closer to point $\mathbf{Z}$ and a smaller weight to those further away. The "distance" between point $\mathbf{Z}$ and a point in the database $\mathbf{Z}(i)$ is defined to be

$$\|\mathbf{Z}-\mathbf{Z}(i)\|=\left\{\sum_{k=1}^{N(N-1)/2}\left[\mathbf{Z}_k-\mathbf{Z}_k(i)\right]^2\right\}^{1/2}. \tag{2-4}$$

$W_{g,i}(\mathbf{Z})$ is expressed in terms of primitive weight functions, $v_i(\mathbf{Z})$, with

$$v_i(\mathbf{Z})=\frac{1}{\|\mathbf{Z}-\mathbf{Z}(i)\|^{2p}}, \tag{2-5}$$

where p is a parameter chosen so as to increase the convergence rate of Equation (2-3) to the exact potential as $N \rightarrow \infty$. Values in the range $p = 9 - 12$ have been employed in MSI calculations. It is obvious that as the distance of an expansion point in the database from point $\mathbf{Z}$ increases, the corresponding primitive weight function, $v_i(\mathbf{Z})$, decreases, as desired. In terms of the $v_i(\mathbf{Z})$, the weight functions are given by

$$W_{g,i}(\mathbf{Z})=\frac{v_i(\mathbf{Z})}{\sum_{g\in G}\sum_{i=1}^{N}v_{g,i}(\mathbf{Z})}. \tag{2-6}$$

Equation (2-6) ensures that the weight functions are normalized.

Once the database has been selected and the energy and derivatives at these points computed, Equations (2-1) through (2-6) define the MSI representation of the PES for molecules containing four atoms or less.

2.2.3. MSI Method for Molecules with More than Four Atoms

When the system of interest contains more than four atoms, some modifications of the above procedure are required because the number of interparticle distances now exceeds the number of required independent internal coordinates. That is, $N(N-1)/2 > 3N-6$, if $N > 4$. When this is the case, the summations in Equation (2-1) become ambiguous. The MSI solution to this problem is to expand the potential at each integration point in terms of $3N$-6 internal coordinates rather than in terms of the reciprocal interparticle distances.

The set of internal coordinates employed in MSI calculations at data point $\mathbf{Z}(i)$ comprises linear combinations of the reciprocal interparticle distances that are guaranteed to be orthonormal at $\mathbf{Z}(i)$ and presumed to be well-behaved in regions near $\mathbf{Z}(i)$. These linear combinations must be determined at each configuration point in the database. That is, the internal coordinates vary with $\mathbf{Z}(i)$. As mentioned at the outset of Section 2.2, this feature of the method leads to the nomenclature "moving" Shepard interpolation.

The sets of internal coordinates employed at each point in the database are obtained by a method analogous to a normal mode calculation in which normal-mode eigenvectors are determined that are linear combinations of the atomic Cartesian displacements coordinates. These eigenvectors are orthonormal and diagonalize the kinetic energy matrix of the system. The MSI procedure[23,25,80,81] employs the Jacobian of the mapping from Cartesian coordinates to reciprocal bond lengths, $\mathbf{B}$. This Jacobian has rank $3N$-6 since reciprocal bond lengths cannot describe the translations of the molecule nor rotations of the molecule about the three principal axes. Thus, only $3N$-6 dimensions of the $3N$-dimensional configuration space of the molecule are described by $\mathbf{B}$ and its rank is, therefore, $3N$-6.

The MSI method employs the singular values decomposition (SVD) of the $\mathbf{B}$ matrix, which is a variant of the Wilson $\mathbf{B}$ matrix,[82] to construct orthogonal, nonredundant, internal coordinates that are linear combinations of the reciprocal bond lengths. Rather than make the set of internal coordinates at point $\mathbf{Z}(i)$ functions of all the R_{ij}^{-1}, it is found that more accurate fitting is obtained if a subset of the internal coordinates are made linear combinations of only the R_{ij}^{-1} associated with the existing molecular fragments at point $Z(i)$ while the remainder are linear combinations of the R_{ij}^{-1} associated with the interfragment distances.

In the MSI method, a "fragment" is defined to be a set of n_i atoms that are all within a chosen distance of each other. For example, consider the abstraction of a hydrogen atom from methane by an incident hydrogen atom. This reaction process can be considered to comprise three distinct regions as shown in Figure 2-1.

Once the fragments present at $\mathbf{Z}(i)$ are identified, it is simple to determine the number of interfragment internal coordinates (degrees of freedom). Let the number of fragments present at point $\mathbf{Z}(i)$ be denoted by f. In each fragment, there will be

$$H + CH_4 \longrightarrow CH_5 \longrightarrow H_2 + CH_3$$

Reactant Channel	Transition State	Product Channel
$f = 2$	$f = 1$	$f = 2$
$q = 3$	$q = 0$	$q = 5$

Figure 2-1 Illustration of the molecular fragments present in various regions for hydrogen-atom abstraction in methane by an incident hydrogen atom. In the figure, f denotes the number of fragments in each region while q gives the number of interfragment degrees of freedom.

n_k $(k = 1, 2, \ldots, f)$ atoms, where these values are easily determined by inspection of the various interatomic distances. The number of intrafragment internal degrees of freedom associated with each fragment is $3n_k - p_k$, where p_k is 6 for a non-linear fragment, 5 for a linear one, and 3 for an atom that can only translate. The total number of internal degrees of freedom for an N-atom system is $3N$-6 since in this section, we have $N > 4$, and the molecular system is highly unlikely to be linear. Therefore, if we represent the number of interfragment degrees of freedom as q, we must have

$$3N - 6 = q + \sum_{k=1}^{f} \left[3n_k - p_k \right]. \tag{2-7}$$

Solving Equation (2-6), we obtain

$$q = \sum_{k=1}^{f} p_k - 6, \tag{2-8}$$

since by conservation of mass, we must have $3N = \sum_{k=1}^{f} 3n_k$. In the illustration shown in Figure 2-1, use of Equation (2-8) shows that in the reactant channel, transition state, and product channel the numbers of interfragment degrees of freedom are 3, 0, and 5, respectively, as indicated in the figure. Thus, in the reactant channel, there will be 3 interfragment internal coordinates and 9 intrafragment internal coordinates. The interfragment internal coordinates will each be linear combinations of the five reciprocal interatomic distances between H and the atoms in CH_4 while the intrafragment internal coordinates will be linear combinations of the 10 R_{ij}^{-1} associated with CH_4.

To obtain the inter- and intrafragment internal coordinates, the MSI method defines two $C(N,2) \times 3N$ matrices denoted by $\mathbf{B}^{int}$ and $\mathbf{B}^{ext}$ at point $X = X_o$, where $C(N,2)$ is the number of combinations of N objects taken 2 at a time, i.e., $C(N,2) = \dfrac{N!}{(N-2)!2!}$. These two matrices are

$$\left. \left(B_{Xo}^{int} \right)_{ij} \right|_{X=Xo} = \begin{cases} \left. \dfrac{\partial Z_i}{\partial X_j} \right|_{X=Xo} & \text{if } Z_i \text{ is an intrafragment reciprocal} \\ & \text{interatomic distance otherwise} \\ 0 \end{cases} \tag{2-9}$$

and

$$\left. \left(B_{Xo}^{ext} \right)_{ij} \right|_{X=Xo} = \begin{cases} \left. \dfrac{\partial Z_i}{\partial X_j} \right|_{X=Xo} & \text{if } Z_i \text{ is an intrafragment reciprocal} \\ & \text{interatomic distance otherwise} \\ 0 \end{cases} \tag{2-10}$$

Equations (2-9) and (2-10) define two $\mathbf{B}$ matrices. B_{Xo}^{int} will have block diagonal form with f blocks. B_{Xo}^{int} will have rank $3N$-6-q while B_{Xo}^{int} has rank q.

The two $\mathbf{B}$ matrices are now employed in the SVD procedure to obtain the desired internal coordinates. In general, an $m \times n$ matrix $\mathbf{B}$ can be written in the form[83,84]

$$\mathbf{B} = \mathbf{U}\boldsymbol{\Lambda}\mathbf{V}^{T}, \tag{2-11}$$

where $\mathbf{U}$ is an $m \times m$ orthogonal matrix, $\mathbf{V}$ is an $n \times n$ orthogonal matrix, and $\boldsymbol{\Lambda}$ is an $m \times n$ diagonal matrix with $\boldsymbol{\Lambda} = \mathrm{diag}(\lambda_1, \lambda_2, \ldots, \lambda_{\mathrm{Rank}}(\mathbf{B}), 0, 0, \ldots, 0)$. In the present application, the columns of $\mathbf{U}$ are the normalized eigenvectors of $\mathbf{BB}^{T}$, the rows of $\mathbf{V}^{T}$ are the normalized eigenvectors of $\mathbf{B}^{T}\mathbf{B}$, and the elements of $\boldsymbol{\Lambda}$, λ_i, are the non-negative square roots of the eigenvalues of $\mathbf{B}^{T}\mathbf{B}$. In this form, the SVD of $B_{X_o}^{\mathrm{int}}$ is

$$B_{X_o}^{\mathrm{int}} = U^{\mathrm{int}} \Lambda^{\mathrm{int}} \left(V^{\mathrm{int}} \right)^{T}. \tag{2-12}$$

The $3N$-6-q columns of $\mathbf{U}^{\mathrm{int}}$ corresponding to the nonzero eigenvalues identify the linear combinations of intrafragment reciprocal interatomic distances that form the intrafragment internal, orthonormal, nonredundant coordinates required for the Taylor series expansions for systems with $N > 4$.

To obtain the interfragment, nonredundant internal coordinates, the MSI procedure employs columns of $\mathbf{V}^{\mathrm{int}}$ corresponding to the zero values of $\boldsymbol{\Lambda}$. A $3N \times (q+6)$ matrix $\tilde{V}^{\mathrm{int}}$ is defined whose columns are those columns of $\mathbf{V}^{\mathrm{int}}$ corresponding to the zero eigenvalues of $\boldsymbol{\Lambda}$. These columns of $\tilde{V}^{\mathrm{int}}$ are orthonormal and identify the internal $q + 6$ dimensional subspace for the interfragment motions of the system. To ensure that the interfragment internal coordinates are independent of and orthogonal to the intrafragment internal coordinates, the SVD of the product matrix $B_{X_o}^{\mathrm{int}} \tilde{V}^{\mathrm{int}}$ is computed rather than that for $B_{X_o}^{\mathrm{ext}}$, i.e.,

$$B_{X_o}^{\mathrm{int}} \tilde{V}^{\mathrm{int}} = \mathbf{U}^{\mathrm{ext}} \boldsymbol{\Lambda}^{\mathrm{ext}} (\mathbf{V}^{\mathrm{ext}})^{\mathrm{T}}. \tag{2-13}$$

The columns of $\mathbf{U}^{\mathrm{ext}}$ corresponding to the nonzero eigenvalues identify nonredundant linear combinations of the interfragment reciprocal interatomic distances.

The fragment local internal coordinates, $\zeta^{(X_o)}(\mathbf{X})$, are obtained by first defining the matrix $\tilde{U}$:

$$\tilde{U}_{ij} = \begin{cases} U_{ij}^{\mathrm{int}}; & j=1,2,\ldots,3N-6-q \\ U_{ik}^{\mathrm{ext}}; & k=1,2,\ldots,q; \quad \text{with } j = 3N-6-q+k \end{cases} \tag{2-14}$$

Once $\tilde{U}$ is defined by Equation (2-14), the desired local, mutually orthogonal, nonredundant, independent internal coordinates are obtained from

$$\zeta^{(X_o)}(\mathbf{X}) = \tilde{U}^{T} \mathbf{Z}. \tag{2-15}$$

The notation used in Equation (2-15) indicates that the set of internal coordinates $\zeta^{(X_o)}(\mathbf{X})$ is valid only at and in the near vicinity of $\mathbf{Z}_o = \mathbf{Z}(X_o)$. In practice, the SVD's indicated in Eqs. (2-9) through (2-15) are computed at every point in the database.

With independent, internal coordinates available at every configuration point in the database, the MSI fitting for systems with $N > 4$ proceeds in essentially the same manner as that previously described in Equations (2-1) and (2-2) for systems with $N \leq 4$, but with $\mathbf{Z}$ being replaced with ζ, i.e., Equation (2-2) is replaced with

$$
\begin{aligned}
\left.\frac{\partial V}{\partial X_n}\right|_{X(i)} &= \sum_{k=1}^{3N-6} \frac{\partial \zeta_k}{\partial X_n} \left.\frac{\partial V}{\partial \zeta_k}\right|_{\zeta(i)}, \\
\left.\frac{\partial^2 V}{\partial X_n \partial X_m}\right|_{X(i)} &= \sum_{k=1}^{3N-6} \sum_{j=1}^{3N-6} \frac{\partial \zeta_k}{\partial X_n} \frac{\partial \zeta_j}{\partial X_m} \left.\frac{\partial^2 V}{\partial \zeta_k \partial \zeta_j}\right|_{\zeta(i)} \\
&+ \sum_{k=1}^{3N-6} \frac{\partial^2 \zeta_k}{\partial X_n \partial X_m} \left.\frac{\partial V}{\partial \zeta_k}\right|_{\zeta(i)}
\end{aligned}
\tag{2-16}
$$

As in the case when $N \leq 4$, Equation (2-16) must be inverted to obtain the required values for the $\dfrac{\partial V}{\partial \zeta_k}$ and the $\dfrac{\partial^2 V}{\partial \zeta_k \partial \zeta_j}$. If the molecular system is planar or close to planar, some of the internal coordinates will become nearly linearly dependent. In such a case, the inversion of Equation (2-16) may be numerically unstable. To avoid this problem, Thompson et al.[80] have developed a procedure for slightly distorting the molecular configuration sufficiently to remove the linear dependence of the internal coordinates and thereby make it possible to accurately invert Equation (2-16). The reader may consult Appendix B of Reference 80 for details.

The Shepard procedure for interpolation is again employed to obtain the values of the potential at point $\mathbf{Z}$

$$
V(\mathbf{Z}) = \sum_{g \in G} \sum_{i=1}^{N} W_{g,i}(\mathbf{Z}) T_{g,i}(\zeta),
\tag{2-17}
$$

with the modification that the Taylor series expansions are now executed into terms of the independent local internal coordinates. That is,

$$
\begin{aligned}
T_i(\zeta) = V[\zeta(i)] &+ \sum_{k=1}^{3N-6} [\zeta_k - \zeta_k(i)] \left.\frac{\partial V}{\partial \zeta_k}\right|_{\zeta=\zeta(i)} \\
&+ \frac{1}{2!} \sum_{k=1}^{3N-6} \sum_{j=1}^{3N-6} [\zeta_k - \zeta_k(i)] \\
&\times [\zeta_j - \zeta_j(i)] \left.\frac{\partial^2 V}{\partial \zeta_k \partial \zeta_j}\right|_{\zeta=\zeta(i)}
\end{aligned}
\tag{2-18}
$$

Initially, Equations (2-4) through (2-6) were employed for the weight functions on the Taylor series expansions. However, Thompson et al.[80] found that the six-atom CH_5 system with reasonable values for p resulted in a PES that changes the relative weighting of the data points too sharply. Empirical investigation showed that a two-part primitive weight function defined by

$$v_i(\mathbf{Z}) = \left\{ \left[\frac{\|\mathbf{Z} - \mathbf{Z}(i)\|}{\mathrm{rad}(i)} \right]^{2q} + \left[\frac{\|\mathbf{Z} - \mathbf{Z}(i)\|}{\mathrm{rad}(i)} \right]^{2p} \right\}^{-1} \tag{2-19}$$

with $p \gg q$ gave much improved results. The empirical determination of good values for the parameters contained in Equation (2-19) has been discussed in detail by Thompson et al.[80] They found no large variations in the interpolation accuracy with q in the range from 2 to 6 and p in the range from 9 to 36; $\mathrm{rad}(i)$ is a type of confidence radius associated with data point $\mathbf{Z}(i)$. In practice, this confidence radius is determined for each point in the database. It is taken to be

$$\mathrm{rad}(i) = \|\mathbf{Z}(j) - \mathbf{Z}(i)\|, \tag{2-20}$$

where $\mathbf{Z}(j)$ is the nth nearest neighbor to $\mathbf{Z}(i)$ in the database. For sufficiently large value of n, the computed dynamics on the PES are insensitive to value taken for n.

2.2.4. MSI Configuration Space Sampling

MSI sampling is done iteratively using a combination of reaction path methods to obtain the starting database and classical trajectory methods to "grow" the database iteratively and thereby improve it. These sampling methods have been described in detail by Jordon et al.[85] and by Thompson and Collins.[86] Therefore, only the essentials of the procedure will be described here.

Step 1: To obtain the initial database to be fitted using MSI methods, a small ensemble (≈ 50) of points spaced along the minimum-energy path (MEP) that connects reactants and products for the process under investigation is selected. Care is taken to include points that sample any stationary points, such as minima and saddle points, that occur along this MEP. These points are then used as a database to which a PES is fitted using MSI methods.

If the reaction of interest is a simple bond rupture or bond formation process, there is very little difficulty in identifying the MEP and selecting the points for the initial database. If, however, the system is complex in that multiple two-, three-, and four-body reactions are occurring simultaneously, the determination of the MEPs for all such processes is often a difficult task. Such systems present a formidable challenge to MSI methods as well as other fitting methods. If finding the approximate MEP for all processes becomes a problem, other sampling methods can be employed to obtain the required initial database without compromising the MSI procedure. Chapter 4 of this monograph presents several powerful alternative methods for accomplishing this.

Step 2: In this step, additional points are added to the database by running classical trajectories on the PES obtained by using MSI methods to fit the initial database.

The initial conditions for these sampling trajectories are chosen such that they are appropriate to the reaction(s) being studied. Such conditions include temperature, internal excitation energy of the molecules, relative velocity distribution, internal rotational and vibrational states, among others. In most MSI applications, the number of trajectories computed is on the order of 10. As the trajectories are integrated, molecular configurations are periodically stored until a set of a few thousand configurations is obtained. In order to increase the probability that the sampling will cover all of the regions of configuration space that are important in the reaction dynamics, the sampling trajectories are generally run in both the forward and reverse directions. That is, some trajectories are initiated in the reactant valley of the MEP while others are started in the product valley.

Step 3: The configurations generated from the sampling trajectories are individually tested to determine which of the configurations will be added to the overall database for the system. Collins and co-workers[86] have examined several possible procedures for performing this selection. We will describe two of these here. For a more comprehensive discussion, the reader is referred to Reference 86.

One selection method is based on the concept that if a configuration point is close to many of those occurring in the sampling trajectories, then it is probably an important point and should be added to the database unless that point is already adequately represented in the present database. The numerical procedure for implementing this concept utilizes the primitive weight function given by Equations (2-4) and (2-5). Let there be N_t total configurations present in the set selected from the sampling trajectories and let $\mathbf{R}(j)$ denote the jth configuration point in this set. A comparative weight, $h[\mathbf{R}(j)]$, is computed for each point in the sampling set as

$$h\big[\mathbf{R}(j)\big]=\frac{1}{N_t-1}-\frac{\displaystyle\sum_{n=1;n\neq j}^{N_t} v_j\big[\mathbf{R}(n)\big]}{\displaystyle\sum_{g\in G}\sum_{k=1}^{N} v_{gk}\big[\mathbf{R}(j)\big]}. \tag{2-21}$$

The numerator of Equation (2-21) is large whenever the trajectory point $\mathbf{R}(j)$ is close to many of the other points in the sampling set. The denominator is large whenever $\mathbf{R}(j)$ is near many of the points already in the database. Consequently, $h[\mathbf{R}(j)]$ will be large when $\mathbf{R}(j)$ is close to many points in the sampling set, but far from points currently in the database. MSI sampling chooses one point sampled from the forward trajectories and one from the reverse trajectories that have the largest value of $h[\mathbf{R}(j)]$.

A second selection criterion chooses the trajectory point whose MSI-predicted energy (using the current database) has the greatest variance from the average of all points in the set with N_t configurations. The predicted average energy of the N_t points is

$$\langle V\rangle=\sum_{i=1}^{N_t} W_i(\mathbf{R})T_i(\mathbf{R}), \tag{2-22}$$

where the summation in Equation (2-22) runs over all configuration points in the set sampled from the trajectories and their symmetry equivalents. The variance of the energy of each configuration in the set from the average is then

$$\sigma_V^2(\mathbf{R}) = \sum_{i=1}^{N_t} W_i(\mathbf{R})\left[T_i(\mathbf{R}) - \langle V \rangle\right]^2. \qquad (2\text{-}23)$$

The configuration associated with the largest value of $\sigma_V^2(\mathbf{R})$ in the points sampled from the forward trajectories and the one with the largest $\sigma_V^2(\mathbf{R})$ among the points sampled in the reverse trajectories are added to the database.

Step 4: Distort the geometry of each selected new configuration from planar, if necessary; compute the coefficients of the Taylor series expansion about that point using the methods previously described; and add the new points to the database. In the nomenclature of MSI, this procedure is said to "grow" the PES. Reevaluate the confidence limits, Equation (2-20), using the points in the new database.

Step 5: Compute the dynamical variables—i.e., reaction cross sections, reaction rates, branching ratios, final energy partitioning, etc.—using classical trajectories computed on the PES existing after Step 4 is completed. If these results are essentially the same as those previously computed prior to the new sampling, the PES is considered to have converged and the growing process is over. On the other hand, if the dynamical variables are found to have changed by an amount larger than some chosen convergence limit, return to Step 2 and iterate until the PES resulting from the database at Step 4 is found to converge.

2.2.5. Applications and Results

Collins and co-workers have applied MSI procedures to obtain PESs for numerous systems containing 3 to 6 atoms. Most of the systems investigated contain four atoms.

To provide a test of accuracy of the method itself, Thompson et al.[80] used MSI methods to fit the analytical surface developed by Jordon and Gilbert[87] for the CH_5 system. After modifying this PES so that all hydrogen-atom exchange symmetry was incorporated, the dynamics of both the forward and reverse reactions $CH_4 + H - CH_3 + H_2$ were investigated.

The initial database used in the MSI calculations was obtained by sampling points along the MEP that were obtained by finding the intrinsic reaction path (IRP) in mass-weighted Cartesian coordinates by following the path of steepest descent from the saddle point into the reactant valley.[87] A second IRP was obtained in the same manner by following the path into the $CH_3 + H_2$ product channel. The initial database was obtained by sampling 50 data points along this MEP.

The PES was grown using trajectories initiated in the reactant $CH_4 + H$ valley. The initial conditions for the trajectories employed to grow the PES as well as those used to compute the reaction dynamics were obtained using the efficient microcanonical sampling (EMS) method developed by Schranz et al.[88] on the Jordan-Gilbert[87] analytical surface. The values of p and q in Equation (2-19) were taken to be 12 and 2, respectively. The value of n used to determine the confidence radius in Equation (2-20) was either $n = 25$, 50, 75, 100, or 125. A PES was obtained for each of these values.

The final PES used to examine the reaction dynamics for the forward reaction was obtained using MSI methods to fit 853 configuration points in the final database. For the reverse reaction, this database was increased to 1,250 configuration points. Convergence of the PES was tested by comparison of the computed dynamics on the MSI surface with those obtained by computing trajectories on the analytic surface.[87] The dynamic properties computed for testing were the distribution of relative translational energies and angular momentum in CH_4 for inelastic collisions and the reaction probabilities for zero-impact parameter collisions for the forward reaction.

For trajectories initiated in the $CH_4 + H$ valley at a total energy of 160 kJ mol^{-1}, the average absolute error in the interpolated potential compared to the "exact" analytic energy was ≈ 1.6 kJ mol^{-1}. The average relative gradient errors were $\approx 8\%$. For trajectories initiated in the $CH_3 + H_2$ valley at the same total energy but using a PES fitted to 1,250 points, the average absolute energy error was about 3.5 kJ mol^{-1}. The relative gradient errors were about 12%–13%.

The MSI methods described above for the CH_5 system have been employed to investigate a wide variety of reactions, most of which are four-body reactions involving only a single reaction channel. Table 2-1 lists the best results reported by the authors for different combinations of fitting parameters and number of points in the database.

2.3. INTERPOLATIVE MOVING LEAST-SQUARES METHODS (IMLS)

2.3.1. General Method

The generalized interpolative moving least squares method as developed by Guo et al.[29] interpolates between the points of a database of molecular energies computed by some appropriate electronic structure method by performing a linear least-squares fitting to this database at the instantaneous position of the system during a trajectory, which we will call the fitting point.

The first step in the execution of an IMLS calculation is the selection of the basis functions in terms of which the PES will be expressed by fitting these basis functions multiplied by coefficients to the database using a least-squares method. Let the set of basis functions be represented by $\mathbf{b}(\mathbf{Z})$, where $\mathbf{Z}$ is a set of coordinates that specify the spatial configuration of the molecular system and $b_i(\mathbf{Z})$ denotes the ith member of the set. The IMLS potential at point $\mathbf{Z}$ is defined to be

$$V_{IMLS}(\mathbf{Z}) = \sum_{i=1}^{M} a_i(\mathbf{Z}) b_i(\mathbf{Z}), \qquad (2\text{-}24)$$

Table 2-1 Examples of fitting accuracy of MSI methods. The table gives the best result reported by the authors if multiple surfaces were developed with different parameter values

System	N(a)	rms or absolute average energy error (kJ mol^{-1})	Gradient Error	Reference(s)
H_3	350	0.48	0.71 [b]	33a and 33b
$C + H_3^+$	406	0.4	4.7% [c]	89
$H_3^+ + O$	727	0.35	(d)	90
$BeH_2 + H$	438	0.35	(d)	91
$OH + H_2$	400	1.	(d)	86
$NH + H_2$	66	12.1	(d)	23
$N + H_3^+$	699	0.05	(d)	92,93
BeH_3	1,300	0.41	5.58% [c]	91,93
CH_3^+	406	0.21	2.29% [c]	86,93
OH_3^+	699	0.36	(d)	93
OH_3	254	0.31	(d)	93
$NeCOH^+$	724	0.62	(d)	93
$ArCOH^+$	569	0.54	(d)	93
$KrCOH^+$	1,086	1.18	(d)	93
$FCOH_2^+$	1,300	1.50	(d)	93
$HOOH$	6,489	1.95	(d)	94
$CH_4 + H \rightarrow CH_3 + H_2$	853	1.6	8% [c]	80
$CH_3 + H_2 \rightarrow CH_4 + H$	1,250	3.5	12–13% [c]	80
$[HFHOC]^+$	1357	1.5	(d)	95
$[H_2OHOC]^+$	3047	3.5	(d)	95
$H_3^+ HD$	728	0.34	(d)	96

(a) N is the number of symmetry-distinct *ab initio* points computed.

(b) Units are kJ mol^{-1} Å^{-1}.

(c) Relative percent error

(d) Not given

where M is the total number of basis functions employed in the fitting process, and the $b_i(\mathbf{Z})$ are polynomials.

The fitting coefficients, $a_i(\mathbf{Z})$, are determined by minimizing the sum of the weighted square deviations of $V_{IMLS}(\mathbf{Z})$ from the *ab initio* results, $V(\mathbf{Z})$, at each of the points in the database. The sum of these square deviations is

$$\sigma^2(\mathbf{Z}) = \sum_{i=1}^{N} w_i(\mathbf{Z}) \left[V_{IMLS}(\mathbf{Z}^{(i)}) - V(\mathbf{Z}^{(i)}) \right]^2, \tag{2-25}$$

where $w_i(\mathbf{Z})$ is the weight assigned to point i in the database that in general will have the property that $w_i(\mathbf{Z})$ will be larger for data points closer to $\mathbf{Z}$ and smaller for the more distant points. N is the total number of configuration points in the database which are located at $\mathbf{Z}^{(1)}$, $\mathbf{Z}^{(2)}$, ... , $\mathbf{Z}^{(N)}$ with corresponding *ab initio* energies $V(\mathbf{Z}^{(1)})$, $V(\mathbf{Z}^{(2)})$, ... , $V(\mathbf{Z}^{(N)})$, respectively. The minimization condition requires that

$$\partial \sigma^2 / \partial a_i = 0 \qquad (2\text{-}26)$$

for all values of the summation index.

Equations (2-25) and (2-26) lead to a set of N linear, homogeneous equations whose unknowns are the fitting coefficients $a_i(\mathbf{Z})$. In matrix form, these equations are

$$\mathbf{B}^T \mathbf{W} \mathbf{B} \mathbf{a} = \mathbf{B}^T \mathbf{W} \mathbf{V}, \qquad (2\text{-}27)$$

where $\mathbf{B}$ is an $N \times M$ matrix

$$B = \begin{pmatrix} b_1\left(\mathbf{Z}^{(1)}\right) & b_2\left(\mathbf{Z}^{(1)}\right) & \cdots & b_M\left(\mathbf{Z}^{(1)}\right) \\ b_1\left(\mathbf{Z}^{(2)}\right) & b_2\left(\mathbf{Z}^{(2)}\right) & \cdots & b_M\left(\mathbf{Z}^{(2)}\right) \\ \vdots & \vdots & \ddots & \vdots \\ b_1\left(\mathbf{Z}^{(N)}\right) & b_2\left(\mathbf{Z}^{(N)}\right) & \cdots & b_M\left(\mathbf{Z}^{(N)}\right) \end{pmatrix}, \qquad (2\text{-}28)$$

and $\mathbf{W}$ is an $N \times N$ diagonal matrix whose nonzero elements are the values of the weight functions, i.e.,

$$W = \mathrm{diag}\left[w_1(\mathbf{Z}), w_2(\mathbf{Z}), \ldots, w_N(\mathbf{Z})\right]. \qquad (2\text{-}29)$$

and $\mathbf{V}$ is a $N \times 1$ column vector whose elements are the computed *ab initio* energies at the N data points. That is, $\mathbf{V} = \mathrm{diag}[V(\mathbf{Z}^{(1)}), V(\mathbf{Z}^{(2)}), \ldots, V(\mathbf{Z}^{(N)})]^T$.

Once the coefficient vector $a(\mathbf{Z})$ is obtained by solving Equation (2-27), the IMLS fitted potential at point $\mathbf{Z}$ is given by Equation (2-24). In practice, Equation (2-27) is solved using the SVD technique that has been described in Section 2.2.3.

The IMLS method is ideally suited to the computation of the first derivatives of the potential that are required in the integration of trajectories. The matrix operations used to solve Equation (2-27) can be used again to obtain the first derivatives of $V_{IMLS}(\mathbf{Z})$ with respect to $\mathbf{Z}$.[28] Since the dependence of the basis functions upon $\mathbf{Z}$ are known, their derivatives present no problem. The required derivatives of $\mathbf{a}$ with respect to $\mathbf{Z}$, $\mathbf{a}'$, can be obtained by straightforward differentiation of Equation (2-27) that leads to

$$\mathbf{B}^T \mathbf{W} \mathbf{B} \mathbf{a}' = (\mathbf{B}^T \mathbf{W} \mathbf{V})' - (\mathbf{B}^T \mathbf{W} \mathbf{V})' \mathbf{a} \qquad (2\text{-}30)$$

Thus, the matrix decomposition of $\mathbf{B}^T\mathbf{WB}$ by the SVD technique used in solving Equation (2-27) can be reused to solve Equation (2-30) for $\mathbf{a}'$.

In principle, the set of coordinates $\{\mathbf{q}\}$ used to specify $\mathbf{Z}$ can be chosen by the user. Guo, et al.[29] have employed both reciprocal bond distances, $Z_i = 1/R_i$ and the internal coordinates generally employed in the $\mathbf{Z}$-matrix specification of molecular geometries in most electronic structure programs. These coordinates are a mix of bond distances, three-body angles, and dihedral angles.

The unnormalized weight function, $v_i(\mathbf{Z})$, generally employed in IMLS fitting has the form

$$v_i(\mathbf{Z}) = \frac{1}{\left[\left(\Delta q\right)^2 + e^2\right]^p}, \tag{2-31}$$

where

$$\left(\Delta q\right)^2 = \sum_{k=1}^{N_R}\left[q_k - q_k^{(i)}\right]^2. \tag{2-32}$$

The weight function $w_i(\mathbf{Z})$ is obtained directly from the $v_i(\mathbf{Z})$ by normalization, i.e.,

$$w_i(\mathbf{Z}) = \frac{v_i(\mathbf{Z})}{\Sigma_k v_k(\mathbf{Z})}, \tag{2-33}$$

where the summation in Equation (2-32) runs over M basis functions. If the set $\{\mathbf{q}\}$ comprises coordinates all of which have the same units, Equation (2-31) will be unit-wise consistent. If the coordinates in the set have different units, then formally each must be multiplied by a unit constant to make the units consistent. A typical example is the set of internal coordinates for a four-body system that usually comprises three bond distances, r_1, r_2, and r_3, two three-body angles θ_1, and θ_2, and one dihedral angle ϕ. In this case, the vector $\mathbf{Z}$ might be taken to be

$$\mathbf{Z} = \left(r_1, r_2, r_3, c_1\theta_1, c_2\theta_2, c_3\cos\phi\right), \tag{2-34}$$

where c_1, c_2, and c_3 are unit constants that have the dimensions of length. Since the dihedral angle is discontinuous at $\phi = \pi$, a continuous trigonometry function is used in its place.

Since the weight function is dependent upon the coordinates of the system in hyperspace, the least squares fitting must be done at every integration point encountered in the computation of a trajectory. This is, by far, the major impediment to the use of IMLS methods. The computational time required to solve Equations (2-24) through (2-34) scales as NM^2. Consequently, as the size and complexity of the chemical processes occurring in a given system increase, the number of basis functions required for the fit will substantially increase as will the number of points in the database needed to properly sample all important regions of configuration space. The strong dependence of the required computational time upon N and M will often lead to unacceptably large computational requirements.

An example of the very large computational time requirements of IMLS methods has been provided by Guo et al.[31] They examined the unimolecular dissociation dynamics of H_2CN to $H + HCN$ products using MD methods on an *ab initio* PES obtained by fitting a database containing 830 points using IMLS methods. The reported computation time required for one H_2CN dissociation trajectory lasting 2.5 ps was about 100 minutes. When the database does not need to be refitted at each integration point of a trajectory, as is the case for MSI and NN methods, four-body decomposition trajectories usually require computation times on the order of a few seconds.

Guo et al.[29] have investigated the most effective values of the parameters p and e in Equation (2-31) using the computed root mean square errors (RMSEs) in both the interpolated energies and gradients at an independent set of 5,000 test points obtained from trajectories run on the IMLS surface. Their findings suggest that as long as e is less than 0.01, the RMSEs for both energy and gradient are nearly independent of its value. Consequently, its recommended value is 0.01. The dependence of the fittings errors on p is more complex. The RMSE for the gradient is a near monotonically increasing function of p. In contrast, the RMSE for the energy exhibits a minimum in the neighborhood of $p = 5$. At $p = 4$, the gradient error is found to be asymptotically approaching a near constant value. Consequently, Guo et al.[29] recommend the use of $p = 4$ in Equation (2-31).

2.3.2. Cutoff Function, Basis Sets, and Data Sampling

In order to reduce the large computational overhead of the IMLS method, it is common practice to use a cutoff function that, in effect, allows points in the database at a distance from the fitting point that exceeds some chosen cutoff distance, r_{cut}, to be ignored in the summation in Equation (2-25). The functional form of this cutoff function, $S(\Delta q)$ is

$$S(\Delta q) = \tanh^3\left[\alpha\left(1 - \frac{(\Delta q)^2}{(r_{cut})^2}\right)\right] \quad \text{if } \Delta q \le r_{cut} \qquad (2\text{-}35a)$$

$$S(\Delta q) = 0 \quad \text{otherwise.} \qquad (2\text{-}35b)$$

In Equation (2-35a), α is an adjustable parameter to which Guo et al.[29] assigned a value of $\alpha = 7$. The power of 3 in Equation (2-35a) ensures that the function, first, and second derivatives will all be continuous at the point $\Delta q = r_{cut}$. The cutoff function described by Equations (2-35a) and (2-35b) is incorporated into the IMLS procedure by replacement of the weight function $w_i(\mathbf{Z})$ with $w_i^{new}(\mathbf{Z})$ where

$$w_i^{new}(\mathbf{Z}) = w_i(\mathbf{Z})S(\Delta q). \tag{2-36}$$

Kawano et al.[30] have introduced two modifications to the nature of the cutoff function described by Equations (2-35a) and (2-35b). One of these is a fixed radius cutoff (FRC) method that only includes points within a hypersphere of fixed radius. The second procedure, termed a density adaptive cutoff (DAC) method, includes points within a hypersphere of variable radius that depends upon the point density in the database. The basic idea of both approaches is to reduce the number of data points that have to be considered in the fitting process and thereby reduce the computational time requirements of the IMLS method.

Guo et al.[29] investigated the variation in the RMSE for the energy and the gradient as a function of r_{cut} for the H_2O_2 system. The results showed that the RMSE for the gradient attained a minimum value at $r_{cut} = 1.5$ Å. At this point, the RMSE for the energy was nearly at its minimum value. Consequently, r_{cut} for H_2O_2 was set to this value. In practice, for new systems, this investigation needs to be repeated and the value of r_{cut} thereby optimized.

In the IMLS method, the number and type of basis functions are determined empirically for each system investigated. The selection criterion for the determination is the size of the expansion coefficients associated with each trial term in the basis set averaged over a number of randomly selected testing points. Initially, a large basis set of terms of the form $[q_i^n q_j^m]^{-1}$ is employed in the fitting and the average value of the expansion coefficients for each term is evaluated over a set of randomly selected configurations. In this expression for the basis terms, i and j denote any two coordinates in $\mathbf{Z}$ where $i \neq j$. By successively discarding basis functions with small absolute values of the expansion coefficients, a near optimum basis set of a tractable size can be obtained.

For example, when Guo et al.[29] employed this procedure in their investigation of the H_2O_2 system, two sets of coordinates for $\mathbf{Z}$ were examined. The first was the set of six interatomic distances. The second was the set of internal coordinates comprising the three bond distances, two three-body angles, and the dihedral angle. Empirical testing showed that for the valence internal coordinates, a good basis set could be obtained using a total of 25 terms that included 19 diagonal terms $(i = j)$ with n values from 1 to 4 and 6 cross terms $(i \neq j)$ involving the O-O bond distance and the two three-body angles. Specifically, these terms were the reciprocals of $r\theta_1$, $r\theta_2$, $r^2\theta_1$, $r^2\theta_2$, $r\theta_1^2$, and $r\theta_2^2$, where r is the O-O bond distance. When interatomic distances were used for the coordinate matrix $\mathbf{Z}$, 25 terms were also used. These were three third-order terms r_i^{-3} $(i = 1, 2, 3)$ for the three bonds in HOOH, and all terms up to second order except six of the second-order cross terms.

Several procedures for sampling configuration space have been employed in IMLS fitting calculations to obtain the *ab initio* database. Generally, these procedures are more ad hoc than those employing trajectory sampling and selection methods such as those described in Section 2.2.4 and in Chapter 4 of this mono-graph. For example, the database for the investigation of the dissociation reactions of HOOH comprised two sets of points. The first set contained 25 symmetry-distinct points with 5 selected along the O-O MEP, 3 points at large O-H separa-tion, and 17 points in the low-energy regions with $E < 40$ kcal mol^{-1}. The second set contained randomly selected points chosen with two criteria: (1) the distance from the existing points is no less than a specified value and (2) the energy of the point is less than a chosen maximum value that was taken to be 85 kcal mol^{-1}, which was the largest energy for with the dissociation rate coefficient was computed.

2.3.3. Applications and Results

Most applications of IMLS methods have been to either three- or four-atom systems with the method being applied to fit an empirical potential function. In most applica-tions, the molecular system has been H_2O_2. Table 2-2 reports some of the results. In most cases, the dependence of the accuracy of IMLS fitting upon the parameters of

Table 2-2 Examples of fitting accuracy of IMLS methods. The table gives the best result reported by the authors if multiple surfaces were developed with different parameter values

System	N(a)	rms or absolute average energy error (kJ mol^{-1})	Gradient Error (kJ mol^{-1} Å^{-1})	Reference(s)
HOOH[b]	152	16.5	268.6	29
HOOH[b]	252	14.0	226.4	29
HOOH[b]	502	8.24	153.1	29
HOOH[c]	152	18.0	258.6	29
HOOH[c]	252	13.5	186.6	29
HOOH[c]	502	10.6	168.6	29
HOOH[d]	889	7.95	201.6	30
HOOH[e]	6489	4.52	(f)	94
HOOH[e]	1350	4.18	(f)	97
HOOH[e]	$>10^4$	2.5	(f)	97
H_2CN[g]	330	1.40	(f)	31
H_2CN[g]	830	0.54	(f)	31

(a) N is the number of symmetry-distinct *ab initio* points computed.

(b) Valence internal coordinates used for Z to fit analytic semi-empirical surface described in Ref. 51.

(c) Interatomic distance coordinates used for Z to fit analytic semi-empirical surface described in Ref. 51.

(d) With FRC and DRC cutoff procedures. Fitted to empirical surface. Ref. 51.

(e) Fitted to the symmetrized analytical potential described in Ref. 51.

(f) Not reported.

(g) Fitted to an *ab initio* database.

the procedure, i.e., number of data points employed, values of the parameters, number of basis functions used, etc., are reported. The results given in Table 2-2 are the most accurate fitting result obtained for the system and the reference cited.

2.4. INVARIANT POLYNOMIAL (IP) AND REPRODUCING KERNEL HILBERT SPACE (RKHS) METHODS

2.4.1. Invariant Polynomial Methods

Bowman, Braams, and co-workers[34–39] have developed a different type of least-squares fitting procedure that utilizes expansions in terms of polynomials of functions of the interatomic distances to fit *ab initio* databases obtained from electronic structure calculations. The expansion polynomials are totally symmetrized so that they are invariant to the exchange of any two identical atoms. For this reason, the procedure is generally called an "invariant polynomial" (IP) method.

In the IP method, a point in configuration space, $\mathbf{Z}$, is generally specified by the complete set of $n(n-1)/2$ interatomic distances, R_{ij}, between atoms i and j, where n is the number of atoms in the molecular system of interest. This choice is made because it facilitates the development of a potential that is invariant with respect to permutation of identical nuclei.

The basis functions used in fitting polynomials are functions of the R_{ij}. In some applications these functions are taken to be simple exponentials, X_{ij}, where

$$X_{ij} = \exp\left(-R_{ij}/a\right). \tag{2-37}$$

The polynomials $p(x)$ employed in the fit are functions of the X_{ij} where it is understood that $p(x)$ is a polynomial in all of the X_{ij}, which is denoted by x, that is invariant with respect to permutation of any two identical atoms.

In some calculations, it was found that the use of Equation (2-37) for the basis functions led to poor fits in regions where the nuclei are in close proximity. In such regions, it was found empirically that basis functions of the form

$$Y_{ij} = \frac{\exp\left(-R_{ij}/a\right)}{R_{ij}} \tag{2-38}$$

provided superior fitting accuracy.

The formal IP expression for the global potential of the system varies somewhat from application to application. For example, Chen et al.[35] in their investigation of the association reaction of $OH + NO_2$ used the form

$$V(\mathbf{X}) = \Sigma_\alpha h_\alpha\left(p_1(\mathbf{X}),\ldots,p_d(\mathbf{X})\right)q_\alpha(\mathbf{X}), \tag{2-39}$$

where the vector $\mathbf{X}$ has dimension d and denotes a set of variables X_{ij} in Equation (2-37). $p_k(\mathbf{X})$ and $q_k(\mathbf{X})$ denote the primary and secondary invariant polynomials, respectively. These polynomials are both homogeneous in the variables X_{ij} and both are invariant under permutation of identical atoms. There are d primary invariant polynomials and h_α is a polynomial of them. In this work, a was taken to have the value 2 Bohrs.

In their study of the $H_5O_2^+$ system, Huang et al.[36] employed the form

$$V(\mathbf{X}) = p(\mathbf{X}) + \sum_{i<j} q_{ij}(\mathbf{X}) Y_{ij},$$

(2-40)

where $\mathbf{X}$ is defined as stated above and the Y_{ij} are given by Equation (2-38). Again, $p(X)$ and $q_{ij}(\mathbf{X})$ are invariant polynomials. In this application, $p(\mathbf{X})$ was taken to be a seventh-degree polynomial in the X_{ij} while the $q_{ij}(\mathbf{X})$ were third-degree polynomials. In this application, the value $a = 3$ Bohrs was used.

Xie et al.[37] used a many-body expansion (see Chapter 6, Section 6.3) to represent the global potential for the H_5^+ system. Formally, the potential had the form

$$V\left(H_5^+\right) = \Sigma v_H^{(1)} + \Sigma v_{H_2}^{(2)} + \Sigma v_{H_3}^{(3)} + \Sigma v_{H_5}^{(5)}.$$

(2-41)

$V_H^{(1)}$ is the constant energy of a single hydrogen atom. $V_{H_2}^{(2)}$ is a two-body term which depends upon the internuclear distance between the two H atoms. The three-body term, $V_{H3}^{(3)}$, goes to zero as any one of the three hydrogen atoms separates. Likewise, the five-body term, $V_{H_5}^{(5)}$, goes to zero as any fragment separates. The functional form chosen for the n-body term is

$$V^{(n)} = P_N(\mathbf{R}) d(\mathbf{R}),$$

(2-42)

where $\mathbf{R}$ denotes the interatomic distance vector and the damping function $d(\mathbf{R})$ has the form

$$d(\mathbf{R}) = \max\left(0, 1 - \frac{\|\mathbf{R}\|_2}{n^{0.5} a}\right)^5.$$

(2-43)

In Equation (2-43), n is the number of interatomic distances in $\mathbf{R}$ and a is a constant equal to 7 bohr. The $P_N(\mathbf{R})$ in Equation (2-42) are polynomials of $\mathbf{R}$ that are invariant to permutation of identical atoms.

The connecting feature of all the functional forms employed to represent the global potential using IP methods is that the fitting polynomials are all made invariant to permutation of identical atoms. The methods used to ensure this invariance have been described in detail by Huang et al.[36]

The sampling of configuration space to obtain the molecular configurations whose energies form the database to which the IP potentials are fitted are obtained by either sampling configurations obtained using direct dynamics (see Chapter 4, Section 4.3) or by iterative sampling of configurations obtained in trajectories computed on IP potentials fitted to the present database (see Section 2.2.4 of this chapter and Section 4.2 of Chapter 4).

If there are m basis functions in the IP expression for the potential and n configurations in the database, the polynomial coefficients are determined by a weighted least-squares procedure of dimension $m \times n$. The solution for the coefficients is obtained using a singular value decomposition (SVD) of the associated matrix. Similar techniques are also employed in both the MSI and IMLS methods. This SVD is accomplished using the routine DGESVD from LAPACK.[98] An energy-based weight function is employed in the fitting. If the kth entry in the database has *ab initio* energy $E(k)$, then the corresponding weight for that point in the SVD is $\delta/[\delta + E(k) - E_{min}]$, where E_{min} denotes the global minimum potential energy over all entries in the database. δ is a parameter that has been assigned the value 0.1 hartree.

2.4.2. Applications and Results of IP Methods

IP fitting methods have been applied to some very challenging systems containing five, six, and seven atoms. In some cases, a variety of chemical reactions are occurring. The results for six such systems are given in Table 2-3. In most cases, the databases contain from 20,000 to over 100,000 points. While this seems large, in terms of the total number of data items that must be available, it is on the same order as that required by MSI methods.

For example, consider the seven-atom $H_5O_2^+$ system where a database of 48,189 points was employed in the IP fitting. If the fitting were done using MSI methods with a database comprising 500 symmetry-distinct points, the total number of data items that would have to be provided to execute the MSI fit would be 68,000 since at each point, we would need to have the energy, 15 first derivatives of the energy with respect to the $3N$-6 = 15 internal coordinates, and 120 mixed and diagonal second derivatives. This is a total of 136 data items at each point.

The RMSE of the IP fits to the potential are generally in the range from 0.4 to 3 kJ mol[-1]. This is better than that usually achieved using IMLS methods and comparable to that obtained using MSI methods. The RMSE reported for the CH_5^+ system was

Table 2-3 Examples of fitting accuracy of IP methods

System	N(a)	No. Polynomials[b]	RMSE (kJ mol[-1])	Gradient Error (kJ mol[-1] Å[-1])	Reference
CH_5^+	4,096	1,912	~ 0.06	(c)	38
$H_5O_2^+$	48,189	7,962	0.42	(c)	36
$H_3O_2^-$	66,965	(c)	0.23	(c)	34
$H + CH_4$	20,728	(c)	3.14	(c)	39
$H_5^{+\,(d)}$	105,888	(c)	0.55	(c)	37
$OHNO_2$	55,471	(c)	2.13	(c)	35

(a) N is the number of symmetry-distinct *ab initio* points computed.

(b) Number of invariant polynomials used.

(c) Not reported.

(d) Potential represented by many-body expansion with the individual terms in the expansion expressed in terms of invariant polynomials.

several wavenumbers.[38] With a database containing only 4,096 points, this seems unreasonably small, particularly since the authors report that the energy range fitted was about 96 kJ mol^{-1}.

2.4.3. Reproducing Kernel Hilbert Space (RKHS)

The RKHS method solves a generalized Fredholm-integral equation of the first kind to obtain an analytic expression for the PES from a discrete set of *ab initio* potential energies obtained from electronic structure calculations. The method is arguably the most mathematically sophisticated of the fitting procedures that have been proposed. Therefore, we shall content ourselves with providing some results obtained using this methods and refer the reader to an excellent review article by Hollebeek et al.[99] for the computational details.

The RKHS methods involve representing the potential energy surface in terms of a linear combination of basis functions, as in

$$V(\mathbf{x}) = \sum \alpha_i \phi_i(\mathbf{x}). \tag{2-44}$$

This general format is common to many different fitting methods. What sets the RKHS methods apart is the determination of the basis functions $\phi_i(\mathbf{x})$. These functions are based on a Mercer kernel function $k(\mathbf{x}, \mathbf{y})$ that must satisfy the following conditions:

1. $k(\mathbf{x}, \mathbf{x}_i)$ is symmetric about $\mathbf{x}_i$, $k(\mathbf{x}, \mathbf{x}_i) = k(\mathbf{x}_i, \mathbf{x})$, and is maximum at $\mathbf{x} = \mathbf{x}_i$.
2. The total volume under $k(\mathbf{x}, \mathbf{x}_i)$ is constant with respect to $\mathbf{x}_i$.

This kernel can be written

$$k(\mathbf{x}, \mathbf{y}) = \sum \lambda_i \phi_i(\mathbf{x}) \phi_i(\mathbf{y}) \tag{2-45}$$

with positive λ_i. The functions $\phi_i(\mathbf{x})$ are called eigenfunctions and the λ_i are the corresponding eigenvalues.

Consider a Mercer kernel $k(\mathbf{x}, \bullet)$ and the space of real-valued functions of $\mathbf{x}$ that is generated by this kernel. (This is a Hilbert space under certain conditions.)[99] Two such functions are

$$f(\mathbf{x}) = \sum_{i=1}^{l} a_i k(\mathbf{x}_i, \mathbf{x}) \tag{2-46}$$

$$g(\mathbf{x}) = \sum_{j=1}^{n} b_j k(\mathbf{x}'_j, \mathbf{x}) \tag{2-47}$$

where the a_i and b_j are scalar coefficients and the $\mathbf{x}_i$ and $\mathbf{x}'_i$ are some vectors in the configuration space. It is possible to define an inner product on the space generated by the kernel:

$$\langle f,g \rangle = \sum_{j=1}^{n}\sum_{i=1}^{l} a_i k\left(\mathbf{x}_i,\mathbf{x}_j'\right) b_j. \tag{2-48}$$

Using the symmetry property of the kernel, this can be rewritten

$$\langle f,g \rangle = \sum_{i=1}^{l} a_i \sum_{j=1}^{n} b_j k\left(\mathbf{x}_j',\mathbf{x}_i\right). \tag{2-49}$$

Note that, using Equation (2-47), we can rewrite Equation (2-49) as

$$\langle f,g \rangle = \sum_{i=1}^{l} a_i g\left(\mathbf{x}_i\right). \tag{2-50}$$

In a similar way, we can show that

$$\langle f,g \rangle = \sum_{j=1}^{n} b_j f\left(\mathbf{x}_j'\right). \tag{2-51}$$

Now, let

$$g(\mathbf{y}) = k(\mathbf{x},\mathbf{y}) \tag{2-52}$$

in Equation (2-50) to obtain

$$\langle f, k(\mathbf{x},\bullet) \rangle = \sum_{i=1}^{l} a_i k\left(\mathbf{x},\mathbf{x}_i\right) = \sum_{i=1}^{l} a_i k\left(\mathbf{x}_i,\mathbf{x}\right) = f(\mathbf{x}) \tag{2-53}$$

using Equation (2-46). This is known as the reproducing property of the Mercer kernel, which leads to the name of the resulting space generated by the kernel—reproducing kernel Hilbert space.

The power of the RKHS rests in the following property: Any function defined in an RKHS can be represented by a linear combination of the Mercer kernel functions:

$$f\left(\mathbf{x}_j\right) = \sum_{i=1}^{l} a_i k\left(\mathbf{x}_i,\mathbf{x}_j\right). \tag{2-54}$$

In addition, the expansion given in Equation (2-54) is the minimizer of the following regularized performance index:

$$J(f) = \frac{1}{2N}\sum_{n=1}^{N}\left(t(n) - f(\mathbf{x}(n))\right)^2 + \alpha\|f\|^2, \tag{2-55}$$

where $\{\mathbf{x}(n),t(n)\}_{n=1}^{N}$ are the inputs (interparticle distances) and target outputs (potential energy) for the training data set. The second term on the right side of

Equation (2-55) is the regularization penalty term, which is needed to produce a smooth response.

To find the coefficients in Equation (2-54) that minimize $J(f)$ in Equation (2-55) we can solve the equation [56]

$$[\mathbf{K}+\alpha\mathbf{I}]\mathbf{a}=\mathbf{t}, \tag{2-56}$$

where the elements of the matrix $\mathbf{K}$ are

$$k_{i,j}=k\left(\mathbf{x}_i,\mathbf{x}_j\right), \tag{2-57}$$

$\mathbf{a}$ is the vector containing the a_i coefficients, and $\mathbf{t}$ is the vector of target outputs. (This is the case where the number of data points is equal to the number of basis functions.) An efficient solution of Equation (2-56) can be obtained with the Cholesky decomposition.[99]

When using the RKHS method, the key step is the choice of the Mercer kernel. One of the most common choices is the Gaussian kernel:

$$k(\mathbf{x},\mathbf{y})=\exp\left(\frac{\|\mathbf{x}-\mathbf{y}\|^2}{\sigma^2}\right), \tag{2-58}$$

which leads to the radial basis network. However, Hollebeek et al.[99] have found that the Reciprocal Power (RP) and Taylor Spline (TS) reproducing kernels are well-suited for potential energy surface representations.

As seen in Table 2-4, the interpolation accuracy of the RKHS method appears to be comparable to the MSI and IP methods, but the lack of systems containing four or more atoms makes more quantitative comparisons difficult.

Table 2-4 Examples of fitting accuracy of RKHS methods

System	N(a)	RMSE (kJ mol^{-1})	Gradient Error (kJ mol^{-1} Å^{-1})	Reference
$N(^2D) + H_2$	1512	1 - 2[b]	(c)	27
$C(^1D) + H_2$[d]	1748	≈ 2.1	(c)	54
$C(^1D) + H_2$[e]	1748	≈ 1.0	(c)	54
$S(^1D) + H_2$	1643	(c)	(c)	100
$O(^1D) + H_2$	1280	1.26	(c)	26

(a) N is the number of symmetry-distinct *ab initio* points computed.

(b) Not reported but estimated to be on the same order as the rms error in Ref. 26.

(c) Not reported.

(d) RKHS method with a HDMR expansion. Total energy range used.

(e) RKHS method with a HDMR expansion for energies below 4184 kJ mol^{-1}

2.5. HYBRID METHODS

2.5.1. Application to H_3 System

Some investigators have examined the effect of combining the MSI and IMLS methods. Ishida and Schatz have tested such a hybrid method on the $H + H_2$ system[33a] and also on the very exothermic $O + H_2$ reaction.[33b] In this combined MSI/IMLS method, an IMLS-type least-squares fit is obtained at each point in the database. Such fitting must be done only once as opposed to every integration point during the computed trajectories, as is the case if only an IMLS method is used. Once these fits are obtained, the first and second derivatives of the potential with respect to the fitting coordinates are computed directly from them at each point in the database. These derivatives are then employed to execute Shepard interpolations[24] during the computation of trajectories.

The advantage of such a hybrid procedure is that it avoids the necessity of using electronic structure methods to compute the first and second derivatives of the potential at the sampling points, which is required if the MSI method alone is employed. The interpolation is effected using only the energies at these points. In addition, since the actual interpolation during the computation of trajectories is done using the Shepard procedure, the MSI/IMLS method also obviates the need to refit the potential at every integration point, which is the most severe computational bottleneck present in the IMLS method.

The study of the H_3 system[33a] employed the analytical PES of Liu-Siegbahn-Truhlar-Horowitz (LSTH)[101-103] as the "exact" surface to be fitted by MSI and MSI/IMLS procedures. The sampling of data points was executed by random sampling.[104] In this procedure, random configuration points are sampled in the neighborhood of the MEP. If the energy of a selected point exceeds some threshold energy, E_{th}, it is rejected. For the H_3 system, a threshold 1.0 eV higher than the $H + H_2$ asymptote was used since the collision energy considered in the trajectory calculations was 1.0 eV. To confine the sampling to regions in the near vicinity of the MEP, the following range was employed: $0.423 \text{ Å} < R(HH) < 4.232 \text{ Å}$ (8 bohr). Finally, the sampling procedure rejects points that are too near previously sampled points. To incorporate permutation symmetry, all symmetry-equivalent points were added to the data base in the manner previously described in Sections 2.2 to 2.4.

Fifty symmetry-distinct points were used to initiate the sampling. These were chosen along or near the IRC. In addition, the saddle point was included and eight more points in the neighborhood of the saddle. Ten more points with $R(^1H–^2H) = 2.5, 3.5, 4.5, 5.5,$ and $6.5 \text{ Å}, R(^2H–^3H) = 0.74 \text{ Å}$ and the three-body angle either $0°$ or $180°$ were included in the database. Finally, 30 additional points were chosen in the neighborhood of these 10 points and a final point was chosen by random sampling. This database was then updated by the addition of more points in increments of 100 that were obtained using random sampling methods.

The quality of the interpolations were assessed at 10,000 randomly sampled points for which $R(^1H-^2H)$ and $R(^2H-^3H)$ were both less than 6 au with the three-body angle being chosen randomly. Points with energy outside the range from 1 to 20 kcal mol^{-1} were excluded. Quantitative measures of the interpolation accuracies for energy and associated gradient were obtained by computation of RMSE(E) and RMSE(g), the root mean square deviations of the energy and the gradients, respectively, from those predicted by the LSTH surface for the 10,000 sampled points, respectively. Table 2-5 shows the results obtained when $p = 3$ was used in the weight function given by Equation (2-31). Ishida and Schatz[33a] report that within the range of parameter values considered, this choice gave the best surface.

Bettens and Collins[93] have reported that MSI interpolation can be made significantly more accurate by using a Bayesian approach for the weight function in Equations (2-4) to (2-6). The best result was obtained with a weight function that incorporated anisotropic effects. This unnormalized weight function is

$$v_i(\mathbf{Z}) = \left[\sum_{k=1}^{n(n-1)/2} a^2 + \left(\frac{\mathbf{Z}_k - \mathbf{Z}_k(i)}{d_k(i)} \right)^2 \right]^{-p} \tag{2-59}$$

with

$$d_n(i) = M^{-1} \sum_{k=1}^{M} \frac{\left\{ \left[\frac{\partial V(\mathbf{Z}^{(k)})}{\partial \mathbf{Z}_n} - \frac{\partial Ti(\mathbf{Z}^{(k)})}{\partial \mathbf{Z}_n} \right] [\mathbf{Z}_n(k) - \mathbf{Z}_n(i)] \right\}^2}{E_{tot}^2 \left\| \mathbf{Z}^{(k)} - \mathbf{Z}(i) \right\|^2}. \tag{2-60}$$

The results obtained when Equations (2-59) and (2-60) are employed for the weight functions in both the MSI/IMLS and pure MSI calculations are also given in Table 2-5.

The results given in Table 2-5 show that without Bayesian weighting of the points, the hybrid MSI/IMLS method exhibits significantly better interpolation accuracy for the H_3 system. When a Bayesian weighting is employed in both methods, the interpolation accuracy for the pure MSI method is slightly better than that for the hybrid procedure, although both results are very good.

2.5.2. Application to the $O(^1D) + H_2$ System

Ishida and Schatz[33b] have also investigated the performance of the hybrid MSI/IMLS method relative to that obtained with the pure MSI method with and without Bayesian weighting of the points for the reaction of $O(^1D)$ with H_2 to form $OH + H$. The analytic potential developed by Ho et al.[26] is used to define the "exact" potential surface so that the differences in the interpolation methods could be accurately assessed.

The $O(^1D) + H_2 \rightarrow H + OH$ reaction is highly exothermic so that the gradients of the potential and their rate of change with configuration is much larger than is the case for the thermochemically neutral $H + H_2$ exchange reaction. As such, this system poses a much more demanding test of the interpolation accuracy achieved by the fitting methods.

Table 2-5 Examples of fitting accuracy of the MSI/IMLS and pure MSI methods for the H_3 system for $p = 3$ in the weight function given in Eq. (2-31). Data are taken from References 33a and 33b

Method	N^a_{uniq}	N^b_T	RMSE(E) (kJ mol⁻¹)	RMSE(g) (kJ mol⁻¹ Å⁻¹)
IMLS/MSI	50	291	8.19	5.19
IMLS/MSI	150	891	1.18	1.44
IMLS/MSI	250	1491	0.81	1.08
IMLS/MSI	350	2091	0.73	0.92
MSI	250	1491	1.61	(c)
IMLS/MSI[d]	350	2091	0.69	(c)
MSI[d]	350	2091	0.48	(c)

(a) Number of symmetry-distinct *ab initio* points computed.

(b) Total number of points in database after incorporation of permutation symmetry.

(c) Not reported.

(d) Eqs. (2-44) and (2-45) used for the weight functions in both MSI/IMLS and pure MSI calculations with $p = 6$ and $a = 0.03$.

The overall procedure is analogous to that described in Section 2.5.1 for the H_3 system. Configuration space sampling was again executed using random sampling.[104] For this reaction, the threshold energy was taken to be either E_c, $1.2\,E_c$, or $1.5\,E_c$, where E_c is the collision energy used in the trajectory calculations that was 5.0, 6.0, and 7.5 kcal mol⁻¹ in separate calculations. The system was confined to regions near the IRC by taking 1.0 au $< R(OH) <$ 7.0 au and 0.80 au $< R(HH) <$ 8.0 au. The sampling procedure rejected points whose distance to the nearest point already in the database was less than 0.01 of the average distance. Permutation symmetry was incorporated into the PES by adding all symmetry-equivalent points to the database.

The initial database contained either 24 or 116 unique points. The set of 24 comprised 20 points in the asymptotic $O + H_2$ and $OH + H$ regions and 4 points in the interaction region. The 116-point set had 20 points in the asymptotic regions and a grid of points in the strong interaction region. The details of this grid have been given by Ishida and Schatz.[33b] Interpolation accuracy was evaluated by computation of RMSE(E) and RMSE(g) relative to the "exact" analytic surface using samples of ≈1000 points.

In general, for this system, the hybrid MSI/IMLS method performs significantly better than pure MSI even when Bayesian weighting (Equations (2-59) and (2-60)) is employed. The fact that the RMSE(E) obtained using 10,116 unique points is still 2.9 kJ/mol⁻¹ with the hybrid method demonstrates how much more difficult this system is to fit than H_3.

In contrast, the RKHS method produces an RMSE(E) interpolation error of 1.26 kJ mol⁻¹ with a database of 1,280 unique symmetry points.[26]

Table 2-6 Examples of fitting accuracy of the MSI/IMLS and pure MSI methods for the $O(1D) + H_2$ system for $p = 3$ in the weight function given in Eq. (2-31). Data are taken from Reference 33b

Method	N^a_{uniq}	RMSE(E) (kJ mol^{-1})	RMSE(g) (kJ mol^{-1} Å^{-1})
MSI/IMLS	150	≈ 26	≈ 0.7
MSI/IMLS	416	16.7	≈ 0.5
MSI	150	≈ 132	≈ 4
MSI	325	≈ 50	≈ 2
MSI/IMLS[c]	416	≈ 15	(b)
MSI[c]	416	≈ 24	(b)
MSI/IMLS	10,116	2.9[d]	(b)

(a) Number of symmetry-distinct *ab initio* points computed.

(b) Not reported.

(c) Eqs. (2-44) and (2-45) used for the weight functions in both MSI/IMLS and pure MSI calculations with $p = 6$ and $a = 0.03$.

(d) Median absolute error.

2.6. NEURAL NETWORKS APPLICATIONS TO REACTION DYNAMICS

In addition to the methods described in the chapter, NNs can play an important role in the development of PESs. Neural Networks have the capacity to sense the underlying correlations that exist between an output and the various variables that lead to that output. This ability to make connections between input and output variables and thereby perform pattern recognition is the source of a network's power. It also gives rise to the name "neural network" in that the ability to make connections and recognize patterns is the principal mode of operation of the human brain. The brain accomplishes its task by an awesomely complex interconnection of millions of nerve fibers called "neurons." The interconnections between the nodes of an NN bear a resemblance to the interconnections of the brain. Hence, in an NN, these interconnections are usually termed "neurons."

In chemical reaction dynamics, the most obvious application of an NN is to sense the underlying correlations between the structural variables of a molecular system and the corresponding potential-energy of the system. That is, an NN can be employed to "fit" a database of configuration energies obtained from *ab initio* electronic structure calculations and thereby provide an analytic representation of the PES that can then be employed in a wide range of molecular dynamics (MD) and Monte Carlo (MC) applications. After a thorough introduction to NNs in Chapter 3, the methods and procedures employed to execute NN fitting of potential surfaces are described in detail in Chapters 4 and 5.

While the most frequent application of NNs in reaction dynamics is fitting an *ab initio* database to obtain an analytic representation of the PES, the real power of NNs extends far beyond this type of application. For example, NNs are highly useful in expanding the power of genetic algorithms (GA) and many-body and high-dimensional model representations (HDMR) of potential surfaces. They can be used effectively to reduce the computational effort required to accurately investigate intramolecular vibrational energy transfer processes (IVR). NN methods are powerful tools that permit the parameters of an empirical PES to be determined from either *ab initio* databases or from experimental IR and Raman spectra. Such methods have been effectively employed to predict the results of MD trajectories without the necessity to actually integrate the classical equations of motion. This ability provides an efficient means of reducing the statistical error that is always present in MD calculations. Most recently, NNs have been employed to significantly reduce the computational effort required to obtain high-level electronic structure energies and to solve the Schrödinger equation to obtain the vibrational energy levels of small molecules. Each of these applications is described in some detail in Chapters 6 through 10 of this monograph.

3

FEEDFORWARD NEURAL NETWORKS

3.1. INTRODUCTION

In this section, we want to give a brief introduction to neural networks (NNs). It is written for readers who are not familiar with neural networks but are curious about how they can be applied to practical problems in chemical reaction dynamics. The field of neural networks covers a very broad area. It is not possible to discuss all types of neural networks. Instead, we will concentrate on the most common neural network architecture, namely, the multilayer perceptron (MLP). We will describe the basics of this architecture, discuss its capabilities, and show how it has been used on several different chemical reaction dynamics problems (for introductions to other types of networks, the reader is referred to References 105-107).

For the purposes of this document, we will look at neural networks as function approximators. As shown in Figure 3-1, we have some unknown function that we wish to approximate. We want to adjust the parameters of the network so that it will produce the same response as the unknown function, if the same input is applied to both systems.

For our applications, the unknown function may correspond to the relationship between the atomic structure variables and the resulting potential energy and forces.

3.2. NEURON MODEL

The multilayer perceptron neural network is built up of simple components. We will begin with a single-input neuron, which we will then extend to multiple inputs. We will next stack these neurons together to produce layers. Finally, we will cascade the layers together to form the network.

A single-input neuron is shown in Figure 3-2. The scalar input p is multiplied by the scalar weight w to form wp, one of the terms that is sent to the summer. The other input, 1, is multiplied by a bias b and then passed to the summer. The summer output

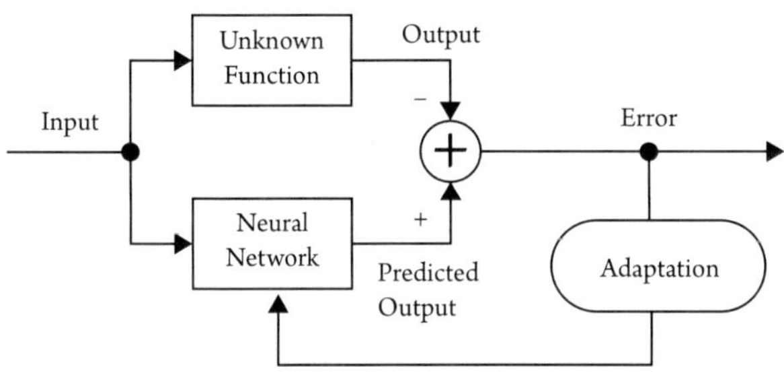

Figure 3-1 Neural network as function approximator.[105]

n, often referred to as the net input, goes into a transfer function f, which produces the scalar neuron output a.

The neuron output is calculated as

$$a = f(wp + b) \tag{3-1}$$

Note that w and b are both adjustable scalar parameters of the neuron. Typically, the transfer function is chosen by the designer, and then the parameters w and b are adjusted by some learning rule so that the neuron input/output relationship meets some specific goal.

The transfer function in Figure 3-2 may be a linear or a nonlinear function of n. One of the most commonly used functions is the *log-sigmoid transfer function*, which is shown in Figure 3-3.

This transfer function takes the input (which may have any value between plus and minus infinity) and squashes the output into the range 0 to 1, according to the expression

$$a = \frac{1}{1 + e^{-n}} \tag{3-2}$$

The log-sigmoid transfer function is commonly used in MLP networks that are trained using the backpropagation algorithm, in part because this function is differentiable.

Typically, a neuron has more than one input. A neuron with R inputs is shown in Figure 3-4. The individual inputs $p_1, p_2, ..., p_R$ are each weighted by corresponding elements $w_{1,1}, w_{1,2}, ..., w_{1,R}$ of the *weight matrix* **W**.

The neuron has a bias b, which is summed with the weighted inputs to form the net input n:

$$n = w_{1,1} p_1 + w_{1,2} p_2 + \cdots + w_{1,R} p_R + b \tag{3-3}$$

This expression can be written in matrix form:

$$n = \mathbf{W}p + b, \tag{3-4}$$

where the matrix **W** for the single neuron case has only one row.

Now the neuron output can be written as

$$a = f\left(\mathbf{W}p + b\right) \tag{3-4}$$

Figure 3-5 represents the neuron in matrix form.

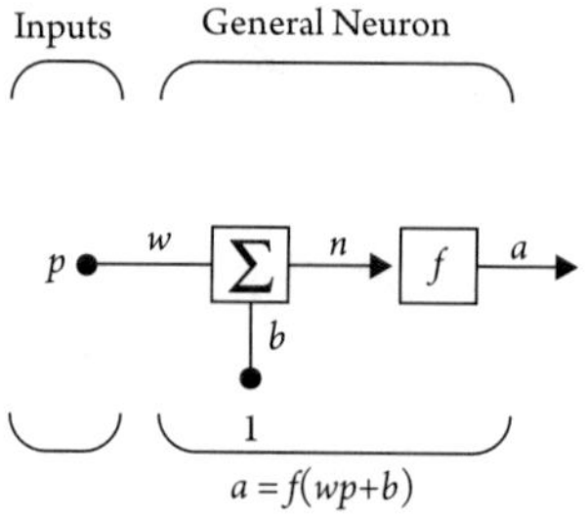

Figure 3-2 Single-input neuron.[105]

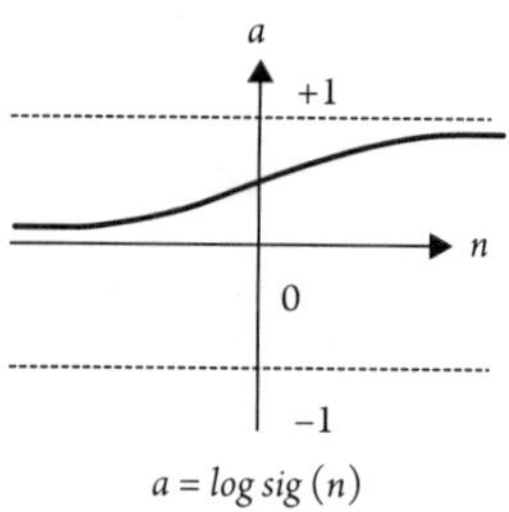

Figure 3-3 Log-sigmoid transfer function.[105]

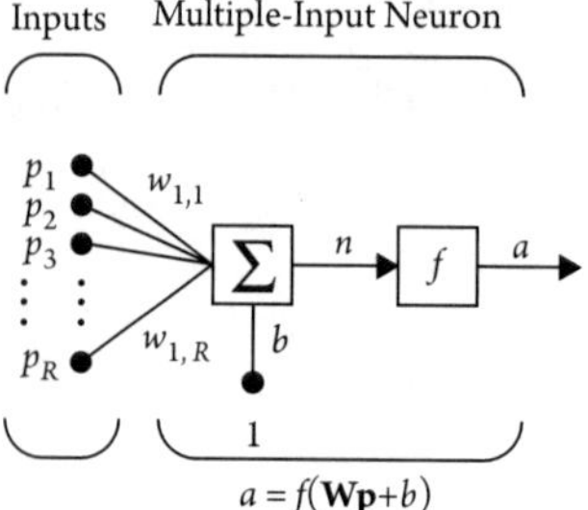

Figure 3-4 Multiple-input neuron.[105]

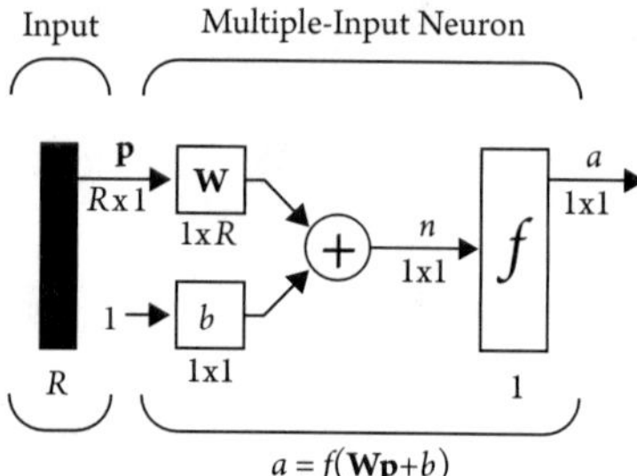

Figure 3-5 Neuron with R inputs, matrix notation.[105]

3.3. NETWORK ARCHITECTURES

Commonly one neuron, even with many inputs, is not sufficient. We might need 5 or 10, operating in parallel, in what is called a layer. A single-layer network of S neurons is shown in Figure 3-6. Note that each of the R inputs is connected to each of the neurons and that the weight matrix now has S rows. The layer includes the weight matrix $\mathbf{W}$, the summers, the bias vector $\mathbf{b}$, the transfer functions, and the output vector $\mathbf{a}$. Some authors refer to the inputs as another layer, but we will not do that here. It is common for the number of inputs to a layer to be different from the number of neurons (i.e., $S \neq R$).

The S-neuron, R-input, one-layer network also can be drawn in matrix notation, as shown in Figure 3-7.

Now consider a network with several layers. Each layer has its own weight matrix $\mathbf{W}$, its own bias vector $\mathbf{b}$, a net input vector $\mathbf{n}$, and an output vector $\mathbf{a}$. We need to introduce some additional notation to distinguish between these layers. We will use superscripts to identify the layers. Thus, the weight matrix for the first layer is written as $\mathbf{W}^1$, and the weight matrix for the second layer is written as $\mathbf{W}^2$. This notation is used in the three-layer network shown in Figure 3-8. As shown, there are R inputs, S^1 neurons in the first layer, S^2 neurons in the second layer, etc. As noted, different layers can have different numbers of neurons.

The outputs of layers 1 and 2 are the inputs for layers 2 and 3, respectively. Thus layer 2 can be viewed as a one-layer network with $R = S^1$ inputs, S^2 neurons, and an

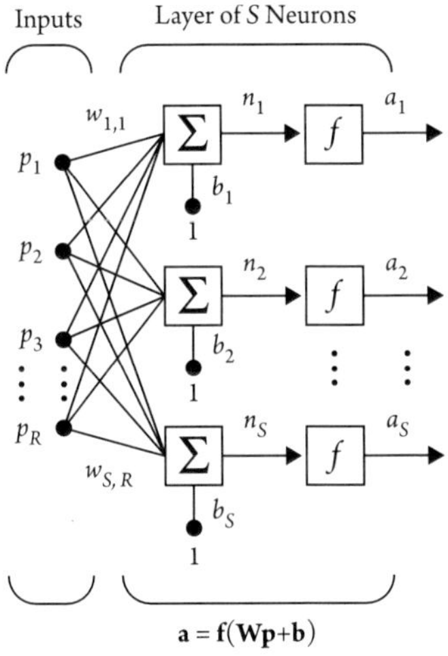

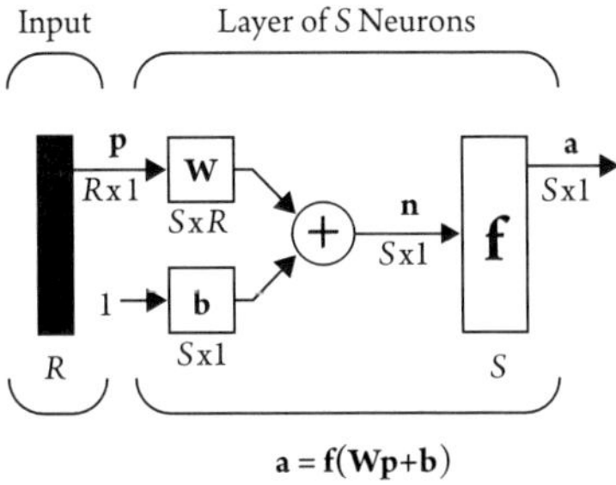

$$\mathbf{a} = \mathbf{f}(\mathbf{Wp}+\mathbf{b})$$

Figure 3-7 Layer of S neurons, matrix notation.[105]

$$\mathbf{a} = \mathbf{f}(\mathbf{Wp}+\mathbf{b})$$

Figure 3-6 Layer of S neurons.[105]

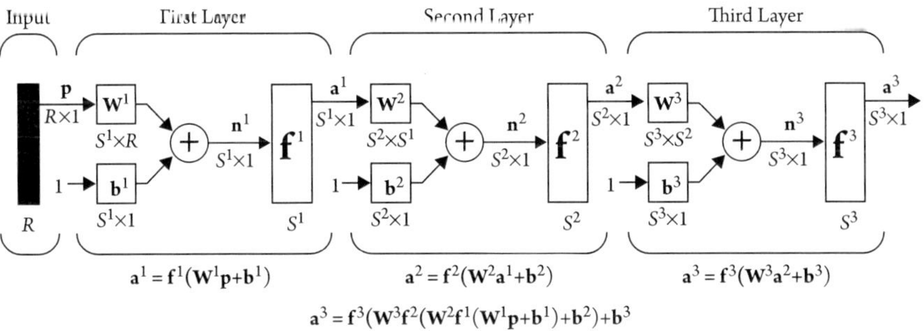

$$a^1 = f^1(W^1p+b^1) \qquad a^2 = f^2(W^2a^1+b^2) \qquad a^3 = f^3(W^3a^2+b^3)$$

$$a^3 = f^3(W^3f^2(W^2f^1(W^1p+b^1)+b^2)+b^3)$$

Figure 3-8 Three-layer MLP network.[105]

$S^2\mathrm{x}S^1$ weight matrix $\mathbf{W}^2$. The input to layer 2 is $\mathbf{a}^1$, and the output is $\mathbf{a}^2$. A layer whose output is the network output is called an output layer. The other layers are called hidden layers. The network shown in Figure 3-8 has an output layer (layer 3) and two hidden layers (layers 1 and 2). Such an NN architecture is commonly denoted by the notation $(R - S^1 - S^2 - S^3)$.

3.4. APPROXIMATION CAPABILITIES OF MULTILAYER NETWORKS

Two-layer networks, with sigmoid transfer functions in the hidden layer and linear transfer functions in the output layer, are universal approximators.[108] A simple example can demonstrate the power of this network for approximation.

Consider the two-layer, (1-2-1) network shown in Figure 3-9. For this example the transfer function for the first layer is log-sigmoid and the transfer function for the second layer is linear. In other words,

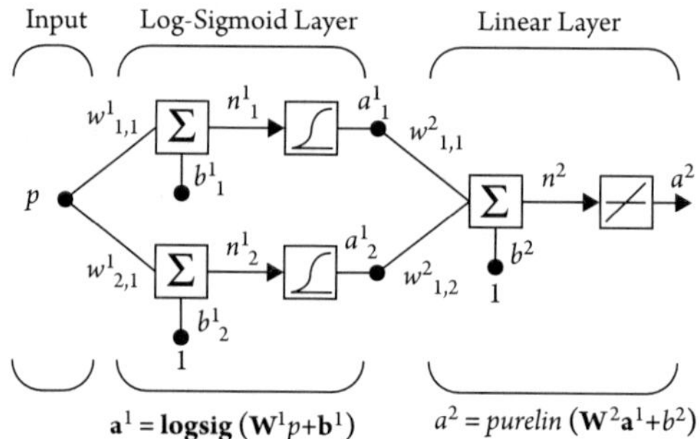

Figure 3-9 Example function approximation network.[105]

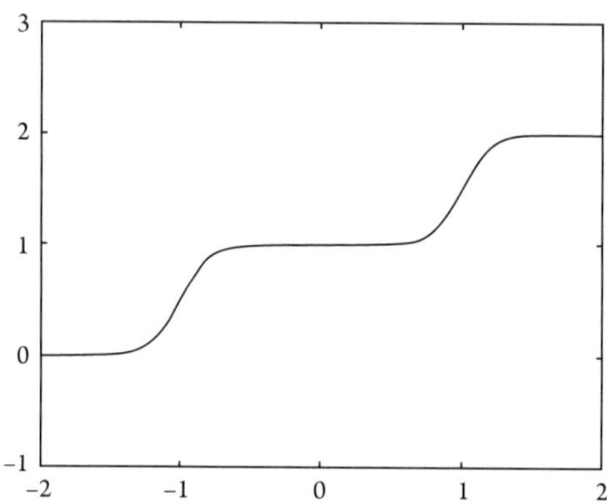

Figure 3-10 Nominal response of network of figure 3-9 (a^2 vs. p).[105]

$$f^1(n) = \frac{1}{1+e^{-n}} \quad \text{and} \quad f^2(n) = n \tag{3-5}$$

Suppose that the nominal values of the weights and biases for this network are

$$w^1_{1,1} = 10, w^1_{2,1} = 10, w^2_{1,1} = 10, w^2_{1,2} = 1, b^1_1 = 10, b^2 = 0.$$

The network response for these parameters is shown in Figure 3-10, which plots the network output a^2 as the input p is varied over the range (–2 to +2). Notice that the response consists of two steps, one for each of the log-sigmoid neurons in the first layer.

By adjusting the network parameters we can change the shape and location of each step, as we will see in the following discussion. The centers of the steps occur where the net input to a neuron in the first layer is zero:

$$n^1_1 = w^1_{1,1}p + b^1_1 = 0 \Rightarrow p = \frac{b^1_1}{w^1_{1,1}} = \frac{-10}{10} = 1 \tag{3-6}$$

$$n^1_1 = w^1_{1,1}p + b^1_1 = 0 \Rightarrow p = \frac{b^1_1}{w^1_{1,1}} = -\frac{10}{10} = -1 \tag{3-7}$$

The steepness of each step can be adjusted by changing the network weights.

Figure 3-11 illustrates the effects of parameter changes on the network response. The nominal response is repeated from Figure 3-10. The other curves correspond to the network response when one parameter at a time is varied over the following ranges:

$$-1 \le w^2_{1,1} \le 1, -1 \le w^2_{1,2} \le 1, 0 \le b^1_2 \le 20, -1 \le b^2 \le 1 \tag{3-8}$$

Figure 3-11(a) shows how the network biases in the first (hidden) layer can be used to locate the position of the steps. Figure 3-11(b) and Figure 3-11(c) illustrate how the weights determine the slope of the steps. The bias in the second (output) layer shifts the entire network response up or down, as can be seen in Figure 3-11(d).

From this example, we can see that the multilayer network is very flexible. It would appear that we could use such networks to approximate almost any function, if we had a sufficient number of neurons in the hidden layer. In fact, it has been shown that two-layer networks, with sigmoid transfer functions in the hidden layer and linear transfer functions in the output layer, can approximate virtually any function of interest to any degree of accuracy, provided sufficiently many hidden units are available. It is beyond the scope of this document to provide detailed discussions of approximation theory, but there are many papers in the literature that can provide a deeper discussion of this field. In Reference 108, Hornik, Stinchcombe, and White present a proof, using the Stone-Weierstrass Theorem,[109] that multilayer perceptron networks are universal approximators. Pinkus gives a more recent review of the approximation capabilities of neural networks in Reference 110. Niyogi and Girosi, in Reference 111, develop bounds on function approximation error when the network is trained on noisy data.

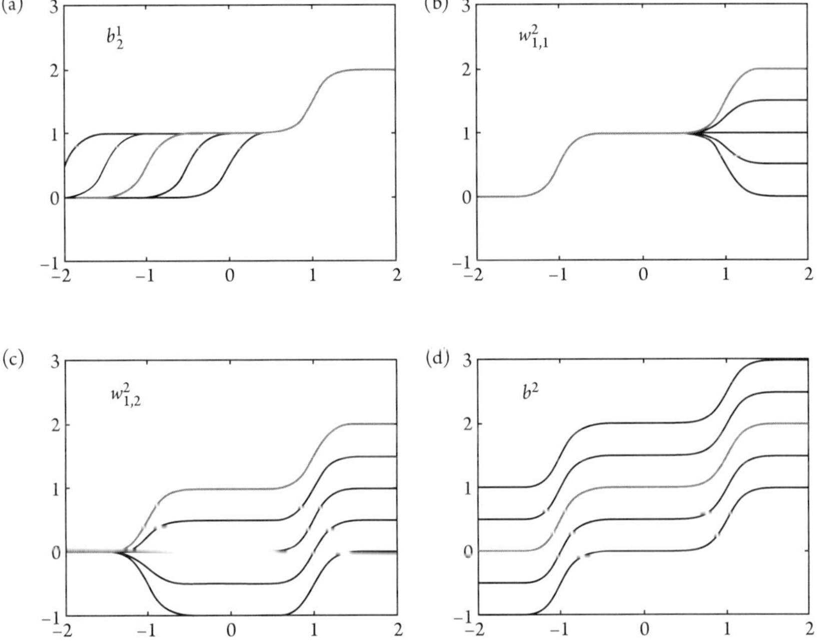

Figure 3-11 Effect of parameter changes on network response (a^2 vs. p).[105]

3.5. TRAINING MULTILAYER NETWORKS

Now that we know multilayer networks are universal approximators, the next step is to determine a procedure for selecting the network parameters (weights and biases) that will best approximate a given function. The procedure for selecting the parameters for a given problem is called training the network. In this section we will outline a training procedure called backpropagation,[112,113] which is based on gradient descent (more efficient algorithms than gradient descent are often used in neural network training).[105]

As we discussed earlier, for multilayer networks the output of one layer becomes the input to the following layer (see Figure 3-8). The equations that describe this operation are

$$\mathbf{a}^{m+1} = f\left(\mathbf{W}^{m+1}\mathbf{a}^m + \mathbf{b}^{m+1}\right) \text{ for } m = 0,1,...,M. \tag{3-9}$$

where M is the number of layers in the network. The neurons in the first layer receive external inputs:

$$\mathbf{a}^0 = \mathbf{p}, \tag{3-10}$$

which provides the starting point for Equation (3-9). The outputs of the neurons in the last layer are considered the network outputs:

$$\mathbf{a} = \mathbf{a}^M. \tag{3-11}$$

The backpropagation algorithm for multilayer networks is a gradient descent optimization procedure in which we minimize a mean square error performance index. The algorithm is provided with a set of examples of proper network behavior:

$$\{\mathbf{p}_1,\mathbf{t}_1\},\{\mathbf{p}_2,\mathbf{t}_2\},...,\{\mathbf{p}_Q,\mathbf{t}_Q\}, \tag{3-12}$$

where $\mathbf{p}_q$ is an input to the network, and $\mathbf{t}_q$ is the corresponding target output. As each input is applied to the network, the network output is compared to the target. The algorithm should adjust the network parameters in order to minimize the mean squared error:

$$F(\mathbf{w}) = \frac{1}{Q}\sum_{q=1}^{Q}\left(e_q\right)^2 = \frac{1}{Q}\sum_{q=1}^{Q}\left(t_q - a_q\right)^2. \tag{3-13}$$

where $\mathbf{w}$ is a vector containing all network weights and biases. If the network has multiple outputs, this generalizes to

$$F(\mathbf{w}) = \frac{1}{Q}\sum_{q=1}^{Q}\mathbf{e}_q^T\mathbf{e}_q = \frac{1}{Q}\sum_{q=1}^{Q}\left(\mathbf{t}_q - \mathbf{a}_q\right)^T\left(\mathbf{t}_q - \mathbf{a}_q\right). \tag{3-14}$$

Using a stochastic approximation, we will replace the mean squared error by the error on the latest target:

$$\hat{F}(\mathbf{w}) = \mathbf{e}^{T}(k)\mathbf{e}(k) = \left(\mathbf{t}(k) - \mathbf{a}(k)\right)^{T}\left(\mathbf{t}(k) - \mathbf{a}(k)\right), \tag{3-15}$$

where the average of the squared error has been replaced by the squared error at iteration k.

The steepest descent algorithm for the approximate mean square error is

$$w_{i,j}^{m}(k+1) = w_{i,j}^{m}(k) - \alpha \frac{\partial \hat{F}}{\partial w_{i,j}^{m}}, \tag{3-16}$$

$$b_{i}^{m}(k+1) = b_{i}^{m}(k) - \alpha \frac{\partial \hat{F}}{\partial b_{i}^{m}}, \tag{3-17}$$

where α is the learning rate.

For a single-layer linear network, these partial derivatives in Equation (3-16) and Equation (3-17) are conveniently computed, since the error can be written as an explicit linear function of the network weights. For the multilayer network, the error is not an explicit function of the weights in the hidden layers; therefore, these derivatives are not computed so easily.

Because the error is an indirect function of the weights in the hidden layers, we will use the chain rule of calculus to calculate the derivatives in Equation (3-16) and Equation (3-17):

$$\frac{\partial \hat{F}}{\partial w_{i,j}^{m}} = \frac{\partial \hat{F}}{\partial n_{i}^{m}} \times \frac{\partial n_{i}^{m}}{\partial w_{i,j}^{m}}, \tag{3-18}$$

$$\frac{\partial \hat{F}}{\partial b_{i}^{m}} = \frac{\partial \hat{F}}{\partial n_{i}^{m}} \times \frac{\partial n_{i}^{m}}{\partial b_{i}^{m}}, \tag{3-19}$$

The second term in each of these equations can be easily computed, since the net input to layer m is an explicit function of the weights and bias in that layer:

$$n_{i}^{m} = \sum_{j=1}^{S^{m-1}} w_{i,j}^{m} a_{j}^{m-1} + b_{i}^{m} \tag{3-20}$$

Therefore

$$\frac{\partial n_{i}^{m}}{\partial w_{i,j}^{m}} = a_{j}^{m-1}, \frac{\partial n_{i}^{m}}{\partial b_{i}^{m}} = 1. \tag{3-21}$$

If we now define

$$s_{i}^{m} \equiv \frac{\partial \check{F}}{\partial n_{i}^{m}}, \tag{3-22}$$

(the sensitivity of $\hat{F}$ to changes in the ith element of the net input at layer), then Equation (3-18) and Equation (3-19) can be simplified to

$$\frac{\partial \hat{F}}{\partial w_{i,j}^m} = s_i^m a_j^{m-1},$$
(3-23)

$$\frac{\partial \hat{F}}{\partial b_i^m} = s_i^m.$$
(3-24)

We can now express the approximate steepest descent algorithm as

$$w_{i,j}^m(k+1) = w_{i,j}^m(k) - \alpha s_i^m a_j^{m-1},$$
(3-25)

$$b_i^m(k+1) = b_i^m(k) - \alpha s_i^m.$$
(3-26)

In matrix form, this becomes:

$$\mathbf{W}^m(k+1) = \mathbf{W}^m(k) - \alpha \mathbf{s}^m \left(\mathbf{a}^{m-1}\right)^T,$$
(3-27)

$$\mathbf{b}^m(k+1) = \mathbf{b}^m(k) - \alpha \mathbf{s}^m,$$
(3-28)

where the individual elements of $\mathbf{s}^m$ are given by Equation (3-22).

It now remains for us to compute the sensitivities $\mathbf{s}^m$, which requires another application of the chain rule. It is this process that gives us the term "backpropagation," because it describes a recurrence relationship in which the sensitivity at layer m is computed from the sensitivity at layer $m+1$:

$$\mathbf{s}^M = -2\dot{\mathbf{F}}^M\left(\mathbf{n}^M\right)(\mathbf{t}-\mathbf{a})^T,$$
(3-29)

$$\mathbf{s}^m = \dot{\mathbf{F}}^m\left(\mathbf{n}^m\right)\left(\mathbf{W}^{m+1}\right)^T \mathbf{s}^{m+1}, \quad m = M-1, M-2, ..., 1$$
(3-30)

where

$$\dot{\mathbf{F}}^m\left(\mathbf{n}^m\right) = \begin{bmatrix} \dot{f}^m\left(n_1^m\right) & 0 & \cdots & 0 \\ 0 & \dot{f}^m\left(n_2^m\right) & \cdots & 0 \\ \vdots & \vdots & \ddots & \vdots \\ 0 & 0 & \cdots & \dot{f}^m\left(n_{S^m}^m\right) \end{bmatrix}.$$
(3-31)

(See Reference 105, Chapter 11, for a derivation of this result.)

In some ways it is unfortunate that the algorithm that is often referred to as backpropagation, given by Equation (3-27) and Equation (3-28), is in fact simply a steepest descent algorithm. There are many other optimization algorithms that can use the backpropagation procedure, in which derivatives are processed from the last layer of the network to the first (as given in Equation (3-30)). For example, conjugate gradient and quasi-Newton algorithms[114-116] are generally more efficient than steepest descent algorithms, and yet they can use the same back-propagation procedure to compute the necessary derivatives. The Levenberg-Marquardt (LM) algorithm is very efficient for training small to medium-size networks, and it uses a backpropagation procedure that is very similar to the one given by Equation (3-30).[117]

3.6. GENERALIZATION (INTERPOLATION AND EXTRAPOLATION)

We now know that multilayer networks are universal approximators, and we have an algorithm with which to train the networks, but we have not discussed how to select the number of neurons and the number of layers necessary to achieve an accurate approximation in a given problem. We have also not discussed how the training data set should be selected. The trick is to use enough neurons to capture the complexity of the underlying function without having the network overfit the training data, in which case it will not generalize to new situations. We also need to have sufficient training data to adequately represent the underlying function.

To illustrate the problems we can have in network training, consider the following general example. Assume that the training data are generated by the following equation:

$$\mathbf{t}_q = \mathbf{g}\left(\mathbf{p}_q\right) + \mathbf{e}_q,\qquad(3\text{-}32)$$

where $\mathbf{p}_q$ is the system input, $\mathbf{g}(.)$ is the underlying function we wish to approximate, $\mathbf{e}_q$ is measurement noise, and $\mathbf{t}_q$ is the system output (network target). Figure 3-12 shows an example of the underlying function $\mathbf{g}(.)$ (thick line), training data target values $\mathbf{t}_q$ (circles), and total trained network response (thin line). The two graphs of Figure 3-12 represent different training strategies.

In the example shown in Figure 3-12(a), a large network was trained to minimize mean squared error (Equation (3-13)) over the 15 points in the training set. We can see that the network response exactly matches the target values for each training point. However, the total network response has failed to capture the underlying function. There are two major problems. First, the network has overfit on the training data. The network response is too complex, because the network has more than enough independent parameters,[118] and they have not been constrained in any way. The second problem is that there are no training data for values of p greater than 0. Neural networks (and other nonlinear black box techniques) cannot be expected to *extrapolate* accurately. If the network receives an input that is outside of the range covered in the training data, then the network response will always be suspect.

While there is little we can do to improve the network performance outside the range of the training data, we can improve its ability to *interpolate* between data

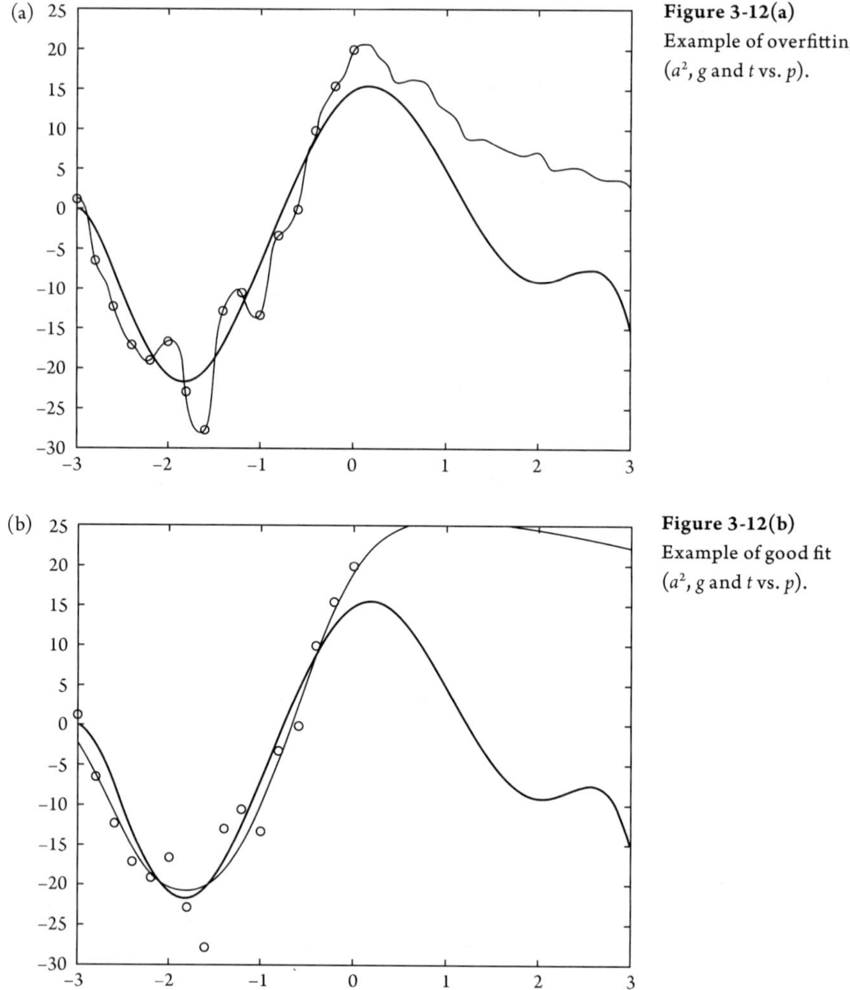

Figure 3-12(a)
Example of overfitting
(a^2, g and t vs. p).

Figure 3-12(b)
Example of good fit
(a^2, g and t vs. p).

points. Improved generalization can be obtained through a variety of techniques. In one method, called early stopping,[118] we place a portion of the training data into a validation data set. The performance of the network on the validation set is monitored during training. During the early stages of training the validation error will come down. When overfitting begins, the validation error will begin to increase, and at this point the training is stopped.

Another technique to improve network generalization is called regularization. With this method the performance index is modified to include a term that penalizes network complexity. The most common penalty term is the sum of squares of the network weights:

$$F(\mathbf{w}) = \frac{1}{Q}\sum_{q=1}^{Q}\mathbf{e}_q^T\mathbf{e}_q + \rho\sum\left(w_{i,j}^m\right)^2. \tag{3-33}$$

This performance index forces the weights to be small, which produces a smoother network response. The trick with this method is to choose the correct regularization parameter ρ. If the value is too large, then the network response will be too smooth and will not accurately approximate the underlying function. If the value is too small, then the network will overfit. There are a number of methods for selecting the optimal ρ. One of the most successful is Bayesian regularization.[119,120] Figure 3-12(b) shows the network response when the network is trained with Bayesian regularization. Notice that the network response no longer exactly matches the training data points, but the overall network response more closely matches the underlying function over the range of the training data.

A complete discussion of generalization and overfitting is beyond the scope of this monograph. The interested reader is referred to References 105, 107, 111, 119, and 120.

3.7. DATA PREPROCESSING

Before a network is trained, there is generally a data preprocessing stage that comes first. The main purpose of the data preprocessing stage is to facilitate network training. Data preprocessing consists of such steps as normalization, nonlinear transformations, feature extraction, coding of discrete inputs/targets, handling of missing data, etc. The idea is to perform preliminary processing of the data to make it easier for the neural network training to extract the relevant information.

For example, in multilayer networks, sigmoid transfer functions are often used in the hidden layers. These functions become essentially saturated when the magnitude of the net input is greater than three, i.e., $[\exp(-3) \sim 0.05]$. We don't want this to happen at the beginning of the training process, because the gradient will then be very small. In the first layer, the net input is a product of the input times the weight plus the bias. If the input is very large, then the weight must be small in order to prevent the transfer function from becoming saturated. It is standard practice to normalize the inputs before applying them to the network. In this way, initializing the network weights to small random values guarantees that the weight-input product will be small. Also, when the input values are normalized, the magnitudes of the weights have a consistent meaning. This is especially important when using regularization. Regularization requires the weight values to be small. However, "small" is a relative term; if the input values are very small, we need large weights to produce a significant net input. Normalizing the inputs clarifies the meaning of "small" weights.

There are two standard methods for normalization. The first method normalizes the data so that they fall into a standard range—typically -1 to $+1$. This can be done with

$$\mathbf{p}^{n} = 2\left(\mathbf{p} - \mathbf{p}^{min}\right)./\left(\mathbf{p}^{max} - \mathbf{p}^{min}\right) - 1 , \qquad (3.34)$$

where $\mathbf{p}^{min}$ is the vector containing the minimum values of each element of the input vectors in the data set, $\mathbf{p}^{max}$ contains the maximum values, ./ represents an element-by-element division of two vectors, and $\mathbf{p}^{n}$ is the resulting normalized input vector.

An alternative normalization procedure is to adjust the data so that they have a specified mean and variance—typically 0 and 1, respectively. This can be done with the transformation

$$\mathbf{p}^{n} = 2\left(\mathbf{p} - \mathbf{p}^{\text{mean}}\right)./\mathbf{p}^{\text{std}} , \tag{3-35}$$

where $\mathbf{p}^{\text{mean}}$ is the average of the input vectors in the data set, and $\mathbf{p}^{\text{std}}$ is the vector containing the standard deviations of each element of the input vectors.

Generally, the normalization step is applied to both the input vectors and the target vectors in the data set.

3.8. PRACTICAL ASPECTS OF NN TRAINING ISSUES

3.8.1. Database, Local Minima, Sampling Bias, Committees, and Derivatives

A very important consideration in network performance is to have enough training points appropriately spaced throughout the configuration space that is important in the process under study. Neural networks do not extrapolate accurately. Therefore, we must have adequate training data in all important regions of configuration space. Achieving this is the basis of the sampling problem, which will be discussed in Chapter 4.

It should be recognized that the random partitioning of the database into training, validation, and testing sets will automatically insert random variations in the fitting accuracy of the NN after training is completed. In addition, the training algorithms that adjust the weight and bias matrices of the NN so as to minimize $F(\mathbf{w})$ in Equation (3-14) may converge to a local minimum rather than the absolute minimum. To handle the local minimum problem, it is customary to train a set of n networks with different choices for the initial values of the weights and biases and then pick the network that provides the lowest fitting error for the testing set. The variations that result from random partitioning are generally handled by sampling m times and using the network that yields the median fitting error of the m networks. The sampling bias may then be estimated by computation of the standard deviation of the errors for all m networks from this median value. The median is usually chosen rather than the average to avoid skewing of the results by an outlier in the m networks.

Since the sampling bias is random, a very effective procedure is to employ all m networks as a "committee" rather than choosing the network with the median error. When a committee is employed, the outputs from all m networks are averaged to obtain the predicted value. Since the sampling bias is random, the output errors will likewise tend to be random. Consequently, the averaging of all m outputs will result in a cancellation of random errors so that it is usually found that the fitting error associated with the average is lower than any of the individual fitting errors of the m networks in the committee. The disadvantage of this procedure is that if m networks are employed as a committee, the computational time consumed in using the committee will be *m* times as great as would be the case if only the network producing the median error were used.

Since feedforward neural networks can be easily differentiated to yield the gradients of the function being fitted, the NN method provides us with a choice of three procedures to execute MD calculations. We can use an NN to fit the computed potentials and then obtain the force fields required by the MD calculations by differentiation of the network function. This is the procedure that has been employed in most applications of NNs to MD studies. Alternatively, we can use a separate neural network to fit the force fields computed by the *ab initio* calculations. This network can then be used in the MD calculations to produce the forces directly. A third option is train one network to fit the computed potentials and at the same time, have the derivatives of the network function with respect to its inputs fit the force fields. The performance index for training this network would be a combination of the squared error on the computed potentials and the squared error on the force fields. To implement a gradient-based algorithm on such a combined performance index, we would need to compute the derivative with respect to the weights of the network function derivative with respect to its inputs. This requires backpropagation operations that are similar to those required for Hessian calculations.[106] The CFDA method[121] recently developed and applied to MD investigations[122] achieves this. This method is described in Section 3.10 of this chapter. The accuracy of the results, which are discussed in Chapter 5, Section 5.3, are the best reported to date using any method.

The fact that NN procedures can be employed to fit the computed derivatives of the potential directly provides a significant advantage over MSI methods.[23-25] To interpolate the force fields directly using Taylor series expansions truncated after the quadratic terms would require the evaluation of mixed third derivatives of the potential with respect to the expansion variables. These derivatives are generally not available from electronic structure calculations.

3.8.2. Input Vector Optimization and Fitting Accuracy

The fitting accuracy of feedforward NNs can be easily increased by employing more hidden layers and additional neurons. This is another major advantage over MSI[23-25] and IMLS[28-33] methods. To increase the accuracy of the Taylor series expansions employed in Shepard interpolations, additional terms must be included in the expansions. These terms require the evaluation of mixed third, fourth, etc. derivatives of the potential function, which are generally not available from electronic structure calculations. To increase the accuracy of the IMLS procedure, the number of basis functions M must be increased and the computational requirements of the method increase with M^2.

In addition to increasing the number of hidden layers and neurons, increased fitting accuracy can often be achieved by appropriate selection of the functional form of the elements of the input vector **p**. The basic principle is to choose functions of the input data that are closely related to the functional dependence of the desired output elements upon the input data. Since NNs do not handle discontinuities well, it is also important that the input elements representing continuous variables exhibit no discontinuities. If proper attention is given to these points, the result is often a significant increase in the fitting accuracy of the NN with only a small increase in the computational effort involved in training and employing the NN.

Malshe et al.[123] have investigated the dependence of NN fitting accuracy upon the functional forms chosen for the input elements when fitting large databases to obtain the PES for molecular systems. The systems chosen for the study were the unimolecular reactions of H_2O_2, HONO, Si_5, and $H_2C = CHBr$. This choice provides a variety of atoms with a widely varying number of electrons in the electronic structure calculations which were carried out using three different levels of accuracy. There is a wide range of chemical bonding present that includes O-O, O-H, N-O, N = O, C-Br, H-Br, H-H, C = C, and three distinct C-H bonds. Moreover, the bonding in small silicon clusters is known to be highly unusual as illustrated by the Si_4 cluster, which is known to be a planar rhombus in its equilibrium configuration. These systems also undergo 10 distinct chemical reactions that include two *cis-trans* isomerizations, six different types of two-center bond ruptures, and two different three-center dissociation reactions.

The H_2O_2 database was obtained by Le et al.[124] It comprised 15,472 configurations whose energies were computed at the MP2 level of theory using a 6-31G* basis set. The energies of the database spanned a range of 6.1053 eV. The HONO and $H_2C = CHBr$ (vinyl bromide) databases were both computed at the UP4(SDQ) level of theory. For HONO, a 6-311(G(d)) basis set was employed to obtain the energies for 21,584 configurations whose energies spanned a range of 4.8816 eV.[125] The vinyl bromide database[126] contained the energies for 68,302 configurations each computed using a 6-331G(d,p) basis set for C and H and Huzinaga's (4333/433/4) basis augmented with split outer s and p orbitals (43321/4321/4) and a polarization f orbital with an exponent of 0.5 for the bromine atom. The energies of this database span a range of 8.0 eV. The Si_5 database[71] contains 10,202 configurations whose energies were obtained from density functional B3LYP calculations using a 6-31G** basis set. These energies span a range of 0.8138 eV.

The input vector required by the NN to fit an *ab initio* database for an N-atom system must contain a minimum of 3N-6 elements. Redundant input vectors, such as the set of interparticle distances, will have dimensions $(M \times 1)$, where $M > 3N-6$. In most cases, the elements of the NN input vector will be chosen to be some combination of interatomic distances between atoms i and j, R_{ij}, three-body angles, θ_{ijk}, and four-body, dihedral angles, ϕ_{ijkl}.

Since the electronic structure calculations to obtain the database are performed prior to NN fitting, the most convenient set of input vectors to specify the molecular configurations is that using the internal coordinates of the Z-matrix[141] that is generally employed in most such calculations. If this choice is made, the individual results of the electronic structure calculations can be transferred without change to the storage array for the database. In this case, each input vector will contain 3N-6 elements that are a combination of distances, three-body angles, and dihedral angles. For notational convenience, we denote this input element as (R_i, θ_j, ϕ_k).

For each of the four systems previously described, Malshe et al.[123] investigated several reasonable choices for the NN input vector. These choices involved various combinations of sine and cosine functions of the angles θ_i and ϕ_i and either the interatomic distances, R_{ij}, their reciprocals, R_{ij}^{-1}, or powers of the reciprocal distances, $R_{ij}^{-n(a,b)}$, where $n(a,b)$ is a parameter that depends on the chemical identity of the two atoms, i and j, forming the interatomic distance R_{ij}. In some cases, $n(a,b)$ was taken to be a constant independent of a and b. In other cases, $n(a,b)$ was taken to be an adjustable parameter that was independent of a and b. In the most general case, $n(a,b)$ was an adjustable parameter dependent upon both a and b.

The NNs employed in the investigation were all two-layer networks that used sigmoid and linear transfer functions in the hidden and output layers, respectively. In each case, the output layer contained a single neuron which provided the predicted potential energy for the system. In order to have a fair comparison of the fitting accuracy of the NN using different input vectors, Malshe et al.[123] chose the number of neurons in the hidden layer to be such that the total number of weight and bias parameters for all NNs used in a given system were either equal or nearly equal.

When the input vector contained no adjustable parameters, the NNs were trained using the usual Levenberg-Marquardt algorithm.[117,127] When some of the elements of the input vector have the form $R_{ij}^{-n(a,b)}$, the usual Jacobian employed by the Levenberg-Marquardt algorithm was modified to include the Jacobian of the computed errors with respect to the $n(a,b)$ parameters so that the procedure simultaneously fits the $n(a,b)$ parameters as well as the weights and biases of the NN.

For each NN with a given input vector, the database for the system was partitioned randomly between training, validation, and testing sets five times. For each partitioning, five NNs were trained using different estimates for the initial weights and biases to attempt to avoid being trapped in a local minima. The NN with the lowest root mean square error (RMSE) was taken to be the global minimum for that particular partitioning of the database. The median of the five lowest RMSE values obtained was taken as the measure of the expected accuracy of the NN for that particular input vector. This procedure, therefore, required the training of 25 NNs for each input vector.

The results for the four molecular systems investigated are shown in Tables 3-1 to 3-4.

For all systems investigated, Malshe et al.[123] found that the best fitting accuracy was obtained by employing input vectors whose elements were functions of the inverse interatomic distances. This is not unexpected since the magnitude of the interaction potentials usually decrease and approach some limiting asymptotic value as the interatomic distances increase. This is the behavior exhibited by inverse powers of R_{ij}, but not by the distances themselves.

When $n(a,b)$ was treated as an adjustable parameter independent of a and b, the optimum value for this parameter was found to lie in the range 2.00 ± 0.38 for each of the four systems studied. Moreover, Malshe et al.[123] found that the rate of change of the RMSE error with respect to $n(a,b)$ is very small for values of $n(a,b)$ near its optimum value. As a result, if a value of $n(a,b)$ at the center of this range, i.e.,

Table 3-1 NN fitting errors to the 10,202 *ab initio* energies for the Si_5 system as a function of the form chosen for the configuration input vector. $N(h)$ and N_T are the number of neurons in the NN hidden layer and the total number of weight and bias parameters, respectively. All data are taken from Ref. 123.

Set	No. Input Elements	$N(h)$	N_T	Input Specification	RMSE (eV)
1	9	45	496	$R_i,\ \theta_j,\ \phi_k$	0.00373
2	9	45	496	$R_i,\ \theta_j,\ \cos(\phi_k)$	0.00302
3	9	45	496	$R_i,\ \cos^2(\theta_j),\ \cos^2(\phi_k)$	0.00200
4	9	45	496	$R_i^{-2},\ \sin^2(\theta_j),\ \sin^2(\phi_k)$	0.00080
5	9	45	496	$R_i^{-2},\ \cos(\theta_j),\ \cos(\phi_k)$	0.000732
6	9	45	496	$R_i^{-1},\ \cos^2(\theta_j),\ \cos^2(\phi_k)$	0.000718
7	9	45	496	$R_i^{-1},\ \cos(\theta_j),\ \cos(\phi_k)$	0.000576
8	10	41	493	R_i	0.000421
9	10	41	493	R_i^{-1}	0.000319
10	10	41	493	R_i^{-2}	0.000165
11	10	41	493	$R_i^{-n}\ (n = 2.38)$	0.000149

$n(a,b) = 2.00$, is used without optimization, the increases in the RMSE for the four systems investigated were all in the range from 0.03% to 18.3% relative to the RMSE values obtained using the optimized value of $n(a,b)$. The results shown in Table 3-2 indicate that little additional fitting accuracy is achieved for vinyl bromide when $n(a,b)$ is optimized individually for each combination of a and b.

The RMSE values for NNs using input vectors that utilize the Z-matrix coordinates[141] employed in the electronic structure calculations were substantially larger than those for NNs in which the input elements have the form $R_{ij}^{-n(a,b)}$. Malshe et al.[123] suggested that the primary reason for this result was the discontinuities present in the dihedral angles. When more than four atoms are present, a dihedral angle of 175° is not structurally or energetically equivalent to a dihedral angle of −175°. Consequently, a dihedral input element will exhibit a discontinuity at +180° when the element changes from a positive to a negative value. NNs are universal approximators for functions in which both the output functions and the input variables are continuous but not for ones containing discontinuities.[108] This problem becomes particularly acute when the structure frequently moves through the region of the discontinuity, as is the case for vinyl bromide which is planar in its equilibrium configuration. Whenever the dihedral angles are replaced with cosine functions of these angles, the discontinuities are removed, and there is an immediate, substantial error reduction, particularly for vinyl bromide.

Table 3-2 Mean absolute NN fitting errors (MAE) to the 68,302 ab initio energies for vinyl bromide as a function of the form chosen for the configuration input vector. $N(h)$ and N_T are the number of neurons in the NN hidden layer and the total number of weight and bias parameters, respectively. All data are taken from Ref. 123.

Set	No. Input Elements	$N(h)$	N_T	Input Specification	MAE (eV)
1	12	170	2381	R_i, θ_j, ϕ_k	0.0984
2	12	170	2381	$R_i, \theta_j, \cos(\phi_k)$	0.0597
3	12	170	2381	$R_i^{-1}, \cos(\theta_j), \cos(\phi_k)$	0.0568
4	12	170	2381	$R_i^{-2}, \cos(\theta_j), \cos(\phi_k)$	0.0492
5	15	140	2381	R_i	0.0590
6	15	140	2381	R_i^{-1}	0.0508
7	15	140	2381	R_i^{-2}	0.0472
8	15	140	2381	$R_i^{-n}\ (n = 2.27)$	0.0451
9	15	140	2381	$R_i^{-n(a,b)}$	0.0448
				$n(C,C) = 2.04$	
				$n(C,H) = 2.36$	
				$n(C,Br) = 2.01$	
				$n(H,H) = 2.96$	
				$n(H,Br) = 2.11$	

Table 3-3 Root mean square NN fitting errors (RMSE) to the 15 472 ab initio energies for H_2O_2 as a function of the form chosen for the configuration input vector. $N(h)$ and N_T are the number of neurons in the NN hidden layer and the total number of weight and bias parameters, respectively. All data are taken from Ref. 123.

Set	No. Input Elements	$N(h)$	N_T	Input Specification	RMSE (eV)
1	6	40	321	R_i	0.0354
2	6	40	321	R_i^{-1}	0.0327
3	6	40	321	R_i^{-2}	0.0214
4	6	40	321	R_i^{-3}	0.0327
5	6	40	321	R_i^{-4}	0.0108
6	6	40	321	R_i^{-n}	0.0181
				$n(\text{optimum})$	
				$= 1.625$	

Table 3-4 Root mean square NN fitting errors (RMSE) to the 21 584 *ab initio* energies for HONO as a function of the form chosen for the configuration input vector. $N(h)$ and N_T are the number of neurons in the NN hidden layer and the total number of weight and bias parameters, respectively. All data are taken from Ref. 123.

Set	No. Input Elements	$N(h)$	N_T	Input Specification	RMSE (eV)
1	6	41	329	R_i	0.023567
2	6	41	329	R_i^{-1}	0.014435
3	6	41	329	R_i^{-2}	0.013886
4	6	41	329	R_i^{-3}	0.018291
5	6	41	329	R_i^{-n} (optimum=1.68)	0.013890

The data in Tables 3-1 to 3-4 show that the RMSE fitting errors are reduced by factors of 8.86 and 1.67 for Si_5 and vinyl bromide, respectively, when the Z-matrix variables[141] are replaced by the interatomic distances as input elements. If the interparticle distances are replaced with input elements of the form R_{ij}^{-n} with n optimized, further reductions of the RMSE between factors of 1.31 to 2.83 for the four systems investigated are obtained. The only computational cost of the method is a small increase in the required NN training time. Malshe et al.[123] reported this increase to be about 8.5%.

3.9. EXAMPLE TRAINING PROCESS (MATLAB)

In this subsection, we will demonstrate how neural network training can be simply done with a commercially available package—the Neural Network Toolbox for MATLAB. In order to simplify the presentation, we will use an artificial problem with a single input and a single output. For the MD problems, there would be multiple inputs, corresponding to interparticle distances and angles, and the output would correspond to potential energy.

To begin, we will create the data set for our artificial problem. (For an MD problem, the data would be generated by *ab initio* electronic structural calculations.) Our artificial function will be a sine wave, with some noise added. This can be generated in MATLAB with the following commands:

```
p = [-1:.05:1];
```

```
t = sin(2*pi*p)+0.1*randn(size(p));
```

The variable p represents the input to the network, and t is the target output.

The next step is to create the network. This can be done as follows:

```
net = newff(p,t,10,{},'trainbr');
```

```
net.divideFcn ='';
```

The first line creates an MLP network with 10 neurons in one hidden layer. The default for this network is to use a sigmoid transfer function in the hidden layer, and a linear output layer. The training function is `trainbr`, which implements the Bayesian regularization algorithm, which was discussed earlier. The second line will tell the training algorithm not to divide the data into training and validation sets. The data should be divided if early stopping is used to prevent overfitting. In this case, the Bayesian regularization technique will prevent overfitting, and it does not need a validation set.

After the network has been created, the network is trained using the following command:

```
[net1, tr1] = train(net,p,t);
```

Figure 3-13 shows the progress of training. For this simple problem, the training has converged within 100 iterations.

After the network has completed training, we want to verify the operation of the network. A useful tool for this purpose is a scatter plot of the network outputs versus the targets. This can be created with the following commands.

```
a = sim(net1,p);

plotregression(a,t);
```

The first command calculates the network response. The second command draws the scatter plot between the targets and the network outputs. It also performs a regression analysis between the two variables, and calculates the R value. The result is shown in Figure 3-14. The data are well centered on the output = target line, with an R value of 0.99525.

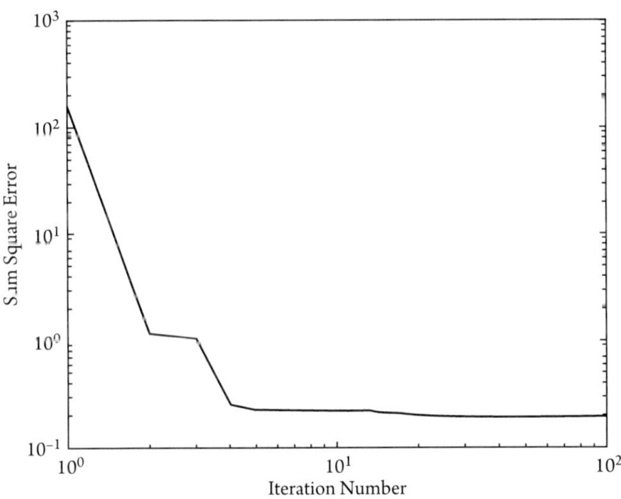

Figure 3-13
Convergence
characteristics.

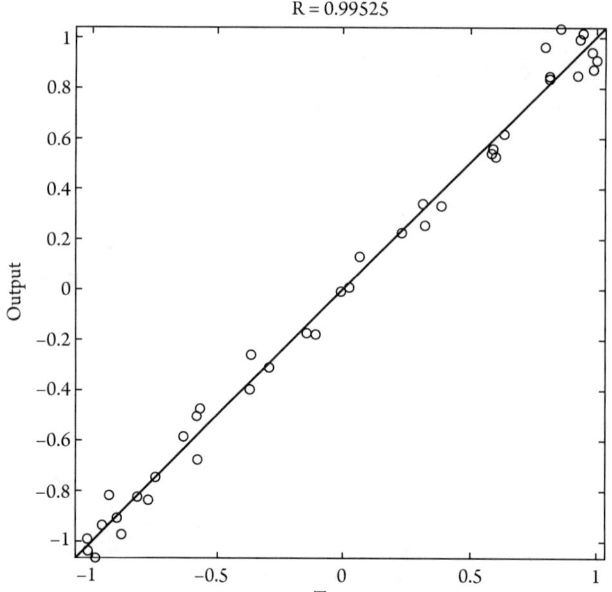

Figure 3-14 Scatter plot of network outputs versus targets.

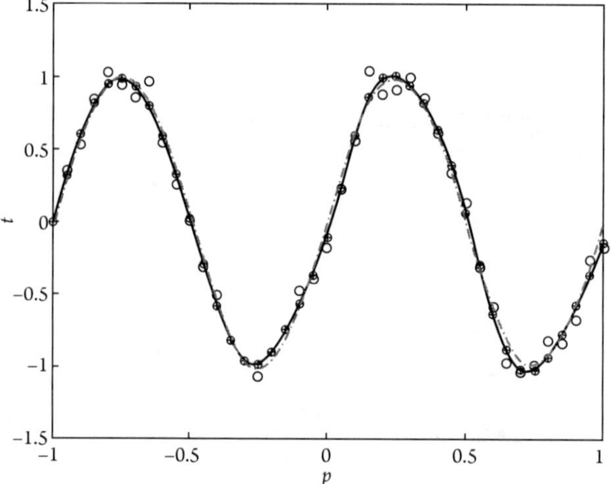

Figure 3-15 Final network fit.

Although this would not be possible in a practical MD problem, which would have many input variables, for this simple problem we can plot the resulting network fit to the underlying sine wave. Figure 3-15 shows the sine wave without noise as the dotted line. The target points (which have noise added) are shown as the large, open circles. The trained network response is shown by the thin continuous line, with small circles and crosses representing the response of the network at the input values of the targets. We can see that the network fits the underlying function, without overfitting to the noise.

3.10. THE COMBINED FUNCTION AND DERIVATIVE APPROXIMATION (CFDA) NN METHOD

The objective of the CFDA is to employ NNs to fit both functions and their derivatives simultaneously.[121] Almost all of the MD studies performed up to the present time have concentrated only on the fitting of the potential energy, although the force field is the key component for the molecular dynamics simulation. This situation is due to two factors. First, the *ab initio* computation of surface gradients requires far more computational effort than just the potential itself. Second, MSI[23-25] and IMLS[28-33] methods are not easily adaptable to fitting surface gradients, and adapting NN methods to the simultaneous fitting of both the potential and its gradients requires significant modification[121] of the usual NN fitting methods.[105-107] Since the forces can be computed by differentiating the potential surface, it has been assumed in the past that an accurate potential surface will yield accurate forces, without specifically fitting the forces. While not unreasonable, this assumption is not always correct.

Hornik et al.[108] showed that feedforward neural networks are capable of approximating arbitrary functions. Pinkus[110] provided additional conditions under which multilayer feedforward neural networks can simultaneously and uniformly approximate a function and various partial derivatives up to order k. These conditions require the transfer functions not to be a polynomial and to be $\mathbf{C}^m$, where $m > k\text{-}1$. Similar conditions were also discussed by Ito[128] and by Li.[129]

An earlier method for approximating both functions and their derivatives was proposed by Ferrari and Stengel[130] in 2005. In a first stage, this method algebraically obtains the network parameters for approximating function values. In a second stage, the parameters are adjusted to improve the derivative approximation. In Basson and Engelbrecht,[131] the derivative approximation was performed by adding an extra output unit for each partial derivative to the regular structure of a feedforward neural network that was used to approximate the function. The standard backpropagation procedure was then used to train the proposed network structure. Hanselmann et al.[132] created an additional network to approximate the derivatives. The derivative approximation obtained from the extra network was then combined with the network output used for function approximation using Taylor's series.

Witkoskie and Doren[133] have recently reported an NN method in which the gradient of a function is utilized to improve the NN fit to the function. The objective of their work was to incorporate the use of gradient information to improve the NN fit to the function with perhaps fewer data and numbers of neurons. However, no data are given concerning the accuracy with which their improved function fit also fits the known gradients. Also, their method does not allow for adjustment of the relative weights to be assigned to fitting the surface and fitting the gradients. Since there is only one potential but many gradients which will span a very large range of values on a typical potential-energy hypersurface, relative weighting can be important in obtaining optimum fitting accuracy.

The CFDA method[121,122] employs a different technique. We do *not* modify the network structure or create an additional network for derivative approximation.

Instead, the same network is used for both function approximation and derivative approximation. In addition, unlike the algorithm developed by Ferrari and Stengel,[130] in the CFDA method, the derivative approximation occurs simultaneously with the function approximation. The output layer contains only a single neuron, which provides the network approximation to the potential. The fitted gradients are obtained directly by differentiation of the network. This feature greatly simplifies the network architecture by reducing the number of neurons and parameters.

Most previous efforts to fit NNs to databases obtained from *ab initio* electronic structure calculations have involved an optimization procedure in which the mean square error between the network output and the desired potential surface was minimized. The typical mean square error is computed as follows:

$$J_f(\mathbf{w}) = \frac{1}{Q} \sum_{q=1}^{Q} \left(V(\mathbf{p}_q) - a(\mathbf{p}_q) \right)^2 \tag{3-36}$$

where V is the potential energy, a is the output of the network, and $\mathbf{p}_q$ is the column vector of the internal coordinates of the qth cluster in the data set currently being used to determine the NN force field.

The basis of the CFDA method is the use of a performance function that explicitly requires the simultaneous fitting of both function and gradient and the incorporation of relative weighting of surface and gradient fitting. In Chapter 5, Section 5.3, we present the results of the application of the method to a realistic, three-body potential-energy hypersurface and demonstrate that MD trajectories computed on the NN resulting from the CFDA method exhibit point-by-point agreement with corresponding trajectories computed on the model analytic potential hypersurface.

The mean square error performance function given by Equation (3-36) does not explicitly penalize errors in the potential gradients, which determine the force field. If the potential is accurately captured by the NN, then we would expect that the gradients would also be reasonably approximated, but this is not assured. We have recently proposed a procedure, which explicitly penalizes both the errors in the potential and in the gradient. The performance index is defined by

$$J(\mathbf{w}) = \frac{1}{Q} \sum_{q=1}^{Q} \left(V(\mathbf{p}_q) - a(\mathbf{p}_q) \right)^2 + \frac{\rho}{QR} \sum_{q=1}^{Q} \sum_{i=1}^{R} \left(\left. \frac{\partial V(\mathbf{p})}{\partial p_i} \right|_{\mathbf{p}=\mathbf{p}_q} - \left. \frac{\partial a(\mathbf{p})}{\partial p_i} \right|_{\mathbf{p}=\mathbf{p}_q} \right)^2, \tag{3-37}$$
$$= J_f(\mathbf{w}) + \rho J_d(\mathbf{w}),$$

where $\mathbf{w}$ is the vector containing all of the weights and biases in the NN, and ρ is a scale factor, which determines the relative importance of potential error and gradient error.[121]

The CFDA algorithm minimizes the performance function defined in Equation (3-37) using standard gradient-based (e.g., the Broyden-Fletcher-Goldfarb-Shanno (BFGS) quasi-Newton method)[114,134] or Jacobian-based (e.g., the Levenberg-Marquardt (LM) method)[117,135] optimization procedures. The difficulty in using Equation (3-37), instead of Equation (3-36), for NN training is that it requires the computation of second derivatives of the NN response. This is in contrast to the standard backpropagation[110] training algorithm for NNs, which is designed for Equation (3-36) and requires only the computation of first derivatives of the NN response.

To employ Equation (3-37) as the performance function in the training of the NN, we must compute the gradient of $J(\mathbf{w})$ with respect to the weights and biases of the network. The gradient of $J_f(\mathbf{w})$ is computed using the standard backpropagation algorithm.[112,113] The following summarizes the gradient calculation for $J_d(\mathbf{w})$. Complete details have been given by Pukrittayakamee et al.[121]

An element of this gradient that corresponds to a network weight will take the form

$$\frac{\partial J_d(\mathbf{w})}{\partial w_{i,j}^m} = \frac{-2}{QR} \sum_{q=1}^{Q} \sum_{i=1}^{R} \left(\left. \frac{\partial V(\mathbf{p})}{\partial p_r} \right|_{\mathbf{p}=\mathbf{p}_q} - \left. \frac{\partial a(\mathbf{p})}{\partial p_r} \right|_{\mathbf{p}=\mathbf{p}_q} \right) \times \left(\frac{\partial}{\partial w_{i,j}^m} \frac{\partial a(\mathbf{p})}{\partial p_r} \right)_{\mathbf{p}=\mathbf{p}_q} . \tag{3-38}$$

If we define the following two terms

$$v_{i,q}^m = \sum_{r=1}^{R} \left(\left(\left. \frac{\partial V(\mathbf{p})}{\partial p_r} \right|_{\mathbf{p}=\mathbf{p}_q} - \left. \frac{\partial a(\mathbf{p})}{\partial p_r} \right|_{\mathbf{p}=\mathbf{p}_q} \right) \times \left(\frac{\partial}{\partial p_r} \frac{\partial a(\mathbf{p})}{\partial n_i^m} \right)_{\mathbf{p}=\mathbf{p}_q} \right) \tag{3-39}$$

$$u_{q_i,r}^m = \left(\left. \frac{\partial V(\mathbf{p})}{\partial p_r} \right|_{\mathbf{p}=\mathbf{p}_q} - \left. \frac{\partial a(\mathbf{p})}{\partial p_r} \right|_{\mathbf{p}=\mathbf{p}_q} \right) \times \left(\frac{\partial a(\mathbf{p})}{\partial n_i^m} \right)_{\mathbf{p}=\mathbf{p}_q} \tag{3-40}$$

we can rewrite the element of the gradient as

$$\frac{\partial J_d(\mathbf{w})}{\partial w_{i,j}^m} = -\frac{2}{QR} \left(\left(\sum_{q=1}^{Q} a_j^{m-1}(\mathbf{p}_q) v_{i,q}^m \right) + \sum_{q=1}^{Q} \sum_{r=1}^{R} \left(\left. \frac{\partial a_j^{m-1}(\mathbf{p})}{\partial p_r} \right|_{\mathbf{p}=\mathbf{p}_q} u_{q_i,r}^m \right) \right) \tag{3-41}$$

.

To use Equation (3-41), we need to compute the following three terms:

$$v_{i,q}^m , u_{q_i,r}^m , \text{ and } \left. \frac{\partial a_j^{m-1}(\mathbf{p})}{\partial p_r} \right|_{\mathbf{p}=\mathbf{p}_q} .$$

The term $\left. \dfrac{\partial a_j^{m-1}(\mathbf{p})}{\partial p_r} \right|_{\mathbf{p}=\mathbf{p}_q}$ is updated forward from layer to layer using

$$\left. \frac{\partial a_j^{m-1}(\mathbf{p})}{\partial p_r} \right|_{\mathbf{p}=\mathbf{p}_q} = \dot{f}^m\left(n_{j,q}^m\right) \sum_{i=1}^{S^{m-1}} w_{j,l}^m \times \left. \frac{\partial a_l^{m-1}(\mathbf{p})}{\partial p_r} \right|_{\mathbf{p}=\mathbf{p}_q} , \tag{3-42}$$

which is initialized at the first layer with

$$\left.\frac{\partial a_j^m(\mathbf{p})}{\partial p_r}\right|_{\mathbf{p}=\mathbf{P}_q} = \begin{cases} 1, & j=r \\ 0, & j\neq r \end{cases} \tag{3-43}$$

The term $u_{q_{i,r}}^m$ is backpropagated from layer to layer using

$$u_{q_{i,r}}^m = \dot{f}^m\left(n_{i,q}^m\right)\sum_{i=1}^{S^{m+1}}\left(w_{l,i}^{m+1}\times u_{ql,r}^{m+1}\right), \tag{3-44}$$

which is initialized at the last layer (layer M) using

$$u_{q_{1,r}}^M = \dot{f}^M\left(n_{i,q}^M\right)\left(\left.\frac{\partial V(\mathbf{p})}{\partial p_r}\right|_{\mathbf{p}=\mathbf{P}_q} - \left.\frac{\partial a(\mathbf{p})}{\partial p_r}\right|_{\mathbf{p}=\mathbf{P}_q}\right). \tag{3-45}$$

The term $v_{i,q}^m$ is backpropagated from layer to layer using

$$v_{i,q}^m = \dot{f}^m\left(n_{i,q}^m\right)\sum_{i=1}^{S^{m+1}}\left(w_{l,i}^{m+1}\times v_{i,q}^{m+1}\right)+\sum_{r=1}^{R}\frac{\partial \dot{f}^m\left(n_{i,q}^m\right)}{\partial p_r}\sum_{l=1}^{S^{m+1}}\left(w_{l,i}^{m+1}\times u_{ql,r}^{m+1}\right), \tag{3-46}$$

which is initialized at the last layer (layer M) using

$$v_{i,q}^m = \sum_{r=1}^{R}\left(\left(\left.\frac{\partial V(\mathbf{p})}{\partial p_r}\right|_{\mathbf{p}=\mathbf{P}_q} - \left.\frac{\partial a(\mathbf{p})}{\partial p_r}\right|_{\mathbf{p}=\mathbf{P}_q}\right)\times\frac{\partial \dot{f}^M\left(n_{i,q}^M\right)}{\partial p_r}\right). \tag{3-47}$$

Therefore, using Equations (3-42), (3-44), and (3-46), we can compute the terms needed in Equation (3-41) to compute the elements of the gradient that correspond to weights in the neural network. For the terms in the gradient that correspond to bias elements, Equation (3-41) is replaced with Equation (3-48):

$$\frac{\partial J_d(\mathbf{w})}{\partial b_i^m} = -\frac{2}{QR}\sum_{q=1}^{Q}v_{i,q}^m. \tag{3-48}$$

This completes the gradient calculation. Once the gradient has been computed, any gradient-based optimization algorithm can be used to optimize the network weights and biases (e.g., conjugate gradient, BFGS quasi-Newton, etc.). For Jacobian-based algorithms (e.g., Gauss-Newton, Levenberg-Marquardt, extended Kalman filter, etc.), a similar procedure to that described above can be used to compute the Jacobian. In our experience, both BFGS quasi-Newton[114,134] and Levenberg-Marquardt[117,135] algorithms have produced excellent results.

There is one other step that is required to use the CFDA algorithm, and that is the selection of ρ. The details of this selection are given in Pukrittayakamee et al.[121] We will provide a summary of the calculations here. The first step was to write ρ as

$$\rho = \frac{\lambda}{\eta^2},\qquad\qquad (3\text{-}49)$$

where η is the calculated ratio of the maximum absolute value of the derivative (force) to the maximum absolute value of the potential energy in the training set. In this way, η will account for the difference in scales between $J_f(\mathbf{w})$ and $J_d(\mathbf{w})$. The same value of the parameter λ can then be used over a variety of problems. This was tested in Pukrittayakamee et al.,[121] and it was found that $\lambda = 10^4$ produced accurate results over a wide variety of fitting problems. That is the value used in our applications of the CFDA method discussed in Chapter 5.

3.11. COMBINED FUNCTION DERIVATIVE APPROXIMATION PRUNING

When fitting both the function and the first derivatives, we expect less overfitting. In the local neighborhoods of training inputs, the neural network approximation should be more accurate since the slopes at the training points are forced to be correct. However, while standard overfitting does not occur with CFDA training, we have observed new types of overfitting. These new types of overfitting cannot be eliminated with standard pruning methods or with early stopping methods. In this section, we will introduce a new pruning procedure that can detect and remove CFDA overfitting. We will focus only on the case of fitting a two-layer network with one output, where the transfer function of the network is hyperbolic tangent sigmoid.

Before describing the pruning method, we first illustrate the difference between standard overfitting (when fitting only the function) and the new CFDA overfitting. In standard overfitting, as shown in Figure 3-16, oscillatory behavior is observed in the network response. The network is very accurate at the training points (for the function approximation, but not for the derivative approximation), but it has significant errors between the training points. However, for CFDA overfitting, as shown in Figure 3-17, the network function and derivative responses are accurate almost everywhere, except over very small regions in the input space. As the overfitting

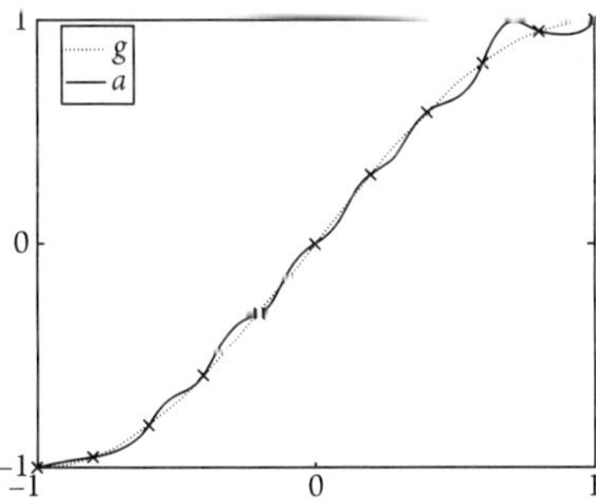

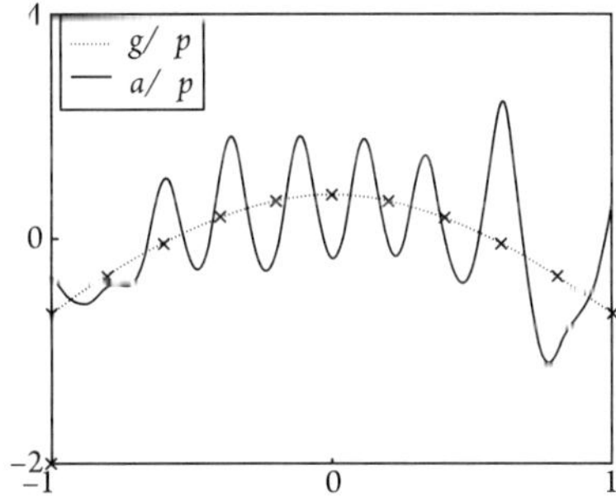

Figure 3-16 Regular overfitting.

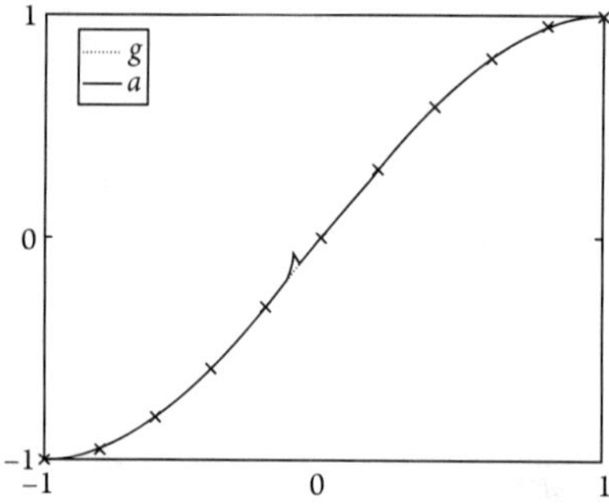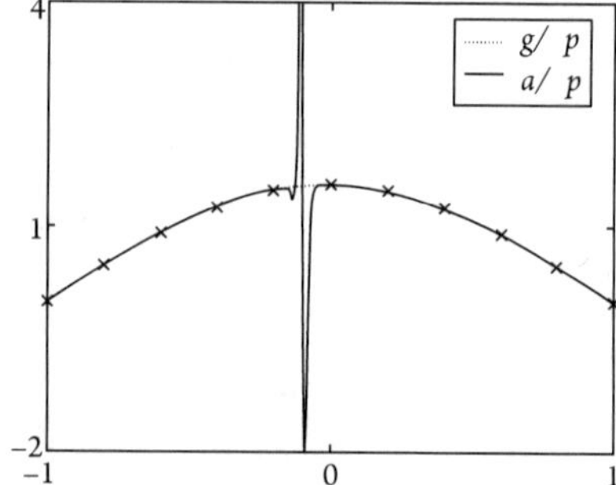

Figure 3-17 CFDA overfitting.

occurs over a very small region, the use of a validation set is not helpful for detecting the new types of overfitting.

In the next sections, we will demonstrate how the network parameters (weights and biases) affect the network response and its derivatives. The insight gained from these demonstrations will lead to the development of our pruning algorithm. Then, we will describe new types of overfitting that occur in networks trained by CFDA algorithms, and we will propose a pruning method to remove these types of overfitting.

3.11.1. Two-layer Network Response

We denote a two-layer network with one output as R-S^1-1. The R-S^1-1 network takes R inputs, with S^1 neurons in the first hidden layer, and produces one output a_1^2. The network parameters consist of the first-layer biases b_i^1, the second-layer bias b^2, the first-layer weights $w_{i,r}^1$, and the second-layer weights $w_{1,i}^2$. The transfer function in the first layer is denoted f^1, while the transfer function at the output layer is linear. The function response of the network can be expressed as

$$a_1^2 = \sum_{i=1}^{S^1} w_{1,i}^2 a_i^1 + b^2, \tag{3-50}$$

where

$$a_i^1 = f^1\left(n_i^1\right) \text{ and } n_i^1 = \sum_{r=1}^{R} w_{i,r}^1 p_r + b_i^1. \tag{3-51}$$

By taking the derivative of Equation (3-50) with respect to the input p_r and using the chain rule of calculus, we obtain the derivative of the network response:

$$\frac{\partial a_1^2}{\partial p_r} = \sum_{i=1}^{S^1} w_{1,i}^2 w_{i,r}^1 \dot{f}\left(n_i^1\right), \tag{3-52}$$

where $\dot{f}^1\left(n_i^1\right)=\partial a_i^1/\partial n_i^1$. For ease of reference, we also introduce the notation

$$a_1^2 = \sum_{i=1}^{S^1} y_i + b^2 \quad \text{and} \quad \frac{\partial y_i}{\partial p_r} \equiv w_{1,i}^2 w_{i,r}^1 \dot{f}\left(n_i^1\right). \tag{3-53}$$

Therefore, Equation (3-50) and Equation (3-52) can also be written as

$$a_1^2 = \sum_{i=1}^{S^1} y_i + b^2 \quad \text{and} \quad \frac{\partial a_1^2}{\partial p_r} = \sum_{i=1}^{S^1} \frac{\partial y_i}{\partial p_r}. \tag{3-54}$$

From Equation (3-53), the terms y_i and $\partial y_i/\partial p_r$ can be viewed as the function and derivative response of the ith neuron. The network's function response a_1^2 and derivative response $\partial a_1^2/\partial p_r$ are linear combinations of the terms y_i and $\partial y_i/\partial p_r$, respectively.

Next, we want to demonstrate how the network parameters affect the function and derivative responses. Knowledge of these effects will help us detect and remove overfitting. It is clear from Equation (3-54) that the term b^2 shifts the entire function response a_1^2 up or down, but it does not contribute to the derivative response $\partial a_1^2/\partial p_r$.

The effects of the other three parameters are more complex. Recall that the terms y_i and $\partial y_i/\partial p_r$ depend on the terms $\dot{f}^1\left(n_i^1\right)$ and $\dot{f}\left(n_i^1\right)$, respectively. With the hyperbolic tangent function, we have

$$\dot{f}\left(n_i^1\right)=1-\left(a_i^1\right)^2. \tag{3-55}$$

The sketches of $\dot{f}^1\left(n_i^1\right)$ and $\dot{f}\left(n_i^1\right)$ are shown in Figure 3-18 for a single-input case.

The first-layer bias b_i^1 controls the center of the neuron response, which occurs at $-b_i^1/w_i^1$. The effect of the first-layer weight w_i^1 on the terms $\dot{f}^1\left(n_i^1\right)$ and $\dot{f}\left(n_i^1\right)$ is

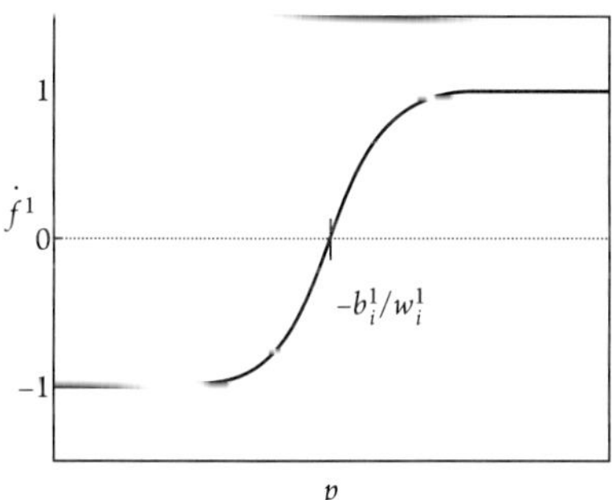

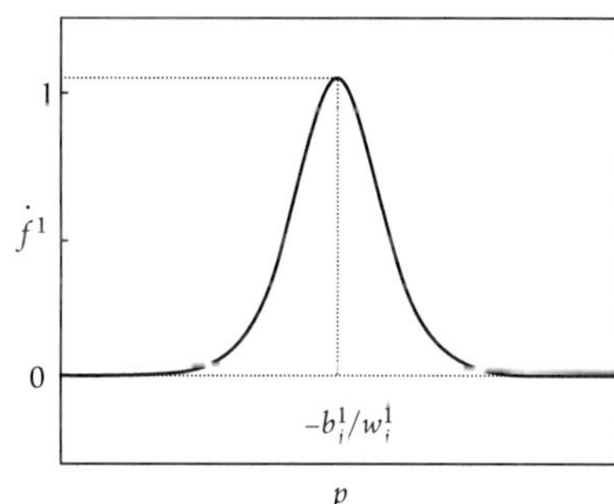

Figure 3-18 Sketches of $\dot{f}^1\left(n_i^1\right)$ and $\dot{f}\left(n_i^1\right)$ [121b] (© 2009 IEEE, reprinted with permission).

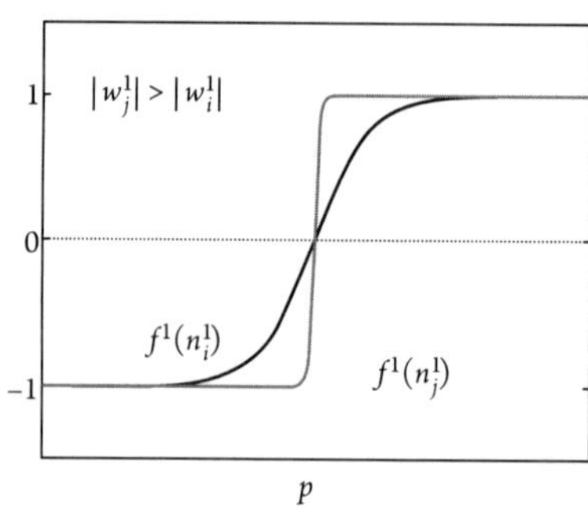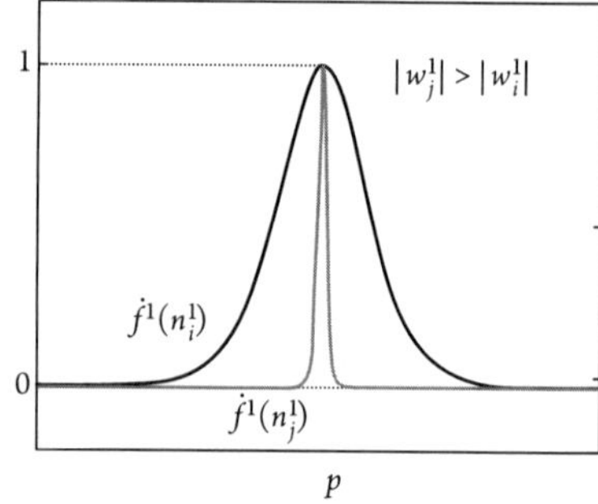

Figure 3-19 Effect of the first layer weight[121b] (© 2009 IEEE, reprinted with permission).

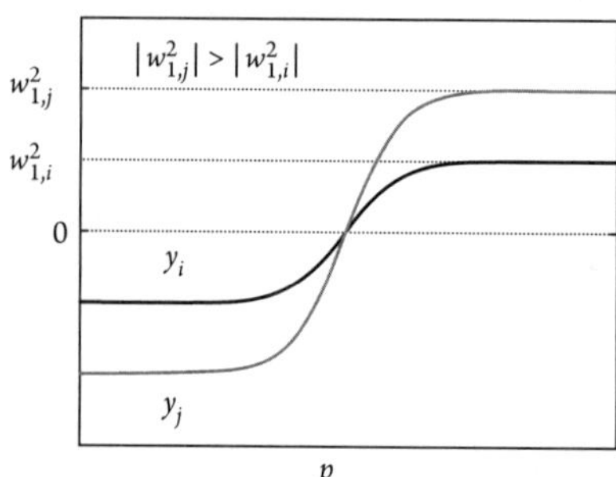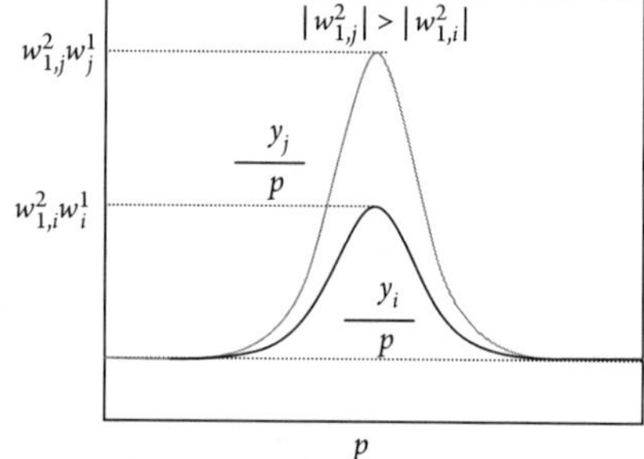

Figure 3-20 Effect of the second layer weight[121b] (© 2009 IEEE, reprinted with permission).

shown in Figure 3-19. We can see that w_i^1 affects the slope of $f^1\left(n_i^1\right)$ and the width of $\dot{f}\left(n_i^1\right)$.

The impacts of the second-layer weight $w_{1,i}^2$ on the terms y_i and $\partial y_i/\partial p_r$ can be seen from Equation (3-53). It scales both $f^1\left(n_i^1\right)$ and $\dot{f}\left(n_i^1\right)$. Figure 3-20 illustrates the impact of $w_{1,i}^2$ when doubling its value.

For the single-input case, we can see from Equation (3-53) and Figure 3-20 that the maximum magnitude of $\partial y_i/\partial p$, which occurs at the neuron center, equals $\left|w_{1,i}^2 w_i^1\right|$. Now, consider the case of multiple inputs. The center of the neuron is governed by the equation

$$n_i^1 = \sum_{r=1}^{R} w_{i,r}^1 p_r + b_i^1 = 0. \tag{3-56}$$

The center of the neuron becomes a line for two dimensional inputs, a plane for three dimensional inputs, and a hyperplane for higher dimensions. The impact of the network parameters on the neuron's responses for multiple inputs is similar to the case for a single input. The magnitude of $\partial y_i/\partial \mathbf{p}$ is

$$\left\|\frac{\partial y_i}{\partial p}\right\| = \sqrt{\sum_{r=1}^{R}\left(\frac{\partial y_i}{\partial p_r}\right)^2} = \left|w_{1,i}^2\right| \times \left\|_i \mathbf{w}^1\right\| \times \left|\dot{f}^1\left(n_i^1\right)\right|. \tag{3-57}$$

Thus, at the neuron center, the magnitude is maximum, and it equals $\left|w_{1,i}^2\right| \times \left\|_i \mathbf{w}^1\right\|$.

3.11.2. Two-Layer Network Response

In this section, we will introduce two types of CFDA overfitting, named Type-A and Type-B. These two types of overfitting occur over very narrow regions of the input space. Therefore, they cannot generally be detected with a validation set.

3.11.2.1. Type-A Overfitting

For the first type of overfitting, the network responses are accurate over the training set. However, at some spaces between training data, the network performs very poorly, causing very large approximation errors on both the function and the first derivatives. The problem comes from a group of neurons (greater than one), whose responses cancel at all training points. Unfortunately, their responses may not cancel at some spaces between training points, thus yielding inaccurate responses at these locations.

To illustrate the idea, consider a $1\text{-}S^1\text{-}1$ network. Assume that two neurons of the network yield the function and derivative responses shown in the first row of Figure 3-21. Note that the symbol X in the figure represents the location of training data. The combined responses of the two neurons are shown in the second row of Figure 3-21.

The responses of the two neurons cancel almost exactly at the training points, but not at the points in between. The remainder after cancellation is Type-A overfitting. If these two neurons are removed, it will not produce a significant change to the original fitting (since their responses yield almost no contribution to the training data). However, it will eliminate the overfitting.

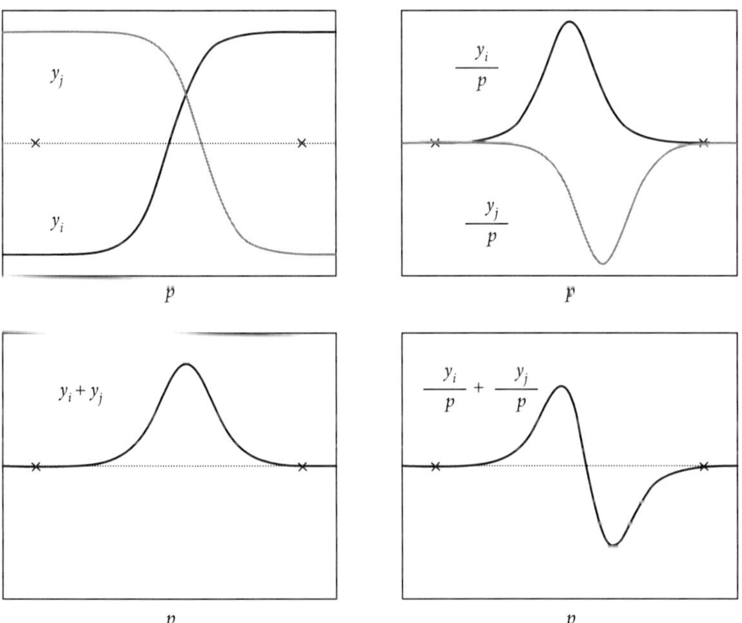

Figure 3-21 Responses of two neurons[121b] (© 2009 IEEE, reprinted with permission).

From Figure 3-21, we can see that the severity of the overfitting in the derivative is associated with the magnitude of the responses of the two neurons. Recall from the previous section that the maximum magnitude of $\partial y_i / \partial \mathbf{p}$ equals the weight product $\left| w_{1,i}^2 \right| \left\| _i \mathbf{x} \right\| \left\| _i \mathbf{w}^1 \right\|$. Therefore, the higher the weight products are, the worse the overfitting could be.

The weight product of a neuron is defined as "high" if it is larger than the actual derivative of the training point closest to the neuron center. The distance between a training point P_q and the ith neuron center (at $p = -b_i^1 / w_i^1$) can be easily computed in the case of single input. For multiple inputs, it requires the calculation illustrated in Figure 3-22.

From geometry, the distance from the training point $\mathbf{P}_q$ to the ith neuron's center (which is $\left(_i \mathbf{w}^1 \right)^T \mathbf{p} + b_i^1 = 0$) is

$$d_{i,q} = \frac{\left| \left(_i \mathbf{w}^1 \right)^T \mathbf{P}_q + b_i^1 \right|}{\left\| _i \mathbf{w}^1 \right\|} = \frac{\left| n_{i,q}^1 \right|}{\left\| _i \mathbf{w}^1 \right\|}. \tag{3-58}$$

Any neuron with high weight product will be selected as a candidate for Type-A overfitting. Once the candidate is chosen, we need to search for other neurons, whose responses will cancel the response of the candidate over the training data. Neurons with centers far from the candidate's center would not provide the response cancellation. Therefore, checking only a set of neurons whose centers are close to the candidate's center would be sufficient. We thus need to define the "neighbors" of the neuron candidate.

From Figure 3-21, we can see that the largest distance between the centers of two neurons that cause overfitting is the distance between the two training points. Therefore, we use the maximum distance between neighboring training points to be the threshold for deciding which neurons should be considered neighbors.

To be a neighbor in the multi-input case requires that a neuron center be close to the candidate center, and that the two centers be close to parallel. We fix the angle tolerance at one degree. To compute the distance between the neuron centers, we

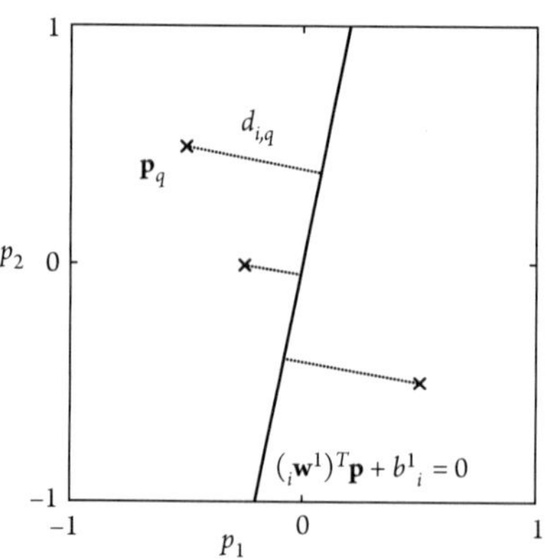

Figure 3-22 Distance between a point and a line[121b] (© 2009 IEEE, reprinted with permission).

assume that the centers are parallel. By geometry, it can be shown that the distance between two parallel hyperplanes, $\left(_j\mathbf{w}^1\right)^T\mathbf{p}+b_j^1=0$ and $\left(_i\mathbf{w}^1\right)^T\mathbf{p}+b_i^1=0$, is

$$d_{i,q}^H=\left|\frac{b_i^1}{\left\|_i\mathbf{w}^1\right\|}-\frac{b_j^1}{\left\|_j\mathbf{w}^1\right\|}\right|,\qquad(3\text{-}59)$$

if $_j\mathbf{w}^1$ and $_i\mathbf{w}^1$ are in the same direction, or

$$d_{i,q}^H=\left|\frac{b_i^1}{\left\|_i\mathbf{w}^1\right\|}+\frac{b_j^1}{\left\|_j\mathbf{w}^1\right\|}\right|.\qquad(3\text{-}60)$$

if $_j\mathbf{w}^1$ and $_i\mathbf{w}^1$ are in opposite directions. We compare the distance between neurons against the distance threshold, specified by the maximum distance between neighboring training points.

Once having a candidate and its neighbors, combinations of these neurons can be tested. We need to verify if cancellation (over the training points) occurs for any of these combinations. If cancellation occurs, the combination of neurons is pruned. Note that there could exist more than one combination yielding an insignificant contribution. In this situation, the combination with the highest number of neurons is pruned.

When cancellation for a combination of neurons is being verified, their combined derivative response is measured over the training set. Then, the maximum magnitude of the derivative response (chosen among the training points) will be compared with the true derivative magnitude evaluated at the training point. If the maximum derivative response from these neurons is much less than the true derivative value, we say that the derivative response from these neurons contributes insignificantly to the fitting. Thus, these neurons are pruned. For all of our tests, we use 10% as the threshold. The exact value is not critical.

3.11.2.2 Type-B Overfitting

For Type-B overfitting, the network produces accurate derivative response over the training set. However, the function response has small errors at some training points. At some locations between training points, the network response changes rapidly. This produces a significant first derivative response and a large second derivative response between training points. This problem is caused by a local minimum in the CFDA performance surface. Although there is a very small function error, the derivative error is minimized. This type of overfitting could be generated by a single neuron, or by multiple neurons.

To demonstrate Type-B overfitting, consider a 1-S^1-1 network. Assume that one neuron of the network yields the function and derivative responses as shown in Figure 3-23.

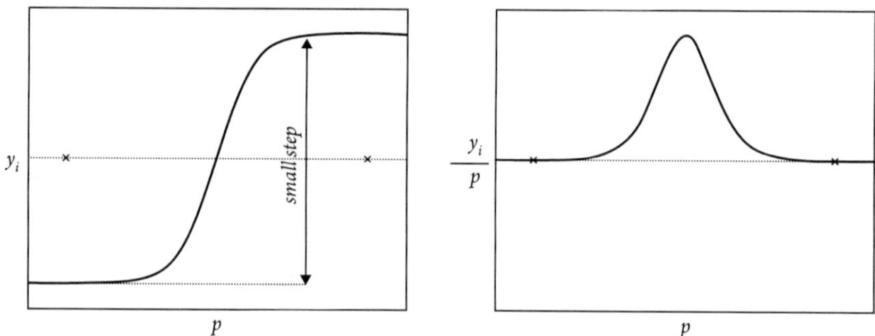

Figure 3-23 Type-B overfitting[121b] (© 2009 IEEE, reprinted with permission).

From Figure 3-23, it should be noted that the step in the function response y_i must be very small. (Otherwise, the function errors at the training points would be large, and the training process could further reduce these function errors.) A small step in y_i corresponds to small magnitude of $\partial y_i / \partial p$, i.e., a small weight product $\left| w_{1,i}^2 w_i^1 \right|$. From Figure 3-23, the response $\partial y_i / \partial p$ does not contribute to the derivative fitting at the training points, but it causes a small fluctuation at points between the training data. If this neuron is eliminated from the network, it will not produce a significant impact on the overall derivative responses at training data. However, it would automatically improve the function response, as well as eliminate the small fluctuation in the derivative response.

To get rid of Type-B overfitting, we want to remove neurons whose derivative response is narrower than the distance between training points. Consider the derivative response in Figure 3-24 and the corresponding buffer zone. If there are no training points within the buffer zone, then that neuron may not contribute to the derivative response at the training points. The neuron can then be further tested for potential removal.

Note that since the combined responses from more than one neuron could also cause Type-B overfitting, the neighbor search procedure (discussed earlier) is also needed. In addition, the process of validating whether the response from a group of neurons significantly contributes to the derivative fitting is the same as when dealing with Type-A overfitting.

In order to proceed with this pruning concept, we first need to define the buffer zone. From Figure 3-19 and Figure 3-24, we can see that the width of the buffer zone depends on the magnitude of the first-layer weight w_i^1 and the term $\dot{f}\left(n_i^1\right)$. To find the relationship, first let's fix the value of $\dot{f}\left(n_i^1\right)$ at x. From Equation (3-55), we have $\dot{f}\left(n_i^1\right)=1-\left(a_i^1\right)^2=x$. Solve for a_i^1 and recall that $a_i^1 =\tanh\left(w_i^1 p+b_i^1\right)$. Thus, we obtain

$$\tanh\left(w_i^1 p+b_i^1\right)=\pm\sqrt{1-x}. \tag{3-61}$$

For multiple inputs, by changing Equation (3-61) from $w_i^1 p+b_i^1$ to $\left(_i\mathbf{w}^1\right)^T \mathbf{p}+b_i^1$ and solving, we obtain two hyperplanes making up the boundary of the buffer zone. These two hyperplanes are parallel to the weight center, and they are

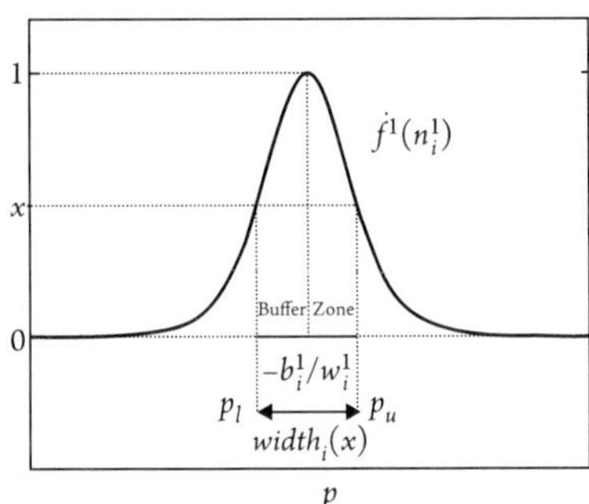

Figure 3-24 Definition of buffer zone[121b] (© 2009 IEEE, reprinted with permission).

$$\left({}_i\mathbf{w}^1\right)^T \mathbf{p}+b_i^1 = a\tanh\left(\sqrt{1-x}\right) \text{ and } \left({}_i\mathbf{w}^1\right)^T \mathbf{p}+b_i^1 = a\tanh\left(-\sqrt{1-x}\right). \tag{3-62}$$

To know the width of the buffer zone, we need to know the distance between the two hyperplanes. Therefore, from Equation (3-59), the distance between the two hyperplanes in Equation (3-62) is

$$width_i(x) = \frac{2a\tanh\left(\sqrt{1-x}\right)}{\left\|{}_i\mathbf{w}^1\right\|}, \quad {}_i\mathbf{w}^1 \neq 0. \tag{3-63}$$

The range of x is $0 \leq x \leq 1$. For pruning purposes, the value of x could be set, for instance, at 0.5. This is the value that we use in all of our tests. The exact value is not critical.

Once we know the width of the buffer zone, we will be able to identify which training points are inside or outside the buffer zone, by comparing with the distance from the training points to the weight center. Then, by counting how many training points satisfy the condition

$$d_{i,q} < \frac{width_i(x)}{2}, \quad \forall q, \tag{3-64}$$

we will be able to tell whether there are training points inside the zone. If there are none in the zone, the neuron becomes a candidate for Type-B overfitting. For all of our tests, we suspect overfitting if less than 1% of the training points lie inside the zone. The algorithm is not sensitive to the exact percentage used.

3.11.3 Pruning Algorithm Summary

In this section, we will provide a summary of the pruning algorithm. Assume we have a neural network trained by a CFDA training method. Once training converges, the pruning method will be executed. The pruning algorithm consists of five steps: (a) Initialization, (b) Candidate Selection, (c) Neighbor Search, (d) Contribution Check, and (e) Pruning and adjusting. After pruning is complete, the pruned network will be retrained by the CFDA training algorithm. This process

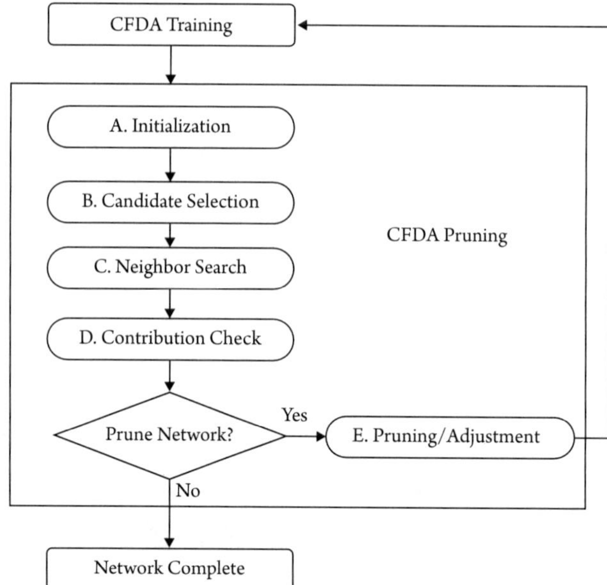

Figure 3-25 Pruning algorithm flow chart[121b] (© 2009 IEEE, reprinted with permission).

will be repeated until the pruning algorithm indicates that no further pruning is required. The flowchart of this process is illustrated in Figure 3-25.

In *Initialization*, we assign values for the various thresholds and set up the procedures for performing the various neighbor search processes. In *Candidate Selection*, we test for neurons with large weight products, which indicates Type-A overfitting, and we test for neurons with a lack of data points inside their buffer zones, which indicates Type-B overfitting. Once a candidate is identified from either of these criteria, we perform a *Neighbor Search* to identify a set of neighboring neurons for each candidate. Once a candidate and its neighbors are identified, we perform a *Contribution Check* to test for the largest combination of these neurons (candidate and neighbors) that does not contribute significantly to the network response at the training points. If such a combination is found, it is removed from the network, and the network second layer bias is adjusted to compensate for the removal.

4

CONFIGURATION SPACE SAMPLING METHODS

4.1. INTRODUCTION

In order to achieve the maximum accuracy in characterizing the PES and the associated force fields for an MD investigation, careful preparation of the database is an essential step in the process. The points that must be addressed include the following:

1. The total volume of configuration space is extremely large, and its size increases as the internal energy of the system rises. For example, consider a four-atom system. For this system, at least six internal coordinates must be specified to determine the spatial configuration of the molecular system. At a given internal energy, each of these six coordinates can span a continuous range of values from some minimum to some maximum. If each variable range is divided into 100 equal increments and the potential energy of the system computed by some *ab initio* method for all possible configurations of the system, a total of 100^6 or 10^{12} electronic structure calculations would need to be executed. This is clearly beyond the computational capabilities of any computational system currently in existence. Grid sampling methods can and have been used effectively for three-atom systems. However, for more complex systems, it is essential that procedures be developed that permit the regions of configuration space that are important in the reaction dynamics to be identified.

2. Sampling methods usually should be optimized to produce a reasonably uniform density of data points in those regions of configuration space that are important in the dynamics. If this is not done and there are regions of very high point density and others with low point density, no fitting technique will function well. The parameters of the method will adjust themselves to fit regions of high density preferentially over those with low density even when the low-density regions may be more important in the dynamics. An exception to the need to have an approximately uniform density of points in the database

occurs in regions where the potential gradient is large. In such regions, the density of points in the database will need to be larger than in regions in which the gradient is small.

3. Methods must be developed that permit the investigator to determine when there is a sufficient number of points in the database to attain the desired convergence of the fitted potential or gradients.

4. Whatever method is employed to sample configuration space, there must be a means to start the process, i.e., the method should be self-starting.

A variety of methods and procedures have been developed to address each of these points. In this chapter, we review some of the more useful of these methods.

4.2. TRAJECTORY AND NOVELTY SAMPLING (NS) METHODS

To achieve the objectives outlined at the beginning of this chapter, a method must be devised that samples the subset of configuration space that is of importance in the reaction dynamics under investigation rather than the complete configuration space. As soon as we direct our attention to the sampling problem, it becomes evident that MD calculations do precisely what we require. Such methods were first introduced by Collins and co-workers.[23,25] The nuclear configurations that are generated in an ensemble of trajectories comprise the subset of configuration space that we wish to sample. All we need to do is to devise efficient procedures for (a) the incorporation of configuration points into the database, and (b) testing for convergence of the potential surface that are well suited to use for complex systems involving multiple reaction channels. It is particularly advantageous if the convergence criteria are independent of the reaction dynamics so that repeated computation of dynamic properties to test for convergence of the surface is not required.

The methods we have frequently employed follow those used by Collins and co-workers.[23,25] We sample the important regions of configuration space in an iterative fashion using trajectories. However, when attempting to execute this process, we are immediately faced with a problem. In order to execute trajectories, we must already have the PES we are seeking to develop. This problem is generally termed the "self-starting problem."

Several procedures have been suggested for handling the self-starting problem. In their original work, Ischtwan and Collins[23] initiated the sampling procedure by generating a series of configurations based on chemical intuition (or chemical knowledge-base) that would be reasonable structures occurring along the reaction coordinate for the process under investigation. These points comprise the initial database. Collins and co-workers[25] have successfully used this technique to investigate several chemical reactions. Once these initial configurations are in-hand, electronic structure calculations at the desired level are executed to obtain the potential energy, gradients, and second derivatives at each point in the initial database. A temporary PES is then generated using moving Shepard interpolation methods.[24] Trajectories are computed on this surface and new points are sampled from these trajectories and added to the database according to a selection procedure based on the local density of points currently in the database. Convergence of the PES resulting from this procedure is tested by computations of relevant dynamic properties of the reaction being studied. These properties include, but are not limited to, reaction probabilities, cross sections, and energy partitioning. When the results of these calculations show that the dynamic properties cease to change significantly when more configuration points are added to the database, the PES is considered to be converged.

When the system under investigation involves only three or four atoms undergoing a single, simple bond rupture reaction, the chemical intuition method for initiating the sampling process works well. However, as the number of atoms present increases and multiple two-, three-, and four-center reactions are occurring simultaneously, the method becomes less likely to be effective.

An alternative method[136] to circumvent the self-starting problem is to initiate the iterative sampling process by employing a semi-empirical potential for the system of interest rather than using Monte Carlo methods or chemical intuition. This is a much more effective starting procedure since it is unlikely that random sampling or intuition will provide an initial sampling of the important regions of configuration space in complex systems with many open reaction channels.

One advantage of employing a semi-empirical potential to initiate the sampling is this: it has been found that the final PES obtained by the iterative sampling of configurations obtained from trajectories integrated on temporary PESs is not sensitive to the choice of the empirical surface employed in the initial step.[136] So long as the choice is reasonable, the iterative procedure will converge. For example, it has been shown[136] that two potentials having different functional forms and different parameters proposed by Tersoff[13] for the Si_5 system lead to essentially the same final PES when each is employed to generate the initial points for iterative trajectory sampling.

When a semi-empirical PES is employed to initialize the sampling, the actual procedure is as follows:

1. Using the empirical potential, a large set of trajectories is computed using standard MD methods[1-7] and the highest energy of interest in the dynamics investigations. The product channels for these trajectories are examined, and a smaller subset is selected such that the trajectories in the subset include examples involving reactions into all the energetically open channels of importance as well as some non-reactive trajectories. The integration of the trajectories in the subset is repeated, and as the trajectories proceed, the nuclear configurations are stored at time intervals determined as described in Step 2. This is continued until we have several thousand such configurations. The energies and

forces present in these configurations are then computed using an appropriately chosen *ab initio* method. This data set comprises our first approximation for the *ab initio* system potential and force field.

2. In the second step, feedforward NNs are fitted to the *ab initio* database for the energies and forces obtained in the first step. The fitting procedure has been described in Chapter 3. A second set of MD trajectories is now computed using the NN force field rather than the original empirical potential. During these calculations, additional nuclear configurations are stored and their energies and forces subsequently computed. As explained below, novelty sampling (NS) algorithms based on the current density of configuration points in the database guide our selection of new configurations.[136] Regions with a low density are preferentially included in the sample. The energies and forces obtained for the newly selected configurations are added to the overall database characterizing the potential-energy hypersurface and force field. New neural networks are trained to fit to this expanded database and the entire procedure repeated. Since the configurations obtained from the MD calculations involve structures in which all of the internal coordinates are simultaneously varying, the computed force fields will properly represent all the coupling terms that influence intramolecular energy transfer and reaction dynamics.

3. The second step is now repeated iteratively. After several iterations, the *ab initio* potential and force field obtained from the trained NN are expected to converge to something close to the experimental potential and force field, at least to the extent that the choice of *ab initio* computational method is capable of yielding energies and forces that approach the experimental values. As described in the following paragraph, the NS[136] itself provides a convenient test for convergence that is independent of the results obtained from MD calculations.

To ensure that the final NN accurately represents the potential and associated force field for the system, it is essential that the training data be spread throughout the subset of configuration space that is important in the dynamics. No interpolation or fitting method, including neural networks, can be expected to extrapolate accurately. While there is little that one can do during training to improve the network performance outside the range of the training data, we can improve its ability to interpolate and fit between data points, and we can work to ensure that the training data cover the relevant range of the input space. To achieve this objective, several methods have been proposed. The two that have been employed in NN studies to date are variable interval sampling and modified NS.

In variable interval sampling, the time interval between selecting configurations during the trajectory sampling for inclusion in the database is varied rather than sampling points at constant time intervals. If configurations are stored at constant time intervals, the database will include many more points from regions where the forces are smaller than in regions characterized by larger forces simply because the atoms will spend less time in regions where the forces are large. To effectively sample configuration space with something close to a uniform density of points over all regions, the time interval between sampling, τ, during the trajectories needs to be a function of the atomic accelerations.

To minimize this effect, a variable interval sampling algorithm[122] has been employed. In this algorithm, τ is given by

$$\tau = \begin{cases} \text{trunc}\left[\alpha\left(a_{\max}\right)^{-1}\right]\Delta t & \text{if trunc}\left[\alpha\left(a_{\max}\right)^{-1}\right] > 0, \text{ and} \\ \text{trunc}\left[\alpha\left(a_{\max}\right)^{-1}\right]\Delta t + \Delta t & \text{if trunc}\left[\alpha\left(a_{\max}\right)^{-1}\right] = 0 \end{cases} \tag{4-1}$$

where

$$a_{\max} = \max\left[a_1, a_2, \ldots, a_i\right], \quad \text{and} \quad a_i = \left(m_i\right)^{-1}\left|P_i\right|. \tag{4-2}$$

In Equations (4-1) and (4-2), a_i is the absolute value of the acceleration of atom i, m_i and P_i are the corresponding atomic mass and momentum vector, respectively, and Δt is the integration step size. $a_{\max}$ is the maximum of the acceleration of the atoms present in the molecule, and the operation trunc(x) yields the integer part of x. The value of τ is periodically updated during the trajectory. Thus, as $a_{\max}$ becomes large when the forces are large, τ will approach Δt, and we sample at every integration point in the trajectory. In regions of small force where $a_{\max}$ is small, τ will assume larger values, and we will sample less frequently. The constant α in the numerator of Equation (4-1) is system dependent and must be determined empirically.

Figures 4-1 and 4-2 show the operation of the time interval algorithm given by Equations (4-1) and (4-2) with $\alpha = 5$ when sampling configuration points using trajectories on a semi-empirical surface for the H_2Br system.[122,137] Each figure shows a histogram of the number of configurations stored as a function of the maximum acceleration for 100 trajectories on the H + HBr analytic surface.[137] The results reported in Figure 4-1 were obtained by sampling at equally spaced time intervals throughout all trajectories. The sampling results shown in Figure 4-2 were obtained using Equations (4-1) and (4-2). When sampling is done at constant time intervals, a preponderance of the sampled points lie in regions where the forces are small. When the weighted sampling described by Equations (4-1) and (4-2) is employed, all regions are sampled much more uniformly.

In modified NS,[136] the selection of new configurations to be added to the database is guided by a modification of the method of novelty detection developed by Pukrittayakamee and Hagan.[138] Let $\mathbf{q}_i$ be a $((3N-6) \times 1)$ column vector of the internal coordinates of the ith cluster in the data set currently being used to determine the NN force field. By computing trajectories using this force field, we generate a new configuration $\mathbf{q}_n$. Qualitatively, we wish to incorporate the forces and energies of this configuration into the database if $\mathbf{q}_n$ lies in a region of configuration space where the density of points in the database is low. However, if this region already contains a high density of configuration points, we wish the probability of incorporating point $\mathbf{q}_n$ to be much lower.

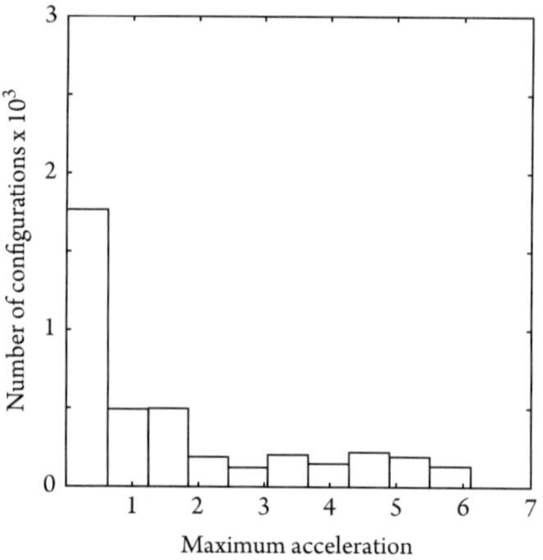

Figure 4-1 Histogram of number of configurations stored and corresponding maximum acceleration when the configurations are stored at equally spaced time intervals. Acceleration is given in units of Å atu^{-2}, where one atomic time unit (atu) is 1.019×10^{-14} s[122] (reprinted with permission from American Institute of Physics).

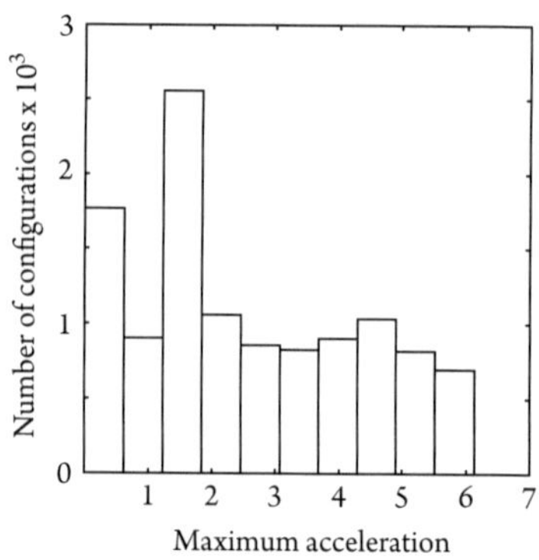

Figure 4-2 Histogram of number of configurations stored and corresponding maximum acceleration when the time interval for sampling is obtained from Eqs. (4-1) and (4-2). Acceleration is given in units of Å atu^{-2}, where one atomic time unit (atu) is 1.019×10^{-14} s[122] (reprinted with permission from American Institute of Physics).

Novelty detection is based on the 2-norm difference between vector $\mathbf{q}_n$ and vector $\mathbf{q}_i$:

$$\left\| \mathbf{q}_i - \mathbf{q}_n \right\| = \left[\left(\mathbf{q}_i - \mathbf{q}_n \right)^T \left(\mathbf{q}_i - \mathbf{q}_n \right) \right]^{1/2} , \tag{4-3}$$

where $\left(\mathbf{q}_i - \mathbf{q}_n \right)^T$ is the transpose of $\left(\mathbf{q}_i - \mathbf{q}_n \right)$. A vector $\mathbf{d}$ is now defined whose elements are all the distances computed using Equation (4-3) for each of the N configuration points in the database. The minimum distance separating point $\mathbf{q}_n$ from other points in the database is the minimum element of $\mathbf{d}$ that we denote as $d_m = \min(\mathbf{d})$. The mean separation distance, $<\mathbf{d}>$, is the average value of the elements of $\mathbf{d}$. The basic concept is to incorporate configuration point $\mathbf{q}_n$ into the database with high probability if d_m or $<\mathbf{d}>$ is large, but low probability if both d_m and $<\mathbf{d}>$ are small.

A selection algorithm based solely upon the magnitude of d_m will often result in the failure to incorporate configuration points that are needed in the database or the incorporation of points that are not critical. This occurs because a minimum distance criterion does not incorporate consideration of the gradient of the function being fitted. In regions where the gradient is large, many more points will be required to obtain an accurate NN fit than is the case in regions characterized by smaller gradients. Unfortunately, the underlying potential function is unknown. Therefore, it is impossible to calculate the gradient at an arbitrary configuration point without executing electronic structure calculations. Pukrittayakamee and Hagan[138] have found that if the minimum distance between a set of modified configuration vectors is used in place of d_m, the effect of the function gradient can be incorporated into the selection algorithm. We create the set of modified vectors, $\boldsymbol{\Gamma}$, using

$$\boldsymbol{\Gamma}_i = [\mathbf{q}_i O_i],\qquad\qquad(4\text{-}4)$$

where O_i is the output potential energy from the NN at configuration point $\mathbf{q}_i$. When the minimum distance, Γ_m, between $\boldsymbol{\Gamma}_i$ and $\boldsymbol{\Gamma}_n$ for all N configuration points in the database is used in place of d_m, the effect of the function gradient is incorporated. This occurs because if the gradient between points $\mathbf{q}_i$ and $\mathbf{q}_n$ is large, the corresponding NN outputs, O_i and O_n, will be very different and Γ_m will increase, as desired.

The first half of the sampling algorithm is based on the normalized probability distribution function of Γ_m values, $P(\Gamma_m)d\Gamma_m$, for all N configuration points in the current database. This distribution function and its maximum value, $P(\Gamma_m)_{\max}$, are computed at the beginning of each iterative cycle in the sampling procedure. As new configuration points are generated during the trajectories computed using the current NN force field, they are added to the database if

$$P\left(\Gamma_m\right) \le T_1 P\left(\Gamma_m\right)_{\max} \quad \text{and } \Gamma_m \ge \Gamma_m^* \quad \text{or, if}$$

$$\frac{P\left(\Gamma_m\right)}{P\left(\Gamma_m\right)_{\max}} \le \xi\left(1 - T_1\right).\qquad\qquad(4\text{-}5)$$

Otherwise, the new configuration point will be rejected by the first test. In Equation (4-5), T_1 is an empirically determined, acceptance threshold in the range ($0 \le T_1 \le 1$) and ξ is a random number whose distribution is uniform on the interval $[0,1]$. Γ_m^* is a small value of Γ_m chosen so as to exclude points that are very near other points in the database. Examination of Equation (4-5) shows that the new point will be accepted if very few points in the database have Γ_m values near that for the new point. It will also be accepted with a probability that decreases as $P(\Gamma_m)$ approaches its maximum value $P(\Gamma_m)_{\max}$.

To be certain that the sampling does not exclude configuration points that are outliers, the first half of the sampling algorithm must be augmented with an additional test that will include outliers. An outlier configuration point is usually defined as one that is inconsistent with the remainder of the database and one that does not follow the majority's underlying correlation.[139]

Suppose there are two configuration points in the database that are far removed from the remainder of the points. This situation is illustrated in Figure 4-3 for the simple case in which the configuration depends upon only two coordinates, $\mathbf{q}_1$ and $\mathbf{q}_2$. The points denoted as 1 and 2 are outliers. Now suppose one of the iterative cycles generates the configuration point denoted by Point 3 in Figure 4-3. Since Γ_m for Point 3 is about the same as that for most of the members of the database, the first selection criterion would very likely result in rejection of this point. However, if configuration space is to be adequately sampled, it is clear that Point 3 needs to be included in the database.

We can ensure that an outlier such as Point 3 will be incorporated into the database by using a second selection criterion. At the beginning of each iterative cycle, we compute the normalized probability distribution function of $<\Gamma>$ values, $P(<\Gamma>)$ $d<\Gamma>_{max}$, for all N configuration points in the current database. From this distribution, we obtain the maximum value, $P(\Gamma)_{max}$. If the first half of the selection algorithm results in rejection of a configuration point, it is still accepted, if

$$P(<\Gamma>) \leq T_2 P(<\Gamma>)_{max} \quad \text{or, if}$$

$$\frac{P(<\Gamma>)}{P(<\Gamma>)_{max}} \leq \xi(1-T_2). \tag{4-6}$$

Proper selection of T_2 will result in the addition of most outlier configurations generated in the MD iterative calculations.

Modified NS not only provides a well-judged rationale for incorporating or rejecting configurations generated by MD trajectories, it also provides a criterion for determining convergence of the PES without requiring the repeated computation of dynamic properties of the reaction under study. The selection criteria described by Equations (4-5) and (4-6) indicate that when the preponderance of new configurations selected for consideration by variable interval sampling or any other suitable method have minimum distances Γ_m that lie beneath the current distribution of minimum distances, $P(\Gamma_m)$, the database is sufficient to yield a converged PES.

The development of a PES for Si_5 clusters that occur in front of and below the cutting tool and in the chip in an MD simulation of a silicon workpiece with an infinitely hard diamond tool serves as an illustrative example of the operation of modified NS to test for convergence of the PES.

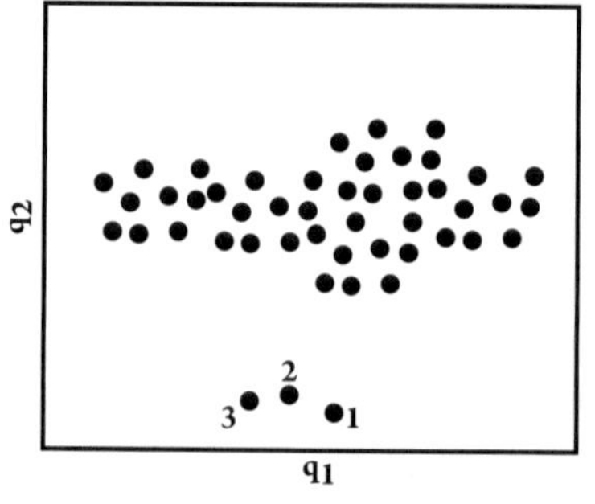

Figure 4-3 Hypothetical distribution of configuration points in a database for a system with two configuration variables, q_1 and q_2. The points labeled 1, 2, and 3 are outliers[36] (reprinted with permission from American Institute of Physics).

A Tersoff potential[13] was employed in the initial step of the iterative process. MD cutting simulations with a $+15°$ rake angle, $10°$ clearance angle, 1 Å cutting depth, 54.3 Å cutting width, at a cutting speed of 491.2 m s^{-1} were carried out.[140] The total number of atoms in the workpiece was 1,675. Throughout the MD simulations, the configurations of the five-atom silicon clusters that are present in front of the tool, in the chip, and within a few unit cell distances in the workpiece beneath the tool were stored. A total of 18,000 Si$_5$ configurations were stored.

These 18,000 configurations were augmented with an additional 10,000 configurations near equilibrium obtained by placing thermal energy corresponding to a temperature of 300 K in the silicon workpiece and then following the vibrational motions of the lattice using molecular dynamics with the same Tersoff[13] potential. Si$_5$ configurations were stored at equally spaced time intervals during the calculations. Subsequent to the MD calculations, the energies and force fields for each of the 28,000 stored configurations were computed using density functional theory (DFT) with a 6-31G** basis set and the B3LYP procedure for incorporating correlation energy.[141]

In the second phase of the procedure, the *ab initio* potential energies and input elements for each of the 28,000 configurations are scaled using Equation (3-34). The scaled database is then divided into training, validation, and testing sets. Approximately 10% of the data are used for the validation set, 10% for the testing set, and the remainder is the training set. A (9-45-1) NN is fitted to the data in the training set using the Levenberg-Marquardt algorithm.[117,135] The NN employed sigmoid transfer functions and linear output functions.

The entire training process for Si$_5$ with a database of 22,400 configurations required about 16 hours of CPU time. It was found[136] that this can be reduced to about eight hours by using a conjugate gradient fitting method[105] initially and then switching to the Levenberg-Marquardt algorithm[117,135] about halfway through the fitting. Comparison of the networks obtained by the two procedures showed that they produce nearly the same interpolated energies and force fields for Si$_5$. Figure 4-4 shows a comparison of the DFT scaled energies and the corresponding NN results for the testing subset of the 28,000 Si$_5$ configurations. If the NN fit were perfect, all points would fall on the $45°$ line shown in the figure. The computed rms deviation of the NN and DFT energies is 0.36 kJ mol^{-1}

The next step in the procedure is to utilize the NN to execute MD calculations during which we employ NS techniques to obtain additional configurations to iteratively improve the potential-hypersurface until an acceptable degree of convergence to the final potential and force fields is obtained. To illustrate the method, Raff et al.[136] examined the vibrational motions of a silicon lattice at 800 K.

Figure 4-5(a) shows the distribution of minimum distances, $P(\Gamma_m)$, for the initial set of configurations. When Equation (4-3) with $T_1 = 0.5$ and $\Gamma_m^* = 0.06$ is applied to the configurations resulting from trajectories computed using the NN fit to the original data set, several thousand additional configurations are identified whose distribution of Γm values from the original database is that shown in Figure 4-5(b). Comparison of this distribution with Figure 4-5(a) for the original configurations

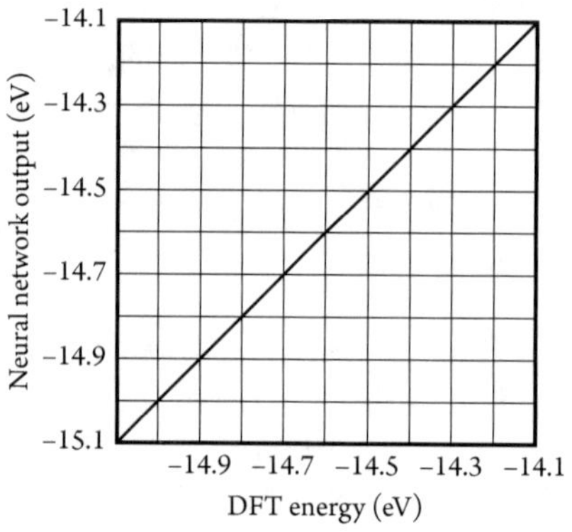

Figure 4-4 Comparison of *ab initio* DFT energies with the predictions of the (9-45-1) NN for the testing subset of the 28,000 Si_5 configurations. If the fit were perfect, all points would fall on the 45° line shown in the figure. The standard deviation of the points from this line corresponds to 0.36 kJ mol^{-1} [136] (reprinted with permission from American Institute of Physics).

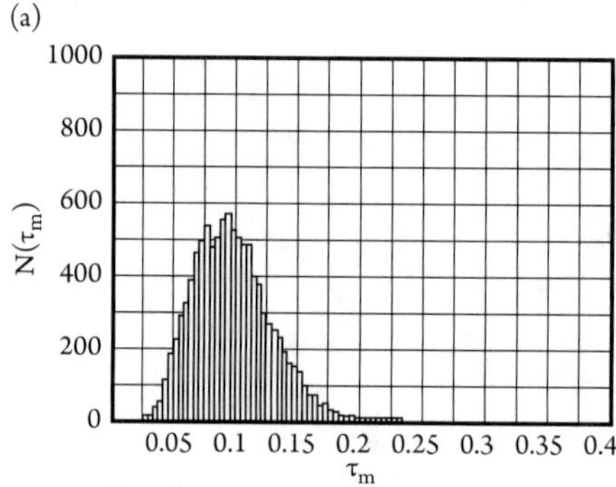

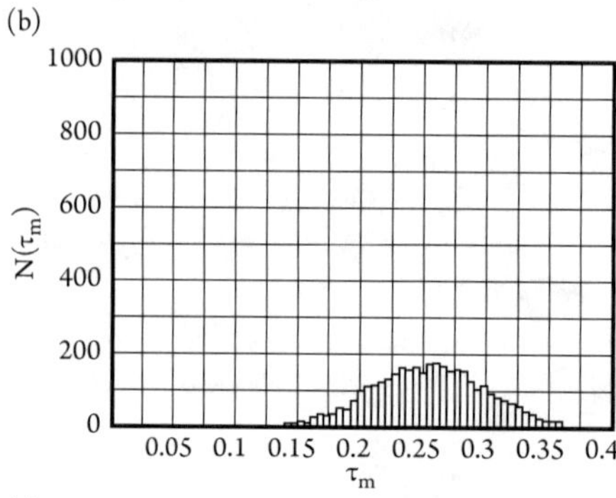

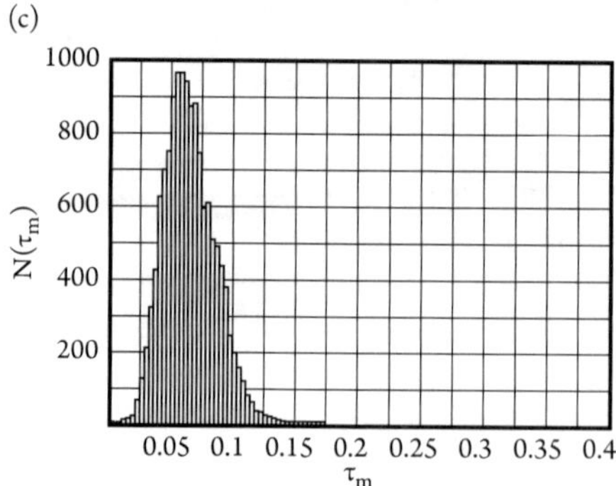

Figure 4-5 (a) Distribution of minimum distances $P(\tau_m)$ for the initial set of 28000 Si_5 configurations. (b) Distribution of τ_m values for the new configurations resulting from the applications of Eqs. (4-5) and (4-6) to Si_5 structures computed using the NN fit to the original Si_5 data set. (c) New distribution of τ_m values for the concatenated set of configurations resulting from the Si_5 structures obtained in the original trajectories using the Tersoff potential and those obtained in the first iteration on the NN surface[136] (reprinted with permission from American Institute of Physics).

shows that the large majority of the new configurations lie in a region of configuration space not adequately sampled by the original data set. These configurations are, therefore, incorporated into the overall database and a new NN is fitted to the computed DFT energies. The new distribution $P(\Gamma_m)$ for the concatenated set of configurations is shown in Figure 4-5(c). During the second iteration, essentially no new configurations are found whose Γ_m values do not lie underneath the distribution shown in Figure 4-5(c). The process has, therefore, converged for the test system. The advantage of this procedure is that it is unnecessary to compute dynamics results in order to test for convergence.

4.3. SELF-STARTING METHOD USING DIRECT DYNAMICS (DD)

The most efficient procedure to obtain the configurations of importance in the dynamics is MD calculations. The nuclear configurations that are generated in an ensemble of trajectories comprise the subset of configuration space that we wish to sample. Such methods were first introduced by Collins and co-workers.[23] There is, however, one significant problem associated with the implementation of such a procedure. The computation of trajectories requires that we already have the PES that we are seeking to develop. Basically, we must have a means to initiate the procedure. This difficulty is usually termed the self-starting problem.

While MSI,[23-25] IMLS,[28-33] and NN methods provide powerful and robust methods for obtaining fits to *ab initio* or other databases once those databases have been obtained, none of these methods provide an efficient, robust, self-starting procedure for sampling the configuration space of the system to obtain that database when the volume of configuration space that must be sampled is extremely large, as will be the case when the system is at high energy and is undergoing several simultaneous two-, three-, and four-center reactions. Since chemical intuition is unlikely to work well in such complex cases and the development of semi empirical potentials for such systems is often tedious and difficult,[16] a more general and robust method for initiating the configuration space sampling is needed.

A sense of the magnitude of the sampling difficulty can be obtained by comparing the size of the databases required to obtain a converged PES for systems involving only a single, two-center bond dissociation reaction with the size required when multiple two-, three-, and four-center reactions are involved. When the reaction of interest is a single, two-center bond dissociation gas-phase reaction in a three- or four-atom system, the database required for obtaining a reasonably accurate PES usually contains between 300 to 6,500 points. Table 4-1 provides a representative sampling of 22 such systems. For these systems, the self-starting problem is usually handled easily using chemical intuition or a relatively simple empirical surface. For single, two-center bond dissociation reactions on surfaces, the situation is the same. In their investigation of H_2 dissociation on Pt(111) and Cu(111) surfaces, Ludwig and Vlachos[142] employed databases containing between 1,000 and 4,000 configurations. Lorenz et al.[143] used a database comprising 659 configurations to

Table 4-1 Comparison of database sizes and interpolation accuracy of *ab initio* energies obtained using various methods for three- and four-body systems.[136] (Reprinted with permission from American Institute of Physics.)

System	N(a)	Method energy error (kJ mol^{-1})	rms or absolute average	Reference
H_3	550	MSI[b]	0.154	144
$C + H_3^+$	406	MSI[b]	0.6	88
$H_3^+ + O$	1,000	MSI[b]	0.5	89
$BeH + H_2$	1,300	MSI[b]	0.35	145
$OH + H_2$	31	MSI[b]	34.2	146
$BeH_2 + H$	438	MSI[b]	0.35	90
$OH + H_2$	400	MSI[b]	1.	86
$NH + H_2$	600	MSI[b]	12.1	23
$N + H_3^+$	699	MSI[b]	0.05	92
BeH_3	500–1,000	MSI[b]	0.03 to 2.39	93
CH_3^+	500–1,000	MSI[b]	0.03 to 2.39	93
NH_3^+	500–1,000	MSI[b]	0.03 to 2.39	93
OH_3^+	500–1,000	MSI[b]	0.03 to 2.39	93
OH_3	500–1,000	MSI[b]	0.03 to 2.39	93
$NeCOH^+$	500–1,000	MSI[b]	0.03 to 2.39	93
$ArCOH^+$	500–1,000	MSI[b]	0.03 to 2.39	93
$KrCOH^+$	500–1,000	MSI[b]	0.03 to 2.39	93
HOOH	6,489	IMLS[b]	4.52	94
HOOH	300	IMLS[c]	8.24–15.6	29
HOOH	6,489	MSI[b]	1.95	94
H_2CN	330	IMLS[c]	1.40	31
H_2CN	830	IMLS[c]	0.53	31

(a) N is the number of *ab initio* points computed.

(b) MSI[b] is the moving Shepard interpolation procedure.

(c) IMLS is the interpolating moving least squares method.

investigate the dissociative adsorption probabilities of H_2 on a (2×2) potassium-covered Pd(100) surface.

In contrast, the study of the simultaneous four-center, *cis-trans* isomerization and two-center N-O bond dissociation reactions of HONO required a database of 21,584 configurations.[125] To study vibrational energies of the five-atom $H_3O_2^-$ system, Huang et al.[34] had to employ a database of 66,965 *ab initio* electronic energies. This large database was necessitated because of the high dimensionality of the five-atom system and also because of the "floppy" nature of the vibrational motions. When Chen et al.[35] investigated the association reactions of OH with NO_2, they had to use a database comprising 55,471 HOONO configurations. To investigate the four distinct, two-center bond dissociations and two, three-center dissociation processes occurring simultaneously in the six-atom vinyl bromide system, the converged database had to contain almost 72,000 points.[35]

Ludwig and Vlachos[142] have reported an investigation of the molecular dynamics of H_2 dissociation on Pt(111) and Cu(111) surfaces using neural networks and NS methods. In their investigation, the database for the NN fitting was obtained by using NS methods to examine the configuration points resulting from running direct dynamics (DD) trajectories in which density functional theory (DFT) was employed at each integration point to obtain the force field required for the solution of Hamilton's equations of motion for the system.

A general, robust solution to the self-starting problem for complex systems can be obtained by extending the direct dynamics (DD) method employed by Ludwig and Vlachos.[142] In DD, trajectories are computed by calculation of the force field at each integration point using some *ab initio* quantum mechanical method. Because of the huge number of computations required for each trajectory, the *ab initio* method chosen must usually be some form of DFT. As a result, only a very limited number of trajectories can be obtained. As the system under investigation increases in size, the computational requirements will quickly overwhelm the available computational resources. The basic problem is that all the information obtained about the potential-energy surface and the corresponding force field during the integration of the DD trajectory is discarded after the completion of the calculation. Therefore, each subsequent trajectory fails to profit from all the computational effort expended in obtaining the previous trajectories. As a result, DD has seen only limited usage.

By combining DD with NS and NN methods,[136,147] we can make both procedures far more powerful. Instead of employing an empirical PES, the sampling is initiated using trajectories computed by DD methods. Such a procedure obviates the need to develop an empirical PES prior to conducting *ab initio* investigations of the dynamics. This DD/MD/NS/NN method, therefore, becomes self-starting as well as efficient because the advantages of the MD/NS/NN procedure also become available.

In Section 4.2, it was noted that the final, converged potential surface is not sensitive to the choice of the particular empirical PES employed in the initial step provided it is reasonably accurate for all reaction channels occurring in the process.[136] As an illustration, we have employed two potentials proposed by Tersoff[13] for the initial step in the investigation of the potential surface for five-atom silicon clusters. These two potentials have different functional forms and different parameters. Figure 4-6 shows a plot of the energies computed using both potentials for a set of several thousand Si_5 configurations generated using MD methods on a silicon workpiece in nanometric cutting.[140] If the two potentials gave the same energies for a given configuration, all points would lie along a 45° line. Obviously, the two empirical potentials predict very different energies. Nevertheless, when neural nets are fitted to each set of *ab initio* energies obtained from the configurations generated by MD calculations using the two empirical potentials, Figure 4-7 shows that the predictions of the two nets for an independent testing set are in excellent accord.

Therefore, if DD samples all the important reaction channels, we expect that the ensemble of configurations generated by the DD trajectories in the initial step will be sufficiently representative of the configuration space important in the dynamics that our NS procedures will lead to convergence of the final database to the desired NN potential surface. Moreover, although the DD trajectories are executed using DFT methods with perhaps a smaller basis set, the electronic structure calculations on the stored configurations obtained in these trajectories can be carried out at any desired level of theory with a substantially larger basis set. In the example given below, this strategy was adopted by running the DD trajectories at DFT level and carrying out the final calculations at MP4(SDQ)/6-311G(g) level.

Proper sampling of configurations associated with reaction channels of minor importance can present difficulties when the DD/MD/NS/NN method is used. When an empirical potential surface is employed to initiate the sampling, thousands of trajectories can be computed in acceptable computational times. Those trajectories reacting in minor reaction channels can then be selected for sampling to ensure that all the important regions of configuration space are included in the database. When DD is used, we do not have the luxury of computing thousands of trajectories. Therefore, a different approach is required. The problem can be addressed in the following ways.

First, one can employ a nonstatistical selection of initial conditions for the DD trajectories. For example, one can employ projection methods[148] to partition the initial excitation energy among the various vibrational and rotational degrees of the system under study. This method permits the initial energy to be preferentially inserted into those modes that chemical intuition suggests will enhance the probability of reaction into selected channels. By combining this procedure with enhanced excitation energy, the DD trajectories can often be forced to sample the important regions of configuration space for minor reaction channels.

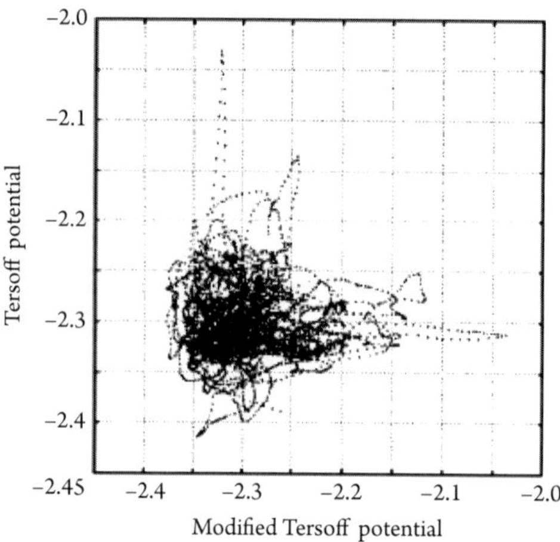

Figure 4-6 Energies (eV) computed using two different Tersoff-type potentials for a set of several thousand Si_5 configurations generated using MD methods in a cutting simulation on a silicon workpiece. If the two potentials gave the same energies for a given configuration, all points would lie along the 45° line[136] (reprinted with permission from American Institute of Physics).

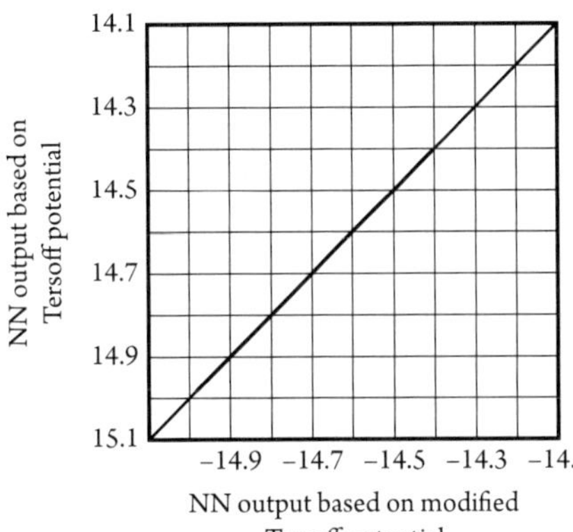

Figure 4-7 Predictions of two different neural nets fitted to *ab initio* energies for Si_5 configurations generated by MD calculations using the two potentials whose results are shown in Fig. 4-6. The plotted points are for an independent testing set. If the two NN produce the same interpolated energies, all results will lie along the 45° line. All energies are in eV[136] (reprinted with permission from American Institute of Physics).

Second, if necessary, the DD trajectories for the reverse reaction can be computed. That is, the DD calculations are initiated by starting in the product channel for the minor reactions of interest. These trajectories are followed until the system reaches the reactant configuration for the process of interest. Since microscopic reversibility ensures that the reaction pathways for both forward and reverse processes will be identical, a proper sampling of the reverse trajectory will provide the configurations required to make the MD/NS/NN method self-starting. It is reasonable to anticipate that this reverse method will be effective since processes that are severely constrained in one direction are often facile in the opposite direction.

Finally, one can initiate the DD trajectories starting at the saddle point on the PES. By proper selection of the initial momenta, the DD trajectory can be forced to proceed either toward the product or the reactant configuration space. Making both selections in turn permits the entire configuration space for the process to be sampled.

It may be noted that the DD procedure will not only produce a new DD/MD/ NS/NN method that is self-starting, it will also serve to greatly enhance the efficiency of DD calculations if one is interested in the dynamics on the *ab initio* surface at the level at which the DD trajectories have been run. No longer will DD suffer from the fact that all the information obtained about the PES and the corresponding force field during the integration of a trajectory is discarded after the completion of the calculation. This information can be retained in a database. An NN is then fitted to this database. After this, the investigation can proceed on the NN surface without the need for further DFT calculations.

To provide an illustrative example of the operation of the DD/MD/NS/NN method, it has been applied to the investigation of the N-O bond dissociation reaction of HONO.[147] With four atoms and both the four-center, *cis-trans* isomerization and two-center, N-O dissociation reaction channels open, this system is sufficiently complex to be a good test of the method.

In this example, attention is focused on the question of whether the final results given by the neural networks obtained by employing (a) direct dynamics and (b) an empirical potential to affect the initial sampling are in agreement. Since we generally initially sample only a few thousand points in the very large configuration space important in complex reactions, it is possible that the two methods will lead to final NN networks that predict very different dynamics. On the other hand, if the two methods lead to very similar NNs that predict similar dynamics, this would suggest that the two initial sampling methods are essentially equivalent. The test system employed is the *cis-trans* isomerization and N-O bond dissociation dynamics of HONO. Since an *ab initio* investigation of the HONO reaction dynamics has recently been reported by Le and Raff[125] using MD/NS/NN methods, we have a good benchmark against which the DD/MD/NS/NN results can be compared.

Le and Raff[125] investigated the isomerization and dissociation dynamics of HONO on an *ab initio* potential surface obtained by fitting the results of electronic structure calculations at 21,584 configurations using previously described MD/NS/NN methods and a semi-empirical PES developed by Guan and Thompson.[149] The electronic structure calculations were executed using GAUSSIAN 98 with a 6-311G(d) basis set at the MP4(SDQ) level of accuracy.[141a] The average absolute interpolation error of the NN fit was found to be 0.017 eV (1.64 kJ mol^{-1}).

To test the DD method, we initiate the sampling of the HONO configuration space by using DD with the "ADMP" command of GAUSSIAN-03[141b] at the B3LYP/6-31G(d) level with a time step of 0.10 fs. To ensure that some trajectories reach the phase space corresponding to the product space (NO + OH), the initial configuration corresponding to bond length r(N-O) for the N-O bond has been taken ~ 2.5 Å while keeping other internal coordinates r(N = O), r(O-H), $\angle$ONO, $\angle$ONH, and torsion angle nearly equal to that of the *cis* or *trans* ground state. The DD trajectories were run for 4,000 integration time steps with the initial nuclear kinetic energy equal to ~ 2.7 eV. One trajectory starting from the *cis* equilibrium as well as one trajectory starting from the *trans* equilibrium state with nuclear kinetic energy equal to 0.14 eV has been run for 500 integration time steps to sample the configurations near the equilibrium states of the system.

Configuration points are selected for inclusion in the database on the basis of distance criteria. A point i occurring during the sampling trajectories is accepted if its distance d_{ij} from each of the previously accepted n points is greater than d_{lim}. Here, d_{ij} is the distance between the points in the six-dimensional hyperspace spanned by the six internuclear distances, $(r_1, r_2, \ldots, r_6)$. It is defined by

$$d_{ij} = \sqrt{\left[\sum \left(r_k^{(i)} - r_k^{(j)} \right)^2 \right]},$$

(4-7)

where the summation runs over all six internuclear distances. The value of d_{lim} is chosen to be 0.1 Å except for the trajectories run near *cis* and *trans* ground states for which d_{lim} is chosen to be 0.01 Å. The initial sampling comprises 6,085 configurations selected from 25 DD trajectories. MP4(SDQ)/6-311G(d) calculations were performed to obtain the potential energy for each of these configurations. This is identical to the level of electronic structure theory employed by Le and Raff[125] in their MD/NS/NN study.

The final phase of the configuration space sampling is executed in a manner analogous to that employed when NS/NN methods are used.[136] A two-layer (6-40-1) NN is employed to fit the *ab initio* 6,085-point database generated from the results of the 25 DD trajectories. The Levenberg-Marquardt algorithm[117,135] is used to affect the training. This NN-fitted potential-energy surface is employed to compute 10,000 trajectories, each with an energy in the range 3.1–3.3 eV. These trajectories are integrated for 5 ps or until N-O bond scission occurs. From these trajectories, 3,827 new configurations have been obtained using novelty sampling procedures[126,136] to guide the selection of new configurations. These new configurations are then added to the already existing 6,085 configurations and a new NN is fitted to this database. This iterative procedure is repeated five times to obtain a database comprising 20,192 configurations. Convergence of the final NN PES is determined using NS criteria. The NN fit to this database is taken as the final HONO potential surface.

Figure 4-8 shows the variation of the potential energy V predicted by the NN with the *ab initio* energy values for all configurations.[147] The difference of V and *ab initio* value of energy is the fitting error. The value of the absolute error averaged over all configurations is found to be 0.022 eV. As expected,[125,126] the error tends to increase as the energy increases. This result is a consequence of the increased size of the configuration space energetically available to the HONO system at higher energies. As the size of the configuration space increases, the sampling becomes more sparse, and the fitting error increases. Figure 4-9 gives the error distribution curve. We note that more than 67% of the configurations have an absolute error less than 0.03 eV.[147]

The results shown in Figure 4-10 allow us to infer that the configuration space covered by the networks obtained by Le and Raff[125] and that obtained by Agrawal et al.[147] are appropriate for the purpose of computing the chemical properties in the range of energy considered. Had there been a significant number of "holes" in the configuration space sampling obtained by the two methods, it would have been reflected in the magnitude of the predicted energies differences between the two networks. The two networks are essentially equivalent.

In addition to the direct comparison between the two networks shown in Figure 4-10, Agrawal et al.[147] have shown that the calculated fundamental HONO vibrational frequencies predicted by the two NNs show an average difference of

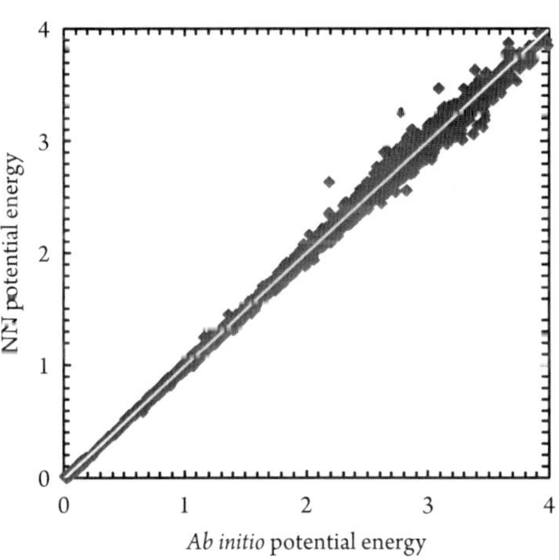

Figure 4-8 Comparison of the potential in eV predicted by the feed forward (6-40-1) NN obtained using novelty sampling of configuration space initiated using DD trajectories with the computed *ab initio* MP4(SDQ)/6-311G(d) energies the final database comprising 20,192 HONO configurations[147] (reprinted with permission from American Chemical Society).

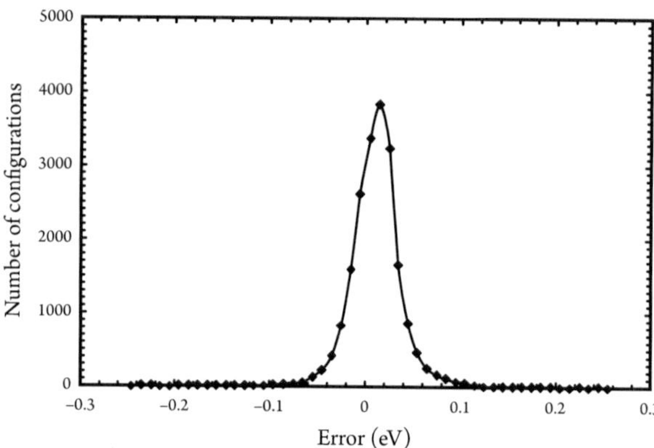

Figure 4-9 Distribution of the differences between (6-40-1) NN obtained using DD to initiate the sampling and the computed *ab initio* MP4(SDQ)/6-311G(d) energies for the final database comprising 20,192 HONO configurations. The average absolute difference is 0.022 eV[147] (reprinted with permission from American Chemical Society).

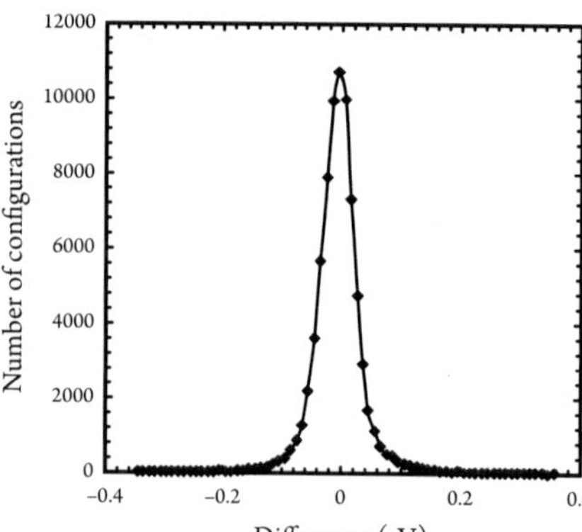

Figure 4-10 Distribution of the differences in energies in eV predicted by the NN obtained using DD to initiate the sampling of configuration space and that reported in Reference 113 where a semi-empirical PES was employed for 75,388 HONO configurations obtained in 1,000 trajectories at an internal energy of 3.1 eV. The average absolute difference is 0.029 eV[147] (reprinted with permission from American Chemical Society).

about 4%. The trajectory-computed N-O dissociation rates at 3.1 and 3.3 eV internal energy with the energy in excess of zero-point energy equally distributed across all vibrational modes differ by about 15%. The rate coefficient deviation expected because of the 0.029 eV average difference in energy between the two NNs is sufficient to account for this 15% variation. Further, the dissociation rates are found to satisfy the RRK equation leading to a frequency factor in agreement with the fundamental frequency of vibration corresponding to the N-O stretching mode.

It is clear that DD trajectories executed at the density functional level of theory with the B3LYP/6-31G(d) procedure for incorporation of the correlation energy provides a robust and powerful self-starting method for initiation of configuration space sampling that obviates the need for a semi-empirical PES.

4.4. CONFIGURATION SAMPLING USING A GRADIENT FITTING METHOD

In this section, we describe a configuration sampling method developed and implemented by Le et al.[124] that does not require trajectory integration which is based primarily on sampling procedures that force an NN fitted to the potential energies to also provide a good fit to the surface gradients. The method produces a more uniform density of configuration points in the final database and more accurate fitting of force field as well as potential energy.

It was noted in Section 4.1 of this chapter that if the density of points in the final database is very non-uniform, the quality of the NN fit will be degraded because too much weight will be placed on fitting regions with a high density of points. When trajectory integration is employed to sample configuration space, variable interval sampling[122] can be employed to ensure a more uniform distribution of sampling points. This method has been described in Section 4.1 of this chapter. When trajectory integration is not employed in the sampling, different methods must be used to produce more uniform distributions of sampling points.

Although most fitting methods used to date have concentrated on fitting the target potential energies, it is the force field that is most important in the execution of MD calculations. The presumption is that if the potentials are accurately fitted, the gradients will also be accurate. However, this is not necessarily the case. The CFDA method[121] described in Chapter 3, Section 3.2, forces the weight and bias matrices of the NN to fit both the function and its gradient.[121,122] Although the method is highly accurate and robust, the standard NN MATLAB[105] toolbox cannot be used to train the network. The sampling method described in this section provides increased accuracy for the gradients of the potential but still permits the network to be trained using the MATLAB toolbox.

The gradient sampling method was originally applied to the development of an NN PES for HOOH.[118] We will use this illustrative example to describe the method.

The configuration coordinates employed to describe the HOOH molecule consist of the three bond distances, the H^1-O^2-O^3 and O^2-O^3-H^4 angles, and one dihedral angle, ϕ. This set of coordinates is used as the input vector for the NN. The *ab initio* potential energy is the NN training target. Le et al.[124] computed the *ab initio* energies at the MP2 level[150–154] with a 6-31G* basis set using the Gaussian suite of programs.[141] Since the products of the HOOH dissociation are two radicals, calculations at this level are not expected to be sufficient to produce chemical accuracy. The PES so obtained is intended to provide a realistic test of the proposed sampling procedure and NN fitting using a large database of *ab initio* energies, not a quantitatively accurate assessment of the reaction dynamics.

The first step of gradient sampling is to construct the initial database of N_o configurations for the system by optimizing the HOOH structures for a series of values of two of the six coordinates required to specify the HOOH configuration using the partial optimization procedure offered by the Gaussian suite of programs.[141] Using this procedure, Le et al.[124] found the minimum HOOH potential energy with one or more z matrix parameters held constant. The remaining parameters were varied in such a way as to minimize the potential energy by using the numerical first and second derivatives of the potential with respect to those parameters. The three molecular bond distances H^1-O^2, O^2-O^3, and O^3-H^4 are denoted as r_1, r_2, and r_3, respectively, while θ_1 and θ_2 are the H^1-O^2-O^3 and O^2-O^3-H^4 angles, respectively, and ϕ is the dihedral angle. In each partial optimization, ϕ and one of the other five

Table 4-2 O-H and O-O bond distances and the O-O-H angle ranges investigated in the initial sampling of the HOOH configuration space. The second column gives the number of configurations added during the calculations. In each sampling, the dihedral angle is varied from zero to 180°, as described in Ref. 124. (Reprinted with permission from American Institute of Physics.)

Varying Parameter	Number of Configurations	Range
O-H bond (r_1)	722	0.71 – 1.91
O-O bond (r_2)	704	1.12 – 2.92
O-O-H angle (θ_1)	722	40.0 – 169.8
Total (N_o)	2148	

parameters were set at certain constant values in energetically reasonable ranges. Subsequently, the four remaining parameters were adjusted to minimize the potential energy. This process was repeated as ϕ is varied from zero to 180° while the other fixed parameter, either r_1, r_2, or θ_1 is varied over ranges given in Table 4-2.

At the conclusion of this initial step, 2,148 HOOH configurations had been selected to provide the initial database. The maximum energy configuration in this ensemble was 4.36 eV relative to the energy of the equilibrium HOOH structure.

In the next phase, Le et al.[124] obtain more data points so as to cover the hyperspace that is energetically important for chemical reactions. The process begins by scaling all input data and the target output using Equation (3-34).

In regions of hyperspace where there are not enough data points to accurately characterize the potential surface, the probability that an NN fit to the current database of N_o points will extrapolate in that region is high. As a result, the NN-predicted first derivatives of the potential energy with respect to bond distances and angles will exhibit significant differences compared to the derivatives predicted by MP2 calculations. This concept allowed Le et al.[124] to identify those regions lacking a sufficient number of data points. To execute the actual search, a temporary NN fit based on the current database is produced. For a single point p of the N_o configurations, the NN derivatives are computed and compared to the MP2 derivatives, which have already been computed. If their difference is higher than 0.03, a search for more data points around point p is executed as follows:

- A unit vector v is randomly generated. A new point is computed as $P_{new} = P+k \cdot grid \cdot v$, where $grid$ is the average distance between the data in the first set. For the present HOOH system where the initial database contained 2,148 points, $grid = 0.119$ in scaled units. k is a scalar random number between 2 and 3.
- If the distance between p_{new} and any present point in the database is smaller than $grid$, the point p_{new} is rejected. In addition, the NN potential at point p_{new} is computed. If the computed potential is greater than 4.36 eV, p_{new} is rejected; otherwise, it is saved.

- Since the HOOH configuration hyperspace is six dimensional, this process is executed for 2^6 times around point p. Once a new point p_{new} is selected, the loop is terminated.

The imposition of the distance criterion in Step 2 assists the production of a database with a more uniform density of points by preventing points very near each other from accumulating in the database. The requirement that the energy at point p_{new} be less than some maximum serves to limit the configuration space volume thereby making the sampling process computationally feasible.

Le et al.[124] perform the derivative analysis for all N_0 points to find those points that are poorly characterized by the current database. The process of finding new points about these points is then executed. Let us assume that this search produces N_1 new points. Le et al.[124] now continue to search for more data points around these N_1 points by execution of the processes previously described to obtain an additional N_2 points. The search procedure about these N_2 points is then executed to obtain N_3 more points. This procedure is terminated when no more configurations can be found or when the number of new configurations reaches 5,000. When termination is reached, a total of $(N_1 + N_2 + N_3 + \cdots)$ new configuration points will have been identified.

Subsequently, *ab initio* calculations of potential energies and derivatives at the desired level (in the present example MP2 level) for the $(N_1 + N_2 + N_3 + \cdots)$ new configurations are executed. Finally, these energies are compared to those predicted by the temporary NN fit. If the percentage difference is less than 1%, the point is discarded, as it is already well described in the database. Inclusion of such points will lead to a non-uniform distribution of points in the database. By rejecting such points, the density of points in the final database will be much more uniform. If the number of rejected points is N_R, the expanded database will contain M_1 points, where $M_1 = N_0 + N_1 + N_2 + N_3 + \cdots - N_R$ configuration points.

This procedure is executed iteratively with N_0 being replaced by M_1 until the number of new points found by the search procedure becomes zero or very small. In

Table 4-3 Number of new configurations identified by the three-step selection process described in the text, total number rejected and accepted because of energies differences between MP2 and NN calculations for each iteration in the selection process. The totals are given in the last row of the table.[124] (Reprinted with permission from American Institute of Physics.)

Iteration #	# Points Identified	# Points Accepted	# Points Rejected
1	4,314	1,803	2,511
2	5,001	3,607	1,394
3	4,997	3,738	1,259
4	1,958	1,508	450
Total	16,270	10,656	5,614

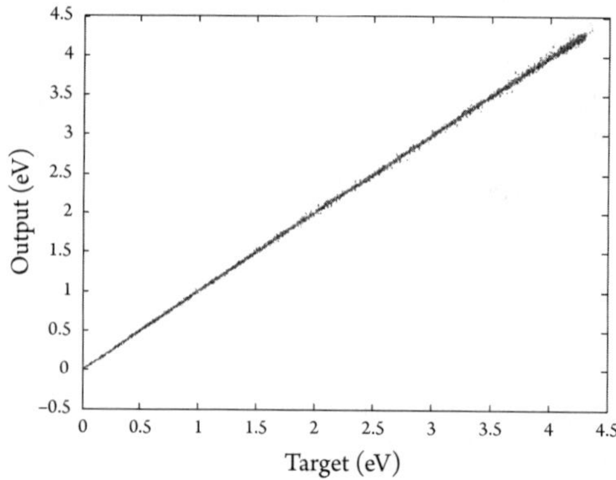

Figure 4-11 The outputs predicted by the NN committee fit are plotted against their corresponding targets for the training set. In the low energy region near equilibrium, the MP2 potential energies are fitted with excellent accuracy. This accuracy results in good agreement between normal mode wave numbers computed from MP2 calculations and from the NN committee. The fitting errors rise in the high energy region due to the increased volume of the HOOH hyperspace[124] (reprinted with permission from American Institute of Physics).

the first iteration for the HOOH system, 1,803 new configurations were obtained. Four iterations were executed before the convergence criteria were satisfied, as discussed above. The numbers of selected new configurations during each iteration, $(N_1 + N_2 + N_3 + \cdots)$, the number of these points rejected, and the number added to the database are given in Table 4-3. A total of 34.5% of the 16,270 points obtained by the selection procedure were rejected in order to produce a more uniform density of configuration points in the database. When combined with 2,148 initial data points, at the termination of the gradient sampling process, 12,804 data points were sampled. The final set contains 25,608 points since the symmetry of hydrogen peroxide permits the duplication of data by exchanging r_1 and r_3, and θ_1 and θ_2. This database is fitted to a five-member NN committee using the procedure described in Chapter 3, Section 3.8. The average potentials and gradients predicted by the committee is the final fit to the H_2O_2 database.

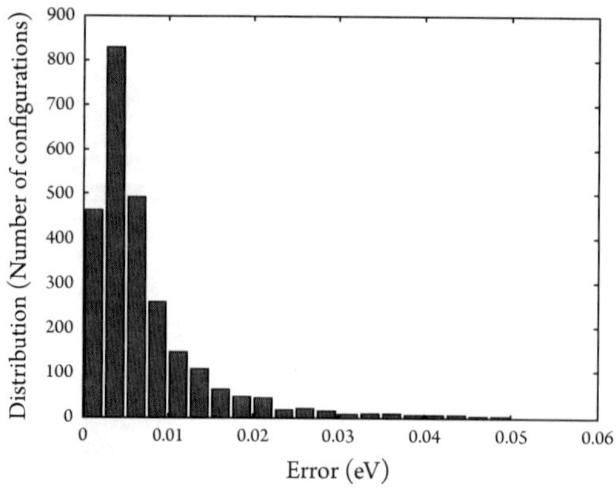

Figure 4-12 Distribution of the absolute testing set errors for the NN committee. The small fitting errors (from 0 to 0.01 eV) clearly dominate the distribution. The average absolute testing set error is 0.0060 eV[124] (reprinted with permission from American Institute of Physics).

Table 4-4 Fundamental vibrational wave numbers resulting from various calculations.[124] (Reprinted with permission from American Institute of Physics.)

	Wave Number (cm⁻¹)			
Mode Description	*NN Predicted*	*MP2/6-31G**	*B3LYP/cc-pVTZ*	*PCPSDE[51]*
Symmetric OH stretching	3723	3740	3766	3778
Antisymmetric OH stretching	3714	3738	3765	3762
Symmetric OOH bending	1456	1463	1438	1453
Antisymmetric OOH bending	1284	1323	1321	1297
O-O stretching	929	928	954	889
Torsion	331	338	366	392

At this point, the quality of the fit is tested by evaluation of the extent of agreement between the normal mode fundamental frequencies obtained from MP2 and NN calculations, and the NN potential energies of random data points obtained from molecular dynamics trajectories with *ab initio* energies. Figures 4-11 and 4-12 show the agreement obtained using the NN fit to the converged database. The normal mode wave numbers given by the final NN fit are shown in Table 4-4. The excellent agreement between the wave numbers predicted by MP2 calculations and those obtained from the NN is apparent.

Because the present sampling method is based on a comparison of the NN-computed gradients at configuration points currently in the database with the target gradients, we are in essence fitting both the potential function and the gradients. As a result, the final NN closely fits both the potential and the gradients. The excellence of the fit is reflected in the very low average absolute testing set error of 0.0060 eV (0.58 kJ mol⁻¹) for the five-member NN committee and by the close agreement of the normal-mode vibrational frequencies predicted by the NN and MP2 calculations. As an added advantage, the standard NN toolbox in MATLAB[133] can be used to affect the NN fitting.

5

APPLICATIONS OF NEURAL NETWORK FITTING OF POTENTIAL-ENERGY SURFACES

5.1. INTRODUCTION

In this chapter, several examples of NN fitting of databases obtained using either *ab initio* electronic structure methods or an empirical potential will be discussed. The objective of this presentation is not to provide a complete and comprehensive review of the field nor is it to acquaint the reader with the details of the reaction dynamics of the particular systems employed as examples. It is rather to provide a clear picture of the power and limitations of NN methods for the investigation of reaction dynamics. We begin with a brief overview of the literature in the field.

Neural networks provide a powerful method to effect the fitting of an ensemble of potential energy points in a database. In 1993, Blank et al.[40,41] employed an NN to fit data derived from an empirical potential model for CO chemisorbed on a Ni(111) surface. Two years later, these same investigators also examined the interaction potential of H_2 on a Si(100)-2 × 1 surface using a data set comprising 750 energies computed using local density functional theory. To the best of our knowledge, these were the first two examples in which NNs were employed to provide the PES for a dynamics study.

Hobday et al.[42,156] have investigated the energies of C-H systems by using a Tersoff potential form[13] in which the three-body term is replaced by an NN comprising five input nodes, one hidden layer with six nodes, and an output layer. In this work, the five input elements are computed by consideration of the bond type, i.e., C-C or C-H, the three-body bond angle θ, which is input to the NN in the form $(1 + \cos \theta)^2$, the connectivity of the local environment, and the second neighbor information. The method was applied to carbon clusters and a wide variety of alkanes, alkenes, alkynes, aromatics, and radicals. Comparison of the atomization energies obtained using the NN potential surfaces with experimental values showed the errors for 12 alkanes, 13 alkenes, 4 alkynes, 7 aromatics, and 12 radicals to lie in the ranges zero to 0.3 eV (alkanes), 0.1 to 1.5 eV (alkenes), 0 to 0.5 eV (alkynes), zero to 1.0 eV (aromatics), and zero to 2.8 eV (radicals).[156]

Tafeit et al.[43] and Gassner et al.[44] have successfully employed NN methods to fit the torsional energies in large biological systems and for the H_2O-Al^{3+}-H_2O system, respectively. Brown et al.[45] have employed a two-layer NN with 32 neurons in the hidden layer to obtain a fit to the PES for the van der Waals systems HF-HF and HF-HCl.

More recently, Lorenz et al.[46] have used an NN to represent the PES for H_2 dissociative adsorption on the (2×2) potassium-covered, Pd(100) surface. These authors conclude that NNs provide an efficient tool to study reaction processes where extensive statistics are required. An even more extensive study of this system by the same investigators was reported in 2006.[157] Here, the authors employed a three-layer, symmetry-adapted (8-50-50-1) NN with sigmoid and linear transfer functions in the hidden and output layers, respectively. The NN was fitted to the analytic PES developed by Gross et al.[158] by fitting DFT calculations reported by Wilke and Scheffler.[159,160] The NN fitting was executed using the extended Kalman filter[161] with 80,685 individually weighted training points taken from the analytic PES. Fitting accuracy was evaluated using 91,665 randomly selected points on the analytic PES. The reported RMS fitting error was 4.4 kJ mol^{-1}. The authors report careful evaluations of the MD-calculated, H_2 sticking probabilities as a function of incident kinetic energy for different sizes of the NN training set and provide comparisons with results obtained using the analytic PES. On the basis of these results, they conclude that if the point sampling is done using grid techniques with equidistant sampling, accurate fitting requires 10^4 to 10^5 points. However, if a more efficient sampling is employed, this number can be reduced by an order of magnitude.

Prudente et al.[162] employed a (3-40-1) NN with exponential transfer functions in the hidden layer and a linear function in the output layer to fit ab initio energies obtained by Dykstra and Swope[163] and by Meyer et al.[164] for the H_3^+ molecular ion. The resulting PES was employed to investigate the vibrational energy levels of the system. The reported rms fitting error was 1.2×10^{-4} au or 26.3 cm^{-1}. The average deviation of the computed vibrational levels from the experimental values was 3.3 cm^{-1}. The same system was investigated more recently by Rocha Filho et al.[165] using a three-layer NN with the architecture (3-12-3-1). The network was trained using the 69 ab initio energies reported by Cencek et al.[166] The average error of the computed vibrational energy levels of H_3^+ was about 2 cm^{-1}.

Prudente and Neto[167] have reported an investigation of the photodissociation process of HCl$^+$ using NNs to fit the $X^2\Pi$ ground state, the $(2)^2\Pi$ electronically excited state, and the dipole moment function for the transition between these two states. In each case, a (3-4-1) network was employed with sigmoid transfer functions in the hidden layer and a linear function for the output layer. The networks were trained using the *ab initio* electronic energies computed by Pradhan et al.[168] The dissociation cross sections were computed as a function of excitation energy. Comparison with the previous results obtained by Prudente et al.[169] using spline methods to obtain the potential surfaces showed good agreement.

Sumpter et al.[73] have shown how neural networks can be employed to predict the temporal details of intramolecular internal energy flow. Their methodology is based on training an NN to recognize the relationships between the phase-space points along a classical trajectory and the associated kinetic mode energies for stretching, bending, and torsional vibrations in both reactive and nonreactive systems. The method is illustrated by application to H_2O_2 using a semi-empirical potential-energy surface. This application of NNs is discussed in more detail in Section 7.2.

Sumpter and Noid[170] have also shown how NNs can be employed to obtain the best parameters for an empirical PES using the measured vibrational spectra of the molecular system. The basis of the method is to train NNs to learn the relationship between the measured spectra and the topology of a multidimensional PES. This work is discussed in more detail in Section 8.2.

Manzhos et al.[47] have employed a nested NN approach to obtain extremely accurate fits to the PES for systems in the neighborhood of their equilibrium configurations. In this approach, one NN is employed to obtain a PES that is reasonably close to the points in the database. A second NN is then developed that fits the difference between predictions of the first NN and the points in the database. Using this approach, the authors report fitting errors on the order of 1 cm^{-1} for H_2O and 2 cm^{-1} for HOOH and H_2CO. To the best of our knowledge, these are the most accurate fitting results reported for the investigation of the vibrational structure of three- and four-atom molecules.

In most applications of two-layer NNs, the transfer functions used between the hidden layer and the output layer are sigmoid functions. Manzhos and Carrington[48] have recently reported an investigation of the use of exponential transfer functions. The potential advantage of such transfer functions lies in the fact that the exponential of a sum is the product of the individual exponentials. This property permits the resulting PES to be written in a sum of products form, which is desirable from the point of view of executing quantum dynamics calculations. Their investigations[48] show that the fitting accuracy of exponential transfer functions is only slightly less than that of sigmoid functions for NNs containing an equal number of neurons.

Raff et al.[136] have developed a novelty sampling—neural network method for sampling multi-dimensional configuration space and then obtaining an accurate fit to the *ab initio* database using neural networks. The novelty sampling provides both the means by which to ensure adequate sampling of the multi-dimensional configuration hyperspace and to test for convergence of the resulting NN potential surface. The details of this method have been discussed in Chapter 4, Section 4.2. To date, the method has been used to investigate four-center, *cis-trans* isomerization and two-center, N-O bond dissociation in HONO[125,147] and the simultaneous dissociation reactions of the six-atom vinyl bromide system that include four, two-center, bond dissociation processes and two, three-center dissociation reactions.[126]

Agrawal et al.[171] have employed NNs to investigate the gas-phase dissociation dynamics of SiO_2 to either $SiO + O$ or $Si + O_2$. These non-adiabatic processes involve singlet, triplet, and pentet potential surfaces.

Pukrittayakamee et al.[122] have reported a study of the exchange and abstraction dynamics for the reactions of hydrogen atoms with molecular HBr using the CFDA NN method described in detail in Chapter 3, Sections 3.10 and 3.11.[121] The NN was fitted to the analytic surface reported by Sudhakaran and Raff[137] using the functional form developed by Kuntz et al.[172] The CFDA was shown to produce nearly exact fits to the analytic surface with an rms deviation of 1.2 cm^{-1} (0.014 kJ mol^{-1}).

Le et al.[124] have reported an MD investigation of the HOOH dissociation dynamics to two OH radicals using an NN fit to an *ab initio* database comprising 25,608 energies computed at the MP2(SDQ) level of theory with a 6-31G* basis set. The selection of configurations for inclusion in the database was executed using a new selection method involving gradient fitting. This method has been described in Chapter 4, Section 4.4. The fitting error was 0.58 kJ mol^{-1} for the five-member NN committee.

Behler and Parrinello[173] (BP) have introduced a generalized NN approach to obtain PESs in large, many-dimensional systems. In their method, the total potential energy of the system is written as the summation of the individual atomic contributions. Neural networks are employed to generate the individual atomic contributions in a manner that incorporates the effects of the local environment about the atom. The procedure automatically incorporates exchange symmetry between identical atoms. The details of the BP method are discussed in Section 5.8.

Lagris et al.,[174] Sugawara,[175] and Nakanishi and Sugawara[176] have employed NNs to solve the vibrational Schrödinger equation for some model one- and three-dimensional systems. In these applications, the vibrational levels of the model systems have been determined one at a time. Most recently Manzhos and Carrington[76,77] have extended this work and demonstrated that by employing radial basis NNs and optimizing a set of neurons for all computed levels, several vibrational levels can be determined simultaneously. They have also presented the first application of this method to a real molecular system, H_2O, where five vibrational levels were determined. This application of NNs is discussed in detail in Chapter 10, Section 10.1.

There are several review articles available that provide succinct summaries of research involving the use of NNs in various aspects of chemical reaction dynamics and structure. Burns and Whitesides[177] have reviewed NN applications in biological sequencing, the interpretation of spectra, and applications related to the analysis of unordered homogeneous data. The review comprises 123 references. A second review by Sumpter et al.[73] describes the use of NN in studies of chemical reactivity, Monte Carlo investigations of molecular dynamics, and PES fitting as well as the same topics as the Burns and Whitesides[177] review. This review cites 299 references. Handley and Popelier[178] have published the most recent review of the field. This review, which cites 52 references, concentrates primarily on the applications of NN to fitting PESs. Zupan and Gasteiger,[179] in their book *Neural Networks in Chemistry and Drug Design,* provide a detailed discussion of different types of neural networks.[80] The 21 chapters contain 215 end-of-chapter references some of which are duplicated.

The remainder of this chapter discusses some of the above systems in more detail to provide the reader with a clear picture of the role NNs can play in the investigation of reaction dynamics.

5.2. NEAR EQUILIBRIUM STRUCTURES— VIBRATIONAL STATE STUDIES

5.2.1. The H_3^+ Molecular Ion

The simplest three-body system from the point of view of electronic structure calculations is the H_3^+ molecular ion. It is unusual in that it is stable relative to dissociation to either H_2^+ and H or to H_2 and H^+. The molecule is readily formed in hydrogen plasmas of medium densities by the four-center exchange reaction

$$H_2 + H_2^+ \rightarrow H_3^+ + H + 167.4 \text{ kJ mol}^{-1}.$$

Because of this, Carney and Porter[180] suggested that it may well be found in the atmosphere of extraterrestrial bodies that are hydrogen-rich. This suggestion has now been realized.

Prudente et al.[162] utilized the *ab initio* electronic structure energies for H_3^+ reported by Dykstra and Swope[163] and the standard NN formalism discussed in Chapter 3 to obtain an analytic PES for the system. The standard NN treatment does not automatically incorportate a consideration of exchange symmetry that a molecular system may possess. Since any pair of hydrogen nuclei can be exchanged in H_3^+ without altering the system potential energy, this feature must be addressed.

There are several methods available for introducing symmetry recognition when an NN is employed to fit *ab initio* data to obtain the PES. The most direct method is to train the NN with additional data obtained by exchanging appropriate coordinates. Since the H_3^+ potential is invariant under permutation operations on the three nuclei, we must have

$$\begin{aligned}
V\left(r_{AB}, r_{BC}, r_{CA}\right) &= V\left(r_{BC}, r_{AB}, r_{CA}\right) \\
&= V\left(r_{CA}, r_{BC}, r_{AB}\right) \\
&= V\left(r_{AB}, r_{CA}, r_{BC}\right) \\
&= V\left(r_{CA}, r_{AB}, r_{BC}\right) \\
&= V\left(r_{BC}, r_{CA}, r_{AB}\right).
\end{aligned}$$

where r_{ij} is the internuclear distance between hydrogen atoms i and j. Using this property, the data set reported by Dykstra and Swope[163] may be augmented using permutation operations on the nuclear coordinates. If the NN is trained using this augmented database, symmetry recognition will be incorporated. Prudente et al.[162] used this method with a (3-40-1) NN and exponential transfer functions to obtain an analytic fit for the H_3^+ potential which they denoted as an unsymmetric neural network (UNN)-PES.

A second method has been employed by Raff et al.[136] when using an NN to fit the *ab initio* database for Si_5 clusters. Instead of augmenting the database with all possible permutation exchanges, the NN is trained for only one permutation and appropriate permutation operations are performed on the input data to obtain the permutation to which the NN has been trained. For example, for the H_3^+ molecular ion, one might train the NN using the A-B-C permutation to obtain $V(r_{AB}, r_{BC}, r_{CA})$. If the potential for the A-C-B permutation is needed, the coordinates of C and B are exchanged prior to presenting the data to the NN. This method has the advantage of allowing the use of a much smaller training database for the NN. It has the disadvantage that nuclear coordinates must be properly permuted before presentation to the NN and after the output is obtained.

Prudente et al.[162] developed a third method to incorporate symmetry recognition into the NN for the H_3^+ system. They first defined a symmetry operator $\varepsilon(ijk)$ by

$$\varepsilon(123) = \varepsilon(132) = \varepsilon(213) = \varepsilon(231) = \varepsilon(312) = \varepsilon(321) = 1$$

$$\text{and } \varepsilon(ijkl) = 0 \text{ for all others.} \tag{5-1}$$

with $y_1^0 = r_{AB}$, $y_2^0 = r_{BC}$, and $y_3^0 = r_{CA}$, they defined symmetric neurons by

$$\tilde{y}_m^1 = \sum_{i=1}^{3} \sum_{j=1}^{3} \sum_{k=1}^{3} \varepsilon_{ijk} \, \phi\left(w_{m1} y_i^0 + w_{m2} y_j^0 + w_{m3} y_k^0 + w_{m0} \right), \tag{5-2}$$

where $\phi(z)$ is the transfer function and the w are the weights. Prudente et al.[162] then used a three-layer NN in which the first hidden layer used symmetric neurons defined by Equation (5-2) while the second hidden layer and the output layer were defined in the usual form. To train this network, they employed backpropagation methods[74,107,114,115] that were suitably modified to account for the different form of the first hidden layer.

Using this symmetric formulation, Prudente et al.[162] fitted the 69 point database reported by Meyer et al.[164] for H_3^+ using a (3-15-1) NN with an exponential transfer function for the hidden layer and a linear function for the output layer. This NN fit was denoted by symmetric neural network (SNN)-PES. The reported rms fitting error was 1.2×10^{-4} au or 26.3 cm^{-1}. The UNN-PES and SNN-PES were utilized with the quantum Monte Carlo method for computation of excited state properties developed by Ceperley and Bernu[181] to obtain the vibrational quantum levels for H_3^+. A total of 33 bound vibrational state energies were reported. Table 5-1 gives the results for the five states where the computed values were compared to experimental results. The average deviation of the computed vibrational levels from the experimental values was 3.3 cm^{-1}.

5.2.2. H_2O, HOOH, and H_2CO

As stated in the overview, Section 5.1, Manhzos et al.[47] have employed a nested NN approach to obtain extremely accurate fits to the PES for systems in the neighborhood of their equilibrium configurations. This approach has been shown to give excellent results for H_2O, for the double-well potential of hydrogen peroxide, H_2O_2, and for formaldehyde, H_2CO.

Table 5-1 Vibrational states of H_3^+ for UNN and SNN potentials obtained using the quantum Monte Carlo method developed by Ceperley and Bernu.[181] The data are given in cm^{-1}.[162]

State	UNN	SNN	Expt[a]
1	2522.9	2519.8	2521.4
2	3175.1	3172.8	3178.4
4	5002.0	4994.4	4997.965
5	5555.1	5547.1	5554.063
9	7499.4	7486.5	7492.91

(a) Refs. 182 and 183

To develop the nested approach in which one NN is utilized to obtain a good fit to the database and a second NN used to refine the results, Manzhos et al.[47] used reference potentials for each of their three test molecules. For H_2O, they employed the analytic potential suggested by Jensen.[184] For HOOH and H_2CO, potentials formulated by Kuhn et al.[51] and Carter et al.,[185] respectively, were used.

Manzhos et al.[47] sample configuration space using a multi-dimensional random selection of points whose energies are less than a cutoff value chosen to restrict the database to points whose energies are important for the purpose of computing low-lying vibrational levels. In practice, they construct their database by selecting configuration points using

$$\frac{E_{cut} - V^0(x_i) + \Delta}{E_{cut} + \Delta} > b_i \qquad (5\text{-}3)$$

where b_i is a random variable uniformly distributed on the interval $[0,1]$. E_{cut} is the energy cutoff value and $\Delta \ll E_{cut}$. For $V^0(x)$, a sum of equilibrium slices is employed. For each coordinate, a set of points is generated by fixing the other coordinates at their equilibrium values and taking 20 equally spaced values (one of which is the equilibrium value) for the coordinate in question. For each of the slice potentials, $V_{1D} < E_{max}$, where Emax is larger than the largest energy level to be computed and is the energy above which fitting is not desired. In general, E_{max} and E_{cut} are close and $E_{max} > E_{cut}$. Excel is used to obtain the 1-D fits to the slices.

The Levenberg-Marquardt algorithm[117,135] was used to train the two-layer NNs which employed sigmoid transfer functions in the hidden layer and linear functions for the output layer. Early stopping with a validation set[118] was used to prevent overfitting. Permutation symmetry for the hydrogen atoms in H_2O and H_2CO and the OH groups in HOOH was incorporated using the first method discussed in Section 5.2.1 of this chapter.

For H_2O, Manzhos et al.[47] used a database of 1,500 points with a second set of 1,500 points included to incorporate the hydrogen-atom symmetry. It was found that an NN with a (3-17-1) architecture was sufficient to produce a potential whose root mean square error (RMSE) for the testing set was 5 cm^{-1}. To decrease the RMSE to 1 cm^{-1}, 23 neurons had to be employed in the hidden layer. The two NNs are denoted as NN1 and NN2. Table 5-2 gives the mean, median, minimum, and maximum errors of the vibrational energy levels computed using the two NN fits to the H$_2$O database.

For hydrogen peroxide, the HOOH configuration was sampled using Equation (5-3) with the dihedral angle lying between 180° and 360°, E_{cut} = 11,000 cm^{-1}, and $\Delta = 0$.[186] This database of 5,000 points was augmented with additional points along equilibrium slices up to E_{max} = 18,000 cm^{-1}. Using a (6-36-1) NN, Manzhos et al.[47] were able to obtain a PES with an RMSE of 20 cm^{-1}. However, when attempting to converge the computed vibrational spectrum by increasing the size of the basis set and the quadrature grid size, holes were discovered in the PES near the outskirts of the fitted region because of the low density of sampling points in the database.

To prevent such holes, Manzhos et al.[47] developed a two-step fitting procedure. In the first step, an approximate, base NN to the full database is obtained. This NN is characterized by using a small number of neurons so as to ensure the absence of overfitting. As an operating principle, the number of neurons was chosen to be such that the total number of parameters in the base NN is at least an order of magnitude less than the total number of points in the database and so that the RMSE of the base NN is at least an order of magnitude greater than the target RMSE for the final PES. In the second step, a second NN is fitted to the difference between the potential predicted by the base NN and the target potential.

For the HOOH system using the two-step procedure with the base NN containing 57 neurons and the difference NN 175 neurons, Manzhos et al.[47] were able to obtain a fit to the HOOH database with an RMSE of 2 cm^{-1}. The mean absolute and median absolute errors were 1.70 and 1.11 cm^{-1}, respectively. The minimum and maximum deviations of the NN PES from the target potentials were 0.00 cm^{-1} and 37.0 cm^{-1}, respectively. The difference between the vibrational eigenvalues obtained with this two-step NN surface and those computed using the reference

Table 5-2 Absolute differences between vibrational energy levels computed on the two NN potential and those computed using the analytic potential reported by Jensen[184] for the HOOH system. Errors are reported in cm^{-1}.[47] (Reprinted with permission from American Institute of Physics and the authors.)

Property	NN1	NN2
Mean error	0.681	0.113
Median error	0.561	0.071
Minimum error	0.000	0.004
Maximum error	4.212	1.180

Table 5-3 Absolute differences between vibrational energy levels computed on the (50/119) NN potential and those computed using the analytic potential reported by Carter et al.[185] for the H_2CO system. Errors are reported in cm^{-1}.[47] (Reprinted with permission from American Institute of Physics and the authors.)

Property	Potential	NN Eigenvalues
mean error	1.05	0.728
median error	0.72	0.440
minimum error	0.00	0.000
maximum error	32.9	8.338

analytic potential reported by Kuhn et al.[51] had mean and median absolute errors of 1.605 cm^{-1} and 0.685 cm^{-1}, respectively. The minimum and maximum absolute deviations were 0.000 cm^{-1} and 11.83 cm^{-1}, respectively.

For formaldehyde, the same two-step NN approach was employed to develop the PES. Based on the quality of the results, Manzhos et al.[47] have suggested that the two-step procedure is probably useful for a large class of molecules.

The sampling for H_2CO was done in a manner very similar to that employed for hydrogen peroxide. Exchange symmetry considerations were incorporated using the first method discussed in Section 5.2.1 of this chapter. Two different two-step sets were examined. The first contained 51 neurons in the base NN and 94 in the difference NN. The second set had 50 neurons in the base NN and 119 in the difference NN. Table 5-3 gives the fitting results for the (50/119) potential and the corresponding computed vibrational eigenvalues relative to the analytic reference potential proposed by Carter et al.[185]

5.3. CFDA FITTING—THE H + H'BR → HBR + H' AND H$_2$ + BR REACTIONS

In Chapter 3, Sections 3.10 and 3.11, a new and improved NN method that simultaneously fits both the potential and its gradient was described. The method is denoted as the Combined Function Derivative Approximation (CFDA). The complete derivation of all relevant equations for implementation of the method have been given by Pukrittayakamee et al.[121] To provide an illustrative example of the method and its quality, Pukrittayakamee et al.[122] have applied it to the inelastic scattering, abstraction, and exchange reactions of the (H, HBr) system. These reactions are

$$H' + HBr \rightarrow H' + HBr^\dagger \quad \text{inelastic scattering,}$$

$$H' + HBr \rightarrow H'\text{-}H + Br \quad \text{abstraction,}$$

and
$$H' + HBr \rightarrow H'Br + H \quad \text{exchange.}$$

The system coordinates used to validate the CFDA training procedure are defined in Figure 5-1.

The system was evaluated on the LEPS-type potential energy surface developed by Kuntz et al.[172]

$$V(r_1,r_1,r_1)=\frac{Q_1}{(1+a)}+\frac{Q_2}{(1+b)}+\frac{Q_3}{(1+c)}$$
$$-\left[\frac{J_1^2}{(1+a)^2}+\frac{J_2^2}{(1+b)^2}+\frac{J_3^2}{(1+c)^2}-\frac{J_1J_2}{(1+a)(1+b)}-\frac{J_2J_3}{(1+b)(1+c)}-\frac{J_1J_3}{(1+a)(1+c)}\right]^{\frac{1}{2}} \quad (5\text{-}4)$$

where

$$\frac{Q_1}{(1+a)}=\frac{1}{2}\left\{{}^1E(r_1)+\left[\frac{(1-a)}{1+a}\right]{}^3E(r_1)\right\}, \quad (5\text{-}5)$$

and

$$\frac{J_1}{(1+a)}=\frac{1}{2}\left\{{}^1E(r_1)-\left[\frac{(1-a)}{1+a}\right]{}^3E(r_1)\right\}, \quad (5\text{-}6)$$

with similar expressions for the other terms. Morse- and "anti-Morse"-type functions are used to represent the singlet and triplet energy states, respectively:

$${}^1E(r)=D\left\{\exp\left[-2\alpha(r-r_c)\right]-2\exp\left[-\alpha(r-r_c)\right]\right\}, \quad (5\text{-}7)$$

and

$${}^3E(r)=\frac{1}{2}D\left\{\exp\left[-2\alpha(r-r_c)\right]+2\exp\left[-\alpha(r-r_c)\right]\right\}, \quad (5\text{-}8)$$

where the Morse parameters, D, α, and r_c are obtained from bond dissociation energies, fundamental vibration frequencies, and equilibrium bond distances in the usual manner. The values for all parameters for the H_2Br system are given for Surface I in Sudhakaran and Raff.[137]

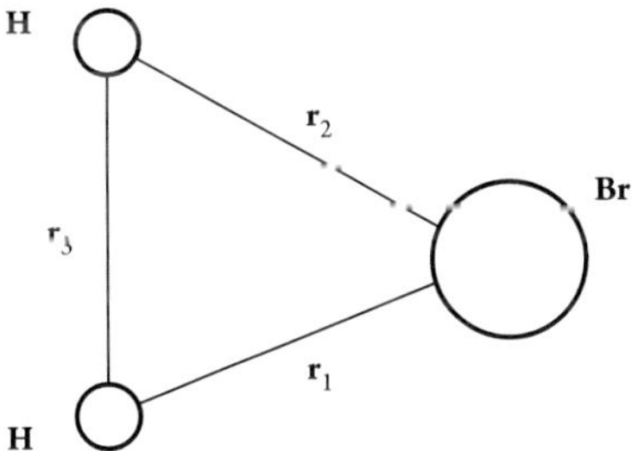

Figure 5-1 H_2Br coordinate system[122] (reprinted with permission from American Institute of Physics)

This LEPS surface has a discontinuity in the gradient. The location of the discontinuity is defined by the following conditions:

$$r_2 = r_3 \text{ and } r_1 = -\frac{1}{\alpha_1}\ln\left(\frac{k_2}{2k_1} \pm \frac{1}{2}\sqrt{\left(\frac{k_2}{k_1}\right)^2 + \frac{8P_2}{k_1}}\right), \tag{5-9}$$

where

$$k_1 = \left\{D_1 - \frac{1}{2}D_1\frac{(1-a)}{(1+a)}\right\}\exp(2\alpha_1 r_c), \quad k_2 = -\left\{2D_1 + D_1\frac{(1-a)}{(1+a)}\right\}\exp(\alpha_1 r_c)$$

$$\text{and } P_2 = \frac{J_2}{(1+b)}.$$

The existence of this discontinuity has caused no problems in many MD investigations that have employed Equations (5-4) to (5-8) for the potential surface because the loci of points on the discontinuity seam fall at such high energies that the discontinuity is never reached for trajectories computed at energies typical of chemical experiments and calculations. Pukrittayakamee et al.[122] ensured that the same was true in their investigations.

The database used for training the NN was obtained by computing 1,000 trajectories for H + HBr using the analytic potential given by Equations (5-4) to (5-8). This method was originally proposed and used by Collins.[25] Each trajectory in a given batch was computed at a constant translational energy. As the trajectories were propagated, H_2Br configurations along with their corresponding potential energies and gradients were stored using variable interval sampling[122] as described in Chapter 4, Section 4.2. Figures 4-1 and 4-2 show typical variable interval sampling results for the H_2Br system. The ensemble of these stored configurations, energies, and gradients comprise 470,000 H_2Br configurations. For the training of each NN, 3,000 of these points were randomly selected to provide the training set. The remaining 467,000 points were used as a testing set. This testing set is far more extensive than that usually employed in most applications of NNs to potential-energy surfaces. It was employed to ensure that the NNs produced by the CFDA method are thoroughly and completely tested in all important regions of the H_2Br configuration space.

To avoid sampling the discontinuity, neural network training data were collected during MD trajectories that were computed at relative translational energies of 0.8, 1.0, and 1.2 eV, and a maximum impact parameter, b_{max}, of 0.2 Å. The impact parameter in individual trajectories was randomly selected over the range from zero to b_{max}. Since the minimum energy at the discontinuity lies 1.65 eV above the reactant H + HBr energy, the sampling trajectories will never reach the discontinuities at the energies used in the sampling.

Figure 5-2 shows the regions where the trajectories were sampled for one random choice of 3,000 points. The location of the discontinuity is indicated by the curve in the figure.

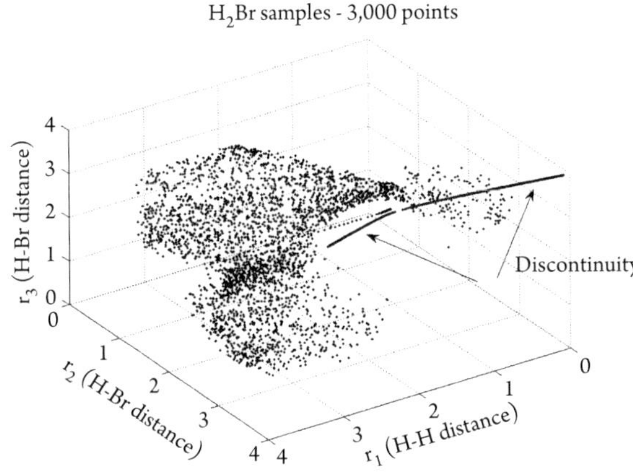

Figure 5-2 Data sampling and location of the discontinuity in the Sudhakaran-Raff analytic potential, denoted as Surface I in Reference 137. The curve in the figure denotes the location of the discontinuity[122] (reprinted with permission from American Institute of Physics).

Because the results can be sensitive to the data selected and to the initial neural network weights, 10 different (3-140-1) neural networks were trained. Each network was trained with a different selection of data points and different initial weights. The median results on the complete 470,000 data points were then used to assess the accuracy of the NN surface. The results for each of the 10 NNs are given in Table 5-4. The median values are given in Table 5-5. As can be seen, all NNs give very close to identical results.

Table 5-5 shows the median rms test set errors for the Bayesian regularization[117] method, where only the potential is fitted, and the CFDA algorithm. In previous work, the Bayesian regularization method has produced the best performance; therefore, it is used as a benchmark. Here, we can see that the CFDA algorithm provides significantly smaller errors on both the potential surface and the force field.

The errors shown in Table 5-5 were obtained with a network trained with only 3,000 data points. The errors were measured over a test set of 467,000 data points. The fact that the test set errors for the CFDA algorithm are lower than those for the Bayesian regularization algorithm (which was designed to prevent overfitting) on such a large test set demonstrates that the CFDA algorithm (without a validation set) does a better job of preventing overfitting than the standard methods. One would not generally have such a large test set available. A large one was employed in this case in order to clearly demonstrate the reduction in overfitting when using CFDA methods (without a validation set). The large test set also permits us to demonstrate that the CFDA algorithm provides an excellent fit over the entire configuration space important in the H$_2$Br reaction dynamics.

Similar results were also reported by Pukrittayakamee et al.[122] for training sets containing 1,500 and 750 points. Since Bayesian regularization[120] is the best standard method for reducing overfitting, this presents a strong case that the CFDA method does a better job of reducing overfitting without a validation set, even when the training set is very small. We also note that the very small errors for the second

Table 5-4 RMS test set errors for Bayesian regularization and CFDA for 10 different (3-140-1) NNs.[122] (Reprinted with permission from American Institute of Physics.)

Bayesian Regularization[a]		CFDA	
Potential (eV)	Forces (eV/Å)	Potential (eV)	Forces (eV/Å)
3.3035e-004	7.6882e-003	1.4525e-004	1.8398e-003
6.5614e-004	9.4262e-003	1.5777e-004	1.6662e-003
3.5913e-004	5.8481e-003	1.5975e-004	1.7043e-003
2.3396e-004	4.3689e-003	1.9860e-004	2.4132e-003
4.2139e-004	7.7385e-003	1.4370e-004	1.2003e-003
3.0202e-004	7.6737e-003	2.3570e-004	3.4086e-003
5.3780e-004	2.3082e-002	1.1207e-004	1.3807e-003
5.0875e-004	7.2271e-003	1.3950e-004	1.4527e-003
1.4652e-003	1.2323e-002	1.4333e-004	2.5245e-003
7.8513e-004	9.3164e-003	2.0005e-004	1.5358e-003

(a) Reference 120

derivatives seen in Table 5-5 indicate that the first derivative cannot be changing rapidly between data points.

In a previously reported study,[121] the CFDA method was employed to fit five other systems that include four analytical potentials and a database for Si_5 obtained from electronic structure calculations. In each case, every test investigated shows that CFDA training (without a validation set) has produced smaller, out-of-sample testing errors than early stopping (with a validation set) or Bayesian regularization (without a validation set). This indicates that CFDA training does a better job of preventing overfitting than the standard methods currently in use.

Figure 5-3 shows a comparison of the Sudhakaran-Raff potential energies[137] and the corresponding NN results for the test set configurations. If the NN fit were perfect, all points would fall on a 45° line. Figure 5-4 shows the distribution of errors

Table 5-5 Median rms test set errors for Bayesian regularization and CFDA. Training set contains 3,000 data points. Test set contains 467,000 data points. The potential errors are expressed in eV and the force errors in eV/Å.[122] (Reprinted with permission from American Institute of Physics.)

Bayesian Regularization[a]			CFDA		
Potential	Force	2nd Deriv.	Potential	Force	2nd Deriv.
4.65E-04	7.71E-03	1.10E-01	1.51E-04	1.68E-03	2.76E-02

(a) Reference 120

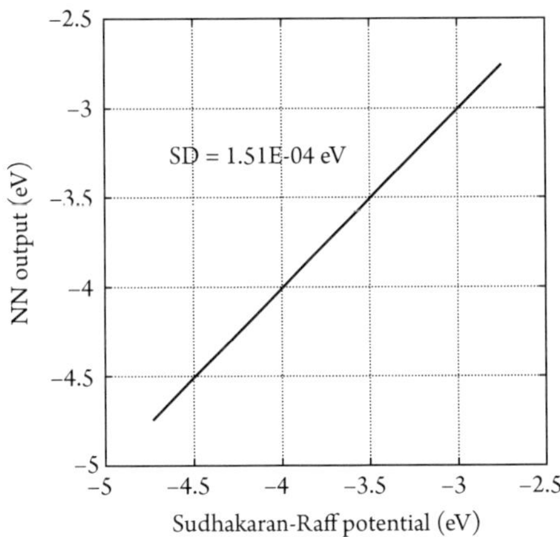

Figure 5-3 Comparison of Sudhakaran-Raff energies with the predictions of the median NN for a test set of 470,000 configurations[122] (reprinted with permission from American Institute of Physics).

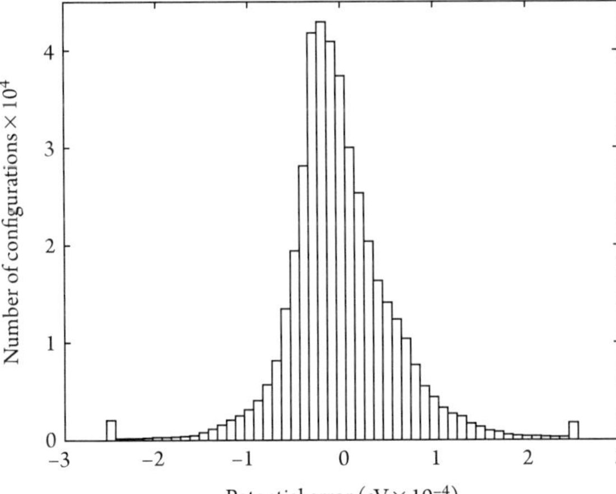

Figure 5-4 Distribution of errors in the potential for the median NN[84] (reprinted with permission from American Institute of Physics).

in the potential for the median network. The median rms deviation of the NN and Sudhakaran-Raff energies is 1.51E-04 eV (1.2 cm^{-1}). Clearly, the fit is nearly exact. Since the energies in the database span a range of about 2 eV, this fitting accuracy represents an average error of about 0.0075%.

Figure 5-5 shows a comparison of the LPS force along r_1 and the corresponding NN results for the test set configurations. Figure 5-6 shows the distribution of errors in the r_1 force for the median network. The median rms deviation of the NN and Sudhakaran-Raff r_1 force is 1.68E-03 eV/Å. Since the forces span a range of about 10.5 eV/Å, this interpolation error represents a percentage error of about 0.016%.

The reaction dynamics of H + HBr were investigated using the median neural network. The reaction channel resulting in the dissociation of all three atoms was energetically closed.

MD trajectories were run using both the median NN surface and the LEPS-type Sudhakaran-Raff potential-energy surface. The velocity Verlet algorithm was employed to integrate the equations of motion during the MD simulations. A total of 1,000 MD trajectories were run at relative translational energies of 0.8 eV, 1.0 eV,

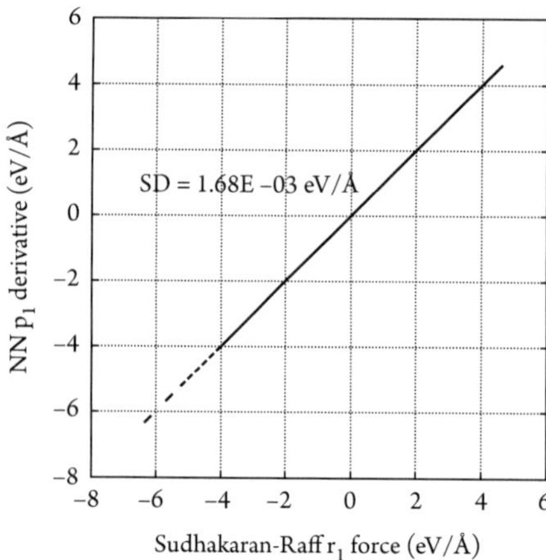

Figure 5-5 Comparison of Sudhakaran-Raff forces (r_1) with the predictions of the median NN for a test set of 470,000 configurations[122] (reprinted with permission from American Institute of Physics).

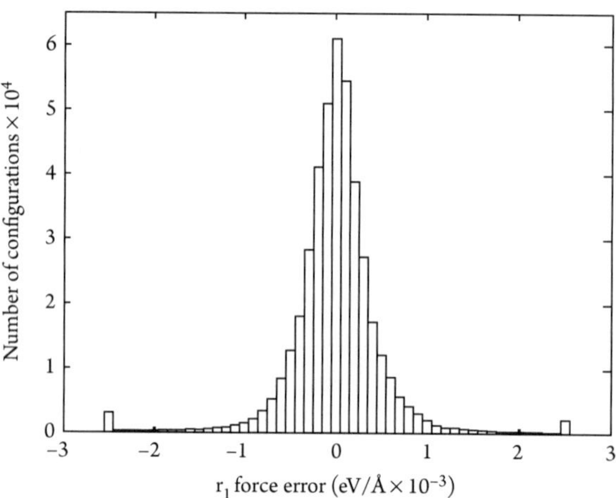

Figure 5-6 Distribution of errors in the force (r_1) for the median NN. These extremely small errors over such an extensive testing set demonstrate that the NN obtained from the CFDA method produces an excellent fit in all regions of configuration space that include the repulsive walls, potential minima, and the flat regions of the potential. The excellent fit also demonstrates that overfitting does not occur in spite of the fact that the CFDA method does not employ a validation set. If overfitting were present, such a small testing set error could not be obtained[122] (reprinted with permission from American Institute of Physics).

and 1.2 eV. None of these trajectories were contained in the set used for sampling the configuration space to obtain the database. The maximum impact parameter and the initial internal HBr energies were selected in the same manner as that used for the sampling trajectories. The impact parameters for the 1,000 MD simulations were selected randomly over the range from zero to b_{max}.

Table 5-6 gives the number of trajectories reacting into each energetically open channel computed using the analytical potential and the NN at a relative

Table 5-6 Reaction yields out of 1,000 trajectories at 1.2 eV.[122] (Reprinted with permission from American Institute of Physics.)

	Analytical Potential[62]	*Median Neural Network*
Exchange	728	728
Abstraction	3	3
No reaction	269	269

translational energy of 1.2 eV. The results for each of the 1,000 trajectories were identical.

Figures 5-7 and 5-8 show the temporal dependence of the difference between the bond distances and corresponding energies, respectively, computed using the analytical potential and the NN during the MD simulation for one randomly chosen trajectory. The figures are both highly magnified to allow appropriate illustration of the very small errors. The maximum absolute deviation for the three bond distances in this trajectory is 0.0003 Å. During most of the trajectory, the deviations are less than 0.0001 Å. The absolute deviation of the total energy lies between 0.0001 eV and 0.0002 eV over most of the trajectory. This particular trajectory shows a momentary increase to 0.0006 eV. Although, Figures 5-7 and 5-8 show results for a single trajectory, these figures are typical of all 1,000 trajectories investigated.

These extremely small differences seen in Figures 5-7 and 5-8 demonstrate that NNs trained using the CFDA method produce essentially a point-by-point match of trajectories computed on the analytic and NN potential surfaces. That is, plots of the temporal variation of r_i ($i = 1, 2, 3$) for a given trajectory on the analytic and NN surfaces are superimposable within the accuracy of the plot. Previous attempts to obtain such point-by-point matching of trajectories on analytic surfaces and numerical fits to such surfaces have not been very successful.[187]

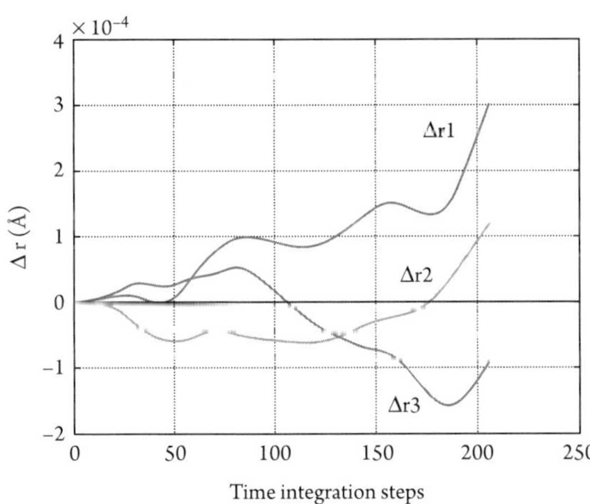

Figure 5-7 Amplified difference between bond distances computed using the analytical potential and the NN during the MD simulation vs. time integration steps of 0.1 fs[122] (reprinted with permission from American Institute of Physics).

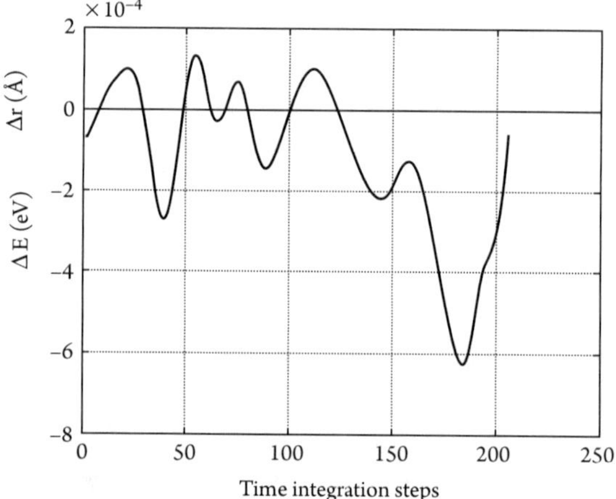

Figure 5-8 Amplified difference between energies computed using the analytical potential and the NN during the MD simulation vs. time integration steps of 0.1 fs[122] (reprinted with permission from American Institute of Physics).

5.4. *CIS-TRANS* ISOMERIZATION AND N-O DISSOCIATION REACTIONS OF HONO

Since 1982, the unimolecular reaction dynamics of HONO have been investigated both theoretically and experimentally. In 1982, McDonald and Shirk studied the HONO *cis-trans* isomerization[188] experimentally in solid N_2 and Ar matrices. The rate of the reaction was found not to be first order due to the collision of HONO molecules with atoms of the matrix. In a latter study, Shirk and Shirk[189] reported that the rate of *cis→trans* isomerization in solid N_2 matrices is about 100 times faster than the rate of the reverse reaction. Khriachtchev et al.[190] have conducted experimental studies of cis-trans isomerization in several matrices (H_2S, NO_2, and Kr). In the gas phase, the reactions (isomerization and dissociation) are usually found to be first order. This was found to be true computationally by Guan et al.[191] in their theoretical study. Also, in later work conducted by Guan and Thompson,[149] the photo-excitation effect was concluded to be mode-specific when different initial excitations were investigated. The rate coefficients for the computed first-order reactions were found to be strongly dependent upon the initial distribution of vibrational energy. Moreover, *cis→trans* isomerization was found to be significantly faster than *trans→cis*. In 2000, a valence bond empirical potential energy surface of HONO was developed by Guo and Thompson[192] which combines the surfaces for *cis* and *trans* HONO into one unique surface.

Le and Raff[125] have reported an *ab initio* study of the unimolecular reactions of HONO using NNs to obtain analytic fits to the *ab initio* database. The complexity and difficulty of developing a neural network (NN) potential surface depends upon the number of atoms involved, the number of open reaction channels, and the internal energy present. In the HONO system, there are two principal open reaction channels. Since HONO has two isomers as shown in Figure 5-9, the first reaction channel to

116

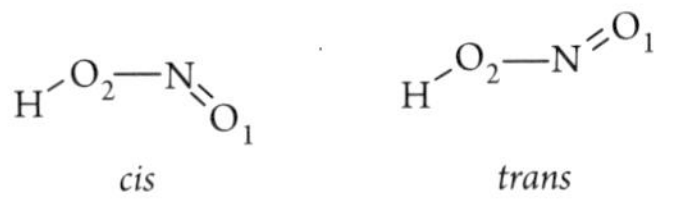

$$H{-}O_2{-}N{=}O_1 \qquad\qquad H{-}O_2{-}N{=}O_1$$

$$cis \qquad\qquad\qquad trans$$

Figure 5-9 The *cis* and *trans* configurations of HONO. The numbers next to the oxygen atoms are used to distinguish one from the other[125] (reprinted with permission from American Institute of Physics).

be considered is *cis→trans* and its reverse *trans→cis* isomerization. The second reaction channel is the dissociation of the N-O bond. With four atoms and two energetically open reaction channels at total energies between 1.70 and 3.30 eV, the chemically important configuration space for HONO is substantially larger than that for a four-atom system undergoing only one two-center bond rupture. Consequently, sampling can be expected to require a larger database.

In this example of a four-body unimolecular reaction in which a four-center, *cis-trans* isomerization and a two-center bond rupture are occurring simultaneously, the MD/NN/NS method described in Chapter 4, Section 4.2, is employed. The *ab initio* energy calculations were executed using the GAUSSIAN[141] suite of programs. The *ab initio* method used in this example is fourth-order, Moller-Plesset perturbation[193] theory with singlet, doublet, and quartet excitations (MP4(SDQ)).

Several basis sets were investigated and compared to find one that seemed suitable in terms of accuracy and computational time requirements. The basis sets investigated included (4-31G),[194–197] 6-31G, 6-31G(d),[198] and 6-311G(d).[199,200] It was found that the computation of the potential energy for many HONO configurations fails to converge when 6-31G and 6-31G(d) basis sets are employed. If an atom is removed too far from its equilibrium position, the optimization of coefficients fails to converge. A 6-311G(d) basis set, on the other hand, has almost no convergence problems for the HONO system. Also, the MP4(SDQ) energy obtained from 6-311G(d) is lower than the energy obtained from using the other basis sets. Therefore, all reported electronic structure calculations were executed with this basis set. An MP4 calculation with this basis set requires 45 to 50 seconds to complete, which means for a set of 21,584 data points, about 280 hours are required using a single processor with a speed of 2.40 GHz.

The data points obtained from MP4(SDQ) calculations with a 6-311G(d) basis set were fitted by using the MATLAB Neural Network software, which is fully discussed by Hagan, Demuth, and Beale.[105] The input vector to the NN comprises the six atomic distances between the four atoms, while the calculated *ab initio* energy values relative to that for HONO in its equilibrium configuration were used as training outputs. Details for operation of the feedforward NN have been described in Chapter 3.

A modified form of novelty sampling[136] of the HONO configuration space is initiated using the empirical potential developed by Guan and Thompson.[149] A series of trajectories with random initial configurations obtained using projection methods[148] are utilized to obtain the initial sampling of configuration space. With a total energy of 2.8 eV, trajectories are run for 5 picoseconds with a step size of 1.018×10^{-16} s, and 8,000 configuration points (4,000 points obtained from running *cis* trajectories and 4,000 points from running *trans* trajectories) are selected and the Cartesian coordinates for each configuration recorded.

To augment this database, a modification of the novelty sampling method suggested by Raff et al.[136] is employed. Empirically, it was found that adding more configurations in regions of configuration space that are already adequately covered by the previous sampling tends to increase the training and testing set errors because the density of sampling configurations becomes highly non-uniform. Therefore, the standard novelty sampling procedures[136] are modified by adding an energy difference criterion before a configuration point is accepted for inclusion in the database. This criterion is that the potential energy of a new configuration must differ by more than 1% from the current neural-network predicted value.

The iterative sampling of configurations using trajectories begins by fitting a feed-forward NN to the initial database of 8,000 configurations. The *cis* and *trans* configurations that have lowest potential energies are obtained, and trajectories are integrated starting at these points to generate more configurations around the area that is believed to contain the *cis* or *trans* equilibrium configuration. In this early iteration, 1,000 additional configurations are obtained using the modified novelty sampling process.

Subsequently, four more iterations are executed to add more configuration points to the database. In the last four iterations, emphasis is placed on adding more configurations related to bond stretching. Various excitation energies are used to ensure that the three bond distances are stretched sufficiently to adequately cover the important region of configuration space. As shown in Table 5-7, additional configurations of 3,314 are added to the database using modified novelty sampling in the second iteration process. After five iterations, a total of 21,584 data points are obtained, and it is recognized that convergence has been obtained using the novelty sampling criteria.[136]

Figure 5-10 shows the non-normalized $P(\Gamma_m)$ (see Chapter 4, Section 4.2, for a definition of the notation) distributions for the original database and for the 3,314 configurations added in the second iteration. Obviously, numerous new configurations are in regions of configuration space far removed from that sampled by the original set. The corresponding distributions after the final iteration are shown in Figure 5-11. Here, very few configurations are found in regions not already sampled. Therefore, the convergence criteria are satisfied.

Table 5-7 Number of configurations added in each iteration of modified novelty sampling.[125] (Reprinted with permission from American Institute of Physics.)

Iteration #	Current # of Data Points	# of Points Added	Total Points
0	0	8,000	8,000
1	8,000	1,000	9,000
2	9,000	3,314	12,314
3	12,314	6,288	18,602
4	18,602	1,996	20,598
5	20,598	986	21,584

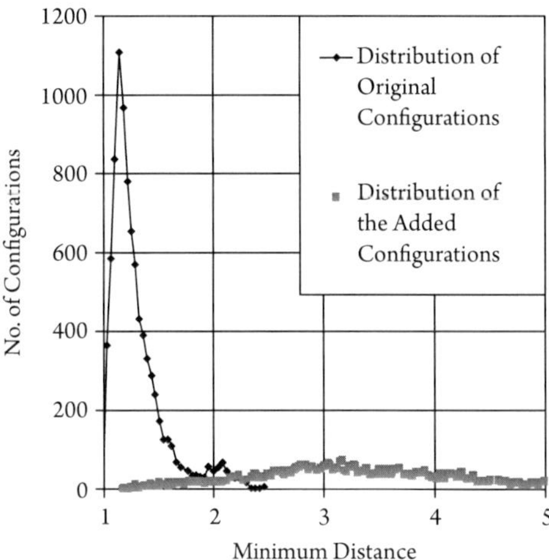

Figure 5-10 Variation of the number of configurations verses minimum distance (first iteration) [125] (reprinted with permission from American Institute of Physics).

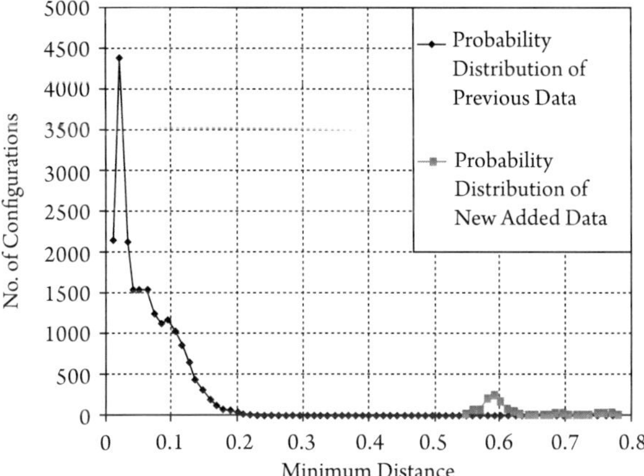

Figure 5-11 Variation of the number of configurations verses minimum distance (final iteration) [125] (reprinted with permission from American Institute of Physics).

The six equilibrium distances predicted by the final NN surface for both the *cis* and *trans* conformations of HONO are generally within 0.01 Å of the results predicted by the empirical potential surface employed by Guan and Thompson.[149] The second derivatives with respect to 12 Cartesian coordinates have been calculated from the final NN potential, and the eigenvalues of the 12×12 second derivative matrix are computed to obtain the normal mode frequencies for the six HONO vibrational modes. The resulting wave numbers for these vibrational modes are reported in Table 5-8. As expected, the MP2 and NN calculated wavenumbers are higher than the experimental values.

As discussed in Chapter 3, Section 3.8.1, more accurate fitting can usually be achieved by employing an NN committee rather than a single NN. Le and Raff[125] employed a five-member NN committee to fit the HONO database. The individual average absolute fitting errors for the five NNs in committee vary from 0.0126 eV to 0.0170 eV. However, when the output is taken as the committee average, this is

Table 5-8 Experimental and calculated wave numbers (cm^{-1}) of *cis* and *trans* HONO.[125] (Reprinted with permission from American Institute of Physics.)

	Vibrational Modes	OH	N=O	HON	O-N	ONO	Torsion
Cis	Experimental[201]	3426	1641	1302	852	609	640
	Calculated by NN surface	3597	1639	1370	968	671	710
	Gaussian (MP2)	3608	1643	1382	958	748	671
Trans	Experimental[201]	3591	1700	1263	790	596	544
	Calculated by NN surface	3799	1738	1348	967	708	549
	Gaussian (MP2)	3796	1694	1338	868	645	616

reduced to 0.0111 eV. That is, the performance of the committee is better than the individual performance of any of its members. Similar averages are taken for the gradients of the surface when performing MD calculations.

Mode specificity of *cis→trans* and *trans→cis* isomerizations has been investigated in both rare-gas matrices, where the reactions are second order,[188] and in the gas phase where the processes are first order.[191,149] All of the theoretical studies to date have been executed using empirical potential surfaces. At a total internal energy of 1.70 eV, the gas-phase studies on empirical surfaces[191,149] generally show the presence of mode specificity with the *cis→trans* rate being higher than the *trans→cis* rate by factors that vary from 2.25 to 22.6 depending upon the vibrational mode initially excited. In *cis-trans* isomerizations, the calculated rate coefficients vary by a factor of 23.8 from 0.063 ps^{-1} to 1.498 ps^{-1} depending upon the vibrational mode excited.[191] For *trans→cis* isomerizations, this factor was found to be about 3.7.[191]

The *cis-trans* and *trans-cis* isomerization rates of HONO were investigated by Le and Raff[125] using the NN fits to the *ab initio* database and the same vibrational excitations as employed in the studies on the empirical HONO surfaces. The *cis-trans* potential barrier on the NN PES is 0.60 eV (13.8 kcal mol^{-1}) as shown in Figure 5-12, while the zero point energy of HONO is 0.525 eV. As reported by Kawana et al.[94] the minimum energy required for isomerization is about 13 kcal mol^{-1}. In the isomerization investigations, the total energy is 1.70 eV, the same as that used by Guan and Thompson.[191,149]

Six different partitionings of the 1.70 eV of internal energy were investigated by Le and Raff.[125] In the first, zero-point energy is inserted into the torsional mode and the remainder of the internal energy is equally partitioned to the remaining five vibrational modes. In the other five initial partitionings, zero-point is given to each vibrational mode, and the remaining internal energy is inserted into one of the five vibrational modes excluding the torsional mode. A study on the torsional mode had been reported by Richter et al.[202] In their investigation, it was found that when the torsional mode is excited, the probability of *cis-trans* isomerizations becomes much higher than for other partitionings of the internal energy with the torsional mode initially in its vibrational ground state.

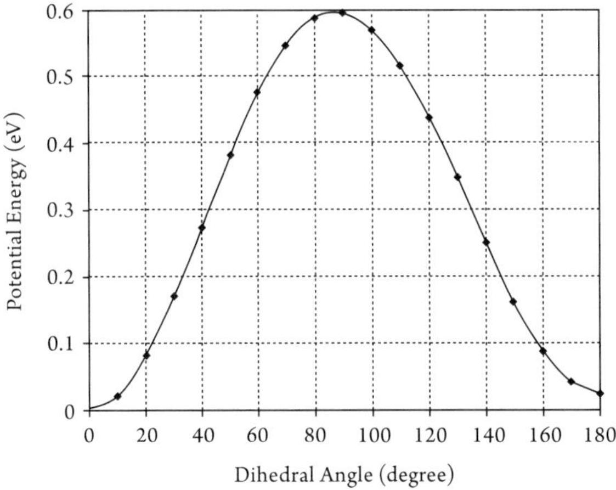

Figure 5-12 Potential barrier of *cis-trans* isomerization optimized by Gaussian 98. The calculated potential energies are given relative to the computed equilibrium potential. In this plot, the dihedral angle is varied in 10° intervals from 0° to 180°. The height of the barrier is about 0.6 eV[125] (reprinted with permission from American Institute of Physics).

The isomerization rate coefficients obtained from the slopes of linear decay curves computed using MD methods[1–7] are given in Table 5-9. At the total energy of 1.70 eV, the rates of *cis-trans* isomerizations are nearly independent of the initial energy partitioning. They vary by a factor of 2.40 from 0.210 ps^{-1} to 0.505 ps^{-1}. This result is in sharp contrast to the factor of 22.8 obtained using an empirical potential surface.[191] The results for *trans→cis* isomerizations are similar. These rates at a total energy of 1.70 eV vary by a factor of 1.82 from 0.199 ps^{-1} when the OH stretch is excited to 0.363 ps^{-1} for excitation of the N-O stretch. Using an empirical surface, Guan and Thompson[149] obtained a factor of 7.0 for this factor.

The MD results clearly indicate that the intramode couplings on the *ab initio* NN potential are much larger than those present in the empirical potential that has previously been employed to examine the HONO dynamics.[191,149] As a result of the larger intramode couplings, the excitation energy spreads to the other modes on a time scale that is large relative to the isomerization rates. Such a result is not surprising since the coupling terms between the vibrational modes were omitted in the potential surface used by Guan et al.[191]

Since the vibrational modes are highly coupled in both the *cis* and *trans* conformations of HONO on the *ab initio* potential, large differences between the *cis→trans* and the *trans→cis* rates are not observed. The (*cis→trans*)/(*trans→cis*) rate coefficient ratios at 1.70 eV total energy for similar mode excitations are found to vary from 0.63 to 1.94. Except for excitation of the HON bend, the *cis→trans* rates always exceed the *trans→cis* rates, but the differences are less than a factor of two.

N-O bond scission reactions of HONO were investigated by Le and Raff[125] at total energies of 3.10 eV and 3.30 eV. It was assumed that if the N-O distance becomes greater than 2.78 Å, the bond can be considered as broken. Therefore, the

Table 5-9 *Cis-trans* isomerization calculated rate constants for the total energy of 1.70 eV (including zero point energy).[125] (Reprinted with permission from American Institute of Physics.)

Reaction	Mode Excited	Rate Coefficient (ps^{-1})	Standard Deviation (ps^{-1})
	OH	0.265	0.005
	N=O	0.491	0.014
	HON	0.210	0.007
cis-trans	O-N	0.405	0.010
	ONO	0.505	0.011
	All[a]	0.401	0.009
	OH	0.199	0.003
	N=O	0.303	0.012
	HON	0.334	0.011
trans-cis	O-N	0.363	0.013
	ONO	0.260	0.010
	All[a]	0.255	0.010

(a) Energy in excess of zero-point energy is equally distributed to all modes using Projection methods.[148]

criterion to stop trajectory integrations is that either the N-O bond length reaches 2.78 Å or 5 ps elapse.

Seven different partitionings of the excitation energy were investigated at each total energy. The *cis* configuration is chosen to be the starting configuration. This choice does not make any significant difference at an excitation energy greater than 3.0 eV since the configuration switches rapidly from *cis* to *trans*, and vice versa.

All N-O bond dissociations were found to be well-described by a first-order rate law. Figure 5-13 shows a typical decay plot at 3.10 eV total energy with the OH vibrational mode excited. The excellent linearity of the result is apparent.

The calculated rate coefficients and their relative standard deviations obtained from the first-order decay plots are given in Table 5-10. As expected, the dissociation rates at 3.10 eV are lower than those at 3.30 eV.

As seen in Table 5-10, excitation of the torsional mode produces the most rapid N-O bond dissociation. Excitation of the N-O stretch is also very effective in promoting N-O bond rupture, as we would expect. The dissociation rates obtained upon excitation of the remaining four vibrational modes at a total energy of 3.10 eV vary from 0.042 ps^{-1} (when the HON bend is excited) to 0.082 ps^{-1} (for excitation of the OH stretch). The ratio of maximum to minimum N-O dissociation rates at 3.10 eV total energy is a factor of 3.9. At 3.30 eV total energy, this factor is 3.3.

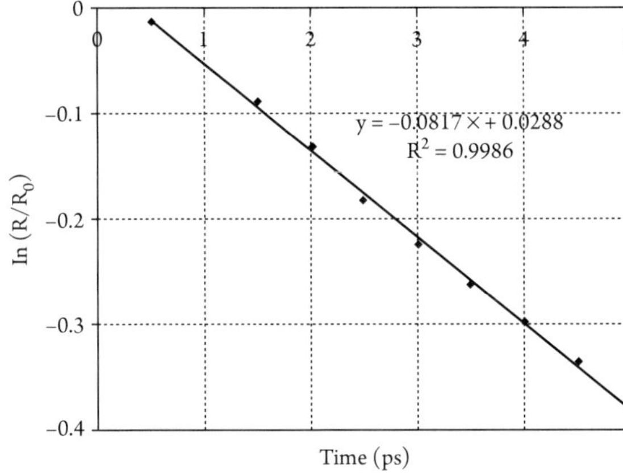

Figure 5-13 First-order decay plot for N-O bong rupture in the NN PES[125] (reprinted with permission from American Institute of Physics).

The extent of mode specific rate enhancement for N-O bond dissociation predicted by the *ab initio* NN surface is significantly less than that obtained by Guan et al.[149] using an empirical surface. For non-rotating HONO, their computed maximum to minimum dissociation rates for the various excitations at a total energy of 2.26 eV was a factor of 69.4 with excitation of the ONO stretch being much more effective than the other possible excitations. The large differences in the extent of

Table 5-10 N-O bond dissociation calculated rate constants at total energies of 3.10 eV and 3.30 eV (including zero point energy).[125] (Reprinted with permission from American Institute of Physics.)

Total Energy	Mode Excited	Rate Coefficient (ps^{-1})	Standard Deviation (ps^{-1})
3.10 eV	OH	0.082	0.001
	N=O	0.075	0.001
	HON	0.042	0.001
	O-N	0.165	0.003
	Torsion	0.136	0.002
	ONO	0.063	0.001
	All	0.092	0.002
3.30 eV	OH	0.151	0.002
	N=O	0.146	0.001
	HON	0.072	0.001
	O-N	0.237	0.008
	Torsion	0.212	0.002
	ONO	0.116	0.002
	All	0.157	0.004

mode specificity on the *ab initio* and empirical surfaces are again the result of the very different intramode couplings present on the two potentials.

5.5. GRADIENT SAMPLING—UNIMOLECULAR DISSOCIATION OF HOOH TO 2 OH

The technique of configuration space sampling using gradient fitting has already been described in detail in Chapter 4, Section 4.4. In that section, the method was described using the hydrogen peroxide system as an illustrative example. In this section, we review the molecular dynamics of H_2O_2 dissociation to two OH radicals obtained using the NN-committee fitted to the database resulting from sampling using a gradient fitting method.

Prior to the execution of MD calculations at different energies, the dissociation barrier of the O-O bond was investigated at HF/6-31G*, MP2/6-31G*, MP4(SDQ)/6-31G*, and density functional theory B3LYP/cc-pVTZ levels of theory. The results are shown in Figure 5-14. In these calculations, the potential energy was minimized for each O-O bond extension from 1.12 Å to 2.92 Å. According to the investigation, the O-O bond is believed to dissociate at a minimum potential energy of 2.57 eV above the equilibrium energy at the MP2 level. The O-O distance at dissociation is approximately 2.65 Å.

There is a good agreement between calculations at MP4(SDQ) and MP2[193] levels of theory with respect to the O-O distance at dissociation. However, the energy barrier resulting from MP4(SDQ) calculations is 2.24 eV, which is 0.34 eV lower than the MP2 barrier. Simple restricted Hartree-Fock (RHF) calculations do not

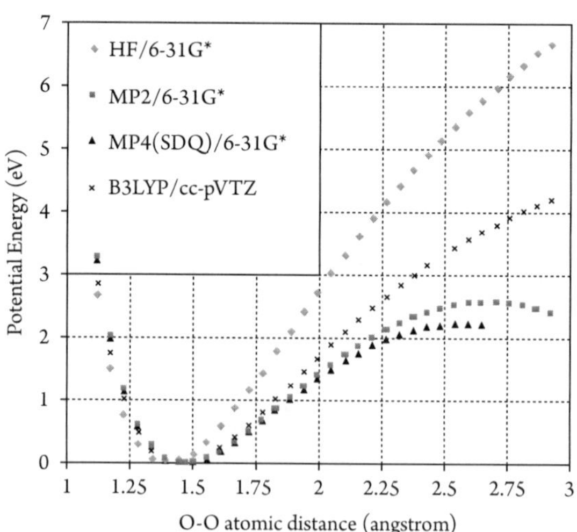

Figure 5-14 The potential energy as a function of the O-O bond distance for HF/6-31G*, MP2/6-31G*, MP4/6-31G*, and DFT/B3LYP-cc-pVTZ levels of electronic structure theory. The energy is minimized with respect to all other coordinates at each point[124] (reprinted with permission from American Institute of Physics).

agree with MP2 and MP4 calculations, and no bond dissociation is found as the potential energy rises to more than 5.5 eV at 2.65 Å. The density functional theory (B3LYP)[203,204] calculations employ a large cc-pTVTZ basis set.[203] The results indicate that the O-O bond is not likely to break at distances less than 2.9 Å as the potential energy continues to increase. In fact, Kuhn et al.[51] have reported that the bond is not broken until it reaches 3.2 Å by executing second-order complete active-space calculations (CASPT2)[205,206] and density functional theory calculations. Since the NN-committee PES is a fit of MP2 calculations, the O-O dissociation distance is taken to be 2.65 Å.

The O-O bond dissociation dynamics of HOOH were investigated by Le et al.[124] using quasi-classical MD.[7] Initially, HOOH was assigned to its equilibrium configuration. Vibrational energy was introduced into each vibrational mode using projection methods.[148] To randomize the vibrational phases, the molecule was allowed to vibrate with no angular momentum for a randomized period of time. Subsequently, excitation energy was introduced into the six vibrational modes equally to bring the total energy to the desired level. The trajectory is integrated with a fixed step size of 1.018×10^{-16} s using a fourth-order Runge-Kutta method,[7] and the O-O bond dissociation was monitored. Trajectories were terminated when 5 picoseconds elapsed or the O-O distance reached 2.65 Å. The rate coefficients k for different energies were obtained from the slope of the first-order decay curves computed from the results of 1,000 trajectories. Figure 5-15 shows a typical result at 3.4 eV internal energy. Table 5-11 gives the rate coefficients obtained at five different internal energies.

The NN PES is converged with respect to the selection criteria previously described in Chapter 4, Section 4.4. Collins and co-workers[23,25] have tested for convergence of the PES by examining the extent to which various dynamics results remain essentially constant when additional points are added to the database used to sample configuration space. This procedure requires repeated computation of whatever property is selected as the testing criterion. However, for three- or four-

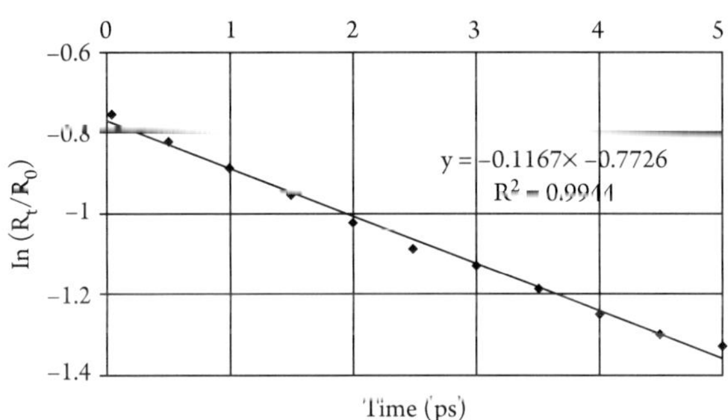

Figure 5-15 First-order decay plot of O-O dissociation reaction when all vibrational modes are equally excited to bring the total energy up to 3.4 eV (including zero point energy). The excellent linearity of the results shows that the statistical accuracy is high and that no angular momentum was present in HOOH. The slope of the least-squares line yields the dissociation rate coefficient[124] (reprinted with permission from American Institute of Physics).

Table 5-11 First-order rate coefficients obtained from decay plots at five different internal energies. The reported standard deviations were obtained from statistical analysis of the linear least-squares fit to the data.[124] (Reprinted with permission from American Institute of Physics.)

Total Energy (eV)	Rate (ps^{-1})	Standard Deviation (ps^{-1})
3.4	0.117	0.003
3.6	0.160	0.004
3.8	0.205	0.005
4.0	0.281	0.007
4.2	0.324	0.009

atom systems undergoing a single reaction, the computational requirements are not large and the method is very effective.

Le et al.[124] have shown that the convergence criterion employed when sampling configuration space using gradient fitting is consistent with the method used by Collins and co-workers.[23,25] This was done by computing first-order dissociation rate coefficients with smaller databases randomly selected from the total database of 25,608 configuration points obtained using gradient fitting for hydrogen peroxide. The dynamics property selected for the convergence test was the computed first-order dissociation rate coefficient for O-O bond rupture at 3.4 eV of internal energy. New NN PESs were obtained using progressively smaller percentages of the total database. The linearity of the resulting decay plots are very similar to that seen in Figure 5-15. The results for the rate coefficients are given in Table 5-12. As can be seen, when the database contains 90% of the points in the total database,

Table 5-12 First-order dissociation rate coefficients for HOOH at a total internal energy of 3.4 eV using NN surfaces trained with smaller databases obtained by random selection of points from the total database.[124] (Reprinted with permission from American Institute of Physics.)

Number of Points in Database	% of Total	k (ps^{-1})
20,486	80	0.094 ± 0.002
23,048	90	0.126 ± 0.002
24,328	95	0.121 ± 0.002
25,608	100	0.117 ± 0.002

the computed rate coefficients are essentially converged to two significant digits. When the percentage reaches 95%, the calculated rate coefficients are within the computed error limits to three significant digits.

The O-O dissociation dynamics of HOOH observed by Le et al.[124] suggest that hydrogen bonds tend to form during the dissociation process. This is seen by the fact that when the O-O bond is stretched, one or both of the hydrogen atoms have a tendency to move closer to the other oxygen. As this occurs, θ_1, or θ_2, or both decrease. As a result, the rate of HOOH dissociation into two OH radicals is slower than would be expected on the basis of the known barrier height, internal energy, and O-O stretching frequency. Figure 5-16 shows several snapshots of the HOOH configuration occurring during a trajectory that eventually results in O-O bond rupture. The snapshots illustrate the formation of hydrogen bonds. Guo et al.[29] obtained a similar result. Harding[207] suggested the presence of such an effect in 1991.

5.6. UNIMOLECULAR DISSOCIATION OF VINYL BROMIDE ($H_2C = CHBR$)

The six-atom vinyl bromide system undergoing dissociation at an internal energy 6.44 eV above zero-point energy, which corresponds to the experimental photolysis energy employed by Liu et al.,[208] presents a formidable challenge to the execution of MD investigations on purely *ab initio* potential-energy surfaces. First, there are eight distinct, energetically open reaction channels that include hydrogen atom dissociation (3 channels), bromine-atom dissociation (1 channel), three-center HBr and H_2 dissociations (2 channels), and four-center H_2 and HBr dissociations (2 channels). If the corresponding back reactions are counted, there are 16 open

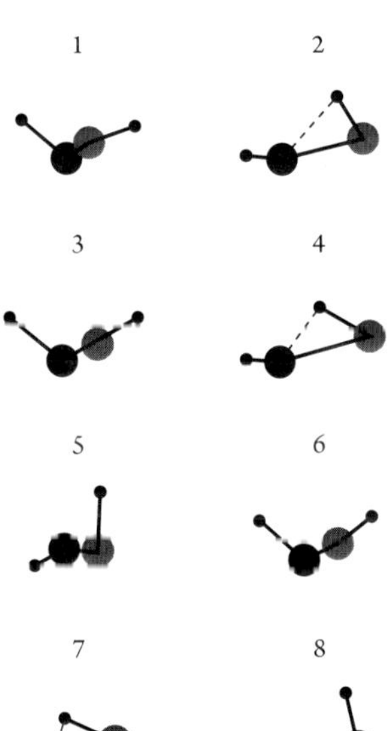

Figure 5-16 Snapshots of HOOH configurations occurring during a trajectory resulting in O-O bond rupture that illustrate the effect of hydrogen bonding on the dissociation processes. In snapshot (1), HOOH configuration is near equilibrium. In snapshots (2)-(4), the O-O bond distance increases from 1.46 Å to 1.85 Å. In snapshot (4), the H-O-O angle has decreased to about 60° forming a hydrogen bond indicated by the dashed line in the snapshot. This interaction prevents O-O bond rupture, and the O-O distance begins to decrease to 1.37 Å shown in snapshot (5). The sequence is repeated again in snapshots (6) and (7), but this time, the hydrogen-bond interaction is not sufficiently strong to prevent the O-O bond rupture shown in snapshot (8)[124] (reprinted with permission from American Institute of Physics).

reaction channels. Second, the presence of six atoms along with the high internal energy makes the total energetically accessible configuration space extraordinarily large. To date, this is the most complex, difficult system whose reaction dynamics have been investigated using purely *ab initio* methods. As such, it serves as a severe test of MD, NN, and NS methods.

Fortunately, previously reported MD investigations using a complex empirical PES[16,209] and the experimental photolysis results[208] indicate that the four-center dissociation channels yielding either H_2 and bromoacetylene or HBr and acetylene play a negligible role in the vinyl bromide unimolecular dissociation dynamics at 6.44 eV of internal energy. This reduces the size of the configuration space that must be sampled to obtain a converged PES for the dissociation reactions of $H_2C = CHBr$. However, even with this simplification, the system still poses a significant and demanding challenge.

The MD/NN/NS method discussed in Chapters 3 and 4 has been employed by Doughan, et al.[209,210] and by Malshe et al.[126] to investigate the vinyl bromide dissociation dynamics. The remainder of this section describes the methods, procedures, and results obtained in their investigations.

Figure 5-17 is the flow chart showing various steps involved in the execution of purely *ab initio* MD studies on a complex system (vinyl bromide) with multiple open reaction channels.

The initial sampling of the vinyl bromide configuration space was executed using the MD/NN/NS method described in Chapter 4, Section 4.2. The initial sampling trajectories were computed using the analytic surface developed by Rahaman and

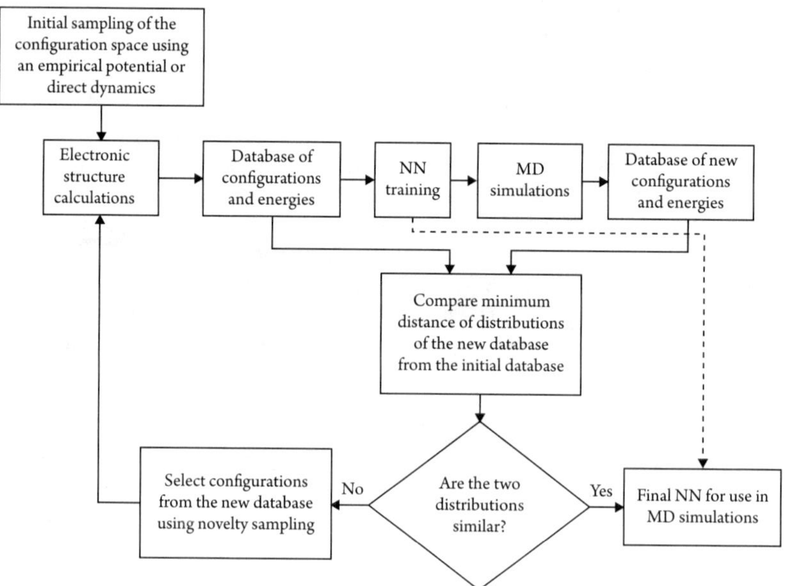

Figure 5-17 Flow chart showing various steps involved in the execution of purely *ab initio* MD studies on a complex system (vinyl bromide) with multiple open reaction channels.

Raff.[16a,b] Projection methods[148] were employed to properly sample over the initial phase space of the system. One thousand trajectories were computed to obtain a subset of 107 trajectories that react via one of the open channels. These trajectories were recomputed. During the recomputation, configurations were stored at equally spaced time intervals. This procedure produced 14,854 configurations. This subset represented the first approximate sampling of the configuration space important in the reaction dynamics for the open reaction channels.

The *ab initio* potentials at each of these 14,854 configurations were computed at the MP4(SDQ) level of accuracy using a 6-31G(d,p) basis set for the carbon and hydrogen atoms and Huzinaga's (4333/433/4) basis set augmented with split outer s and p (43321/4321/4) orbitals for the bromine atom. A polarization f orbital is also added with an exponent of 0.5 to provide a more accurate description. With a CPU clock speed of 1.66 GHz, each potential required about 132 s of computation time.

Examination of the MP4(SDQ) energies shows that the perturbation method begins to fail when the distances between the dissociating atoms becomes too large. For this reason, the sampling of configuration space was restricted to those structures for which the C(2)-Br distances were less than 4.0 Å and the C(1)-H(4), C(1)-H(5), and C(2)-H(3) distances were less than 5.0 Å. The atom numbers are given in Figure 5-18.

In the second step, a (15-140-1) feedforward NN was fitted to the *ab initio* database for the energies obtained in the first step. A second set of MD trajectories was now computed using the NN force field rather than the original empirical potential. During these calculations, additional nuclear configurations were stored and their energies and forces subsequently computed. To compensate for the greatly increased configuration space that occurs upon dissociation, configurations were sampled at five times the frequency whenever any bonded C-H distance in vinyl bromide becomes larger than 3 Å. NS algorithms described in Chapter 4, Section 4.2, were employed to determine whether the sampled configurations were added to the database. If so, the MP4(SDQ) energies for those configurations were computed and added to the overall database characterizing the potential-energy hypersurface. New (15-140-1) neural networks were trained to fit to this expanded database and the entire procedure repeated until convergence was obtained as signaled by the absence of new configurations in unsampled regions of configuration space using the modified distance criteria described in Chapter 4.

The NS convergence criterion is based on two distributions. The first is the distribution of modified minimum distances, $P(\Gamma_m)\, d\Gamma_m$ described in Chapter 4,

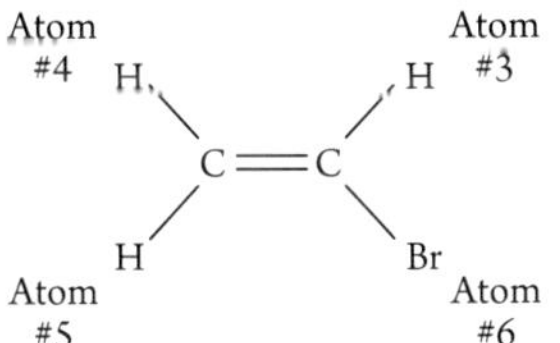

Figure 5-18 Atom numbering used for vinyl bromide[126] (reprinted with permission from American Institute of Physics).

129

Section 4.2, of each configuration point from the remainder of points in the current database. The second is the distribution of modified minimum distances of each *new* configuration point from the points in the original database for that iteration. These two distributions are illustrated in Figures 5-19 and 5-20. The shaded distribution in Figure 5-19 is the modified minimum distance distribution of the 14,854 points obtained from the trajectories on the empirical potential surface. The unshaded distribution is that for the 19,821 new configurations obtained in the first iteration using a feedforward NN fitted to the *ab initio* MP4(SDQ) energies of the original 14,854 configurations. It is clear that many of the configurations in the unshaded distribution are in configuration regions far removed from those present in the initial database. Therefore, many of these configurations need to be incorporated into the database. NS achieves this objective.

The situation after the fifth iteration is very different, as shown in Figure 5-20. Now the shaded and unshaded distributions cover nearly the same minimum distance configurations. Therefore, under novelty sampling criteria, the sampling has converged at iteration 4, and the new configurations obtained in iteration 5 were not included in the database.

A two-layer, 15-140-1 neural network with hyperbolic tangent sigmoid and linear transfer functions in the hidden and output layers, respectively, was employed in each iterative cycle of the novelty sampling process as well as for the final database. The 71,969 *ab initio* energies were divided into training and testing sets of 64,773 and 7,196 configurations, respectively. The accuracy of the NN fit is shown in Figures 5-21 and 5-22. Figure 5-21 shows a plot of the NN output versus the input *ab initio* energies for the testing set. As can be seen, the error increases steadily as the energy increases. This is a consequence of the greatly expanded size of the configuration space available to the system at high energy compared to that available at lower energies. This is the effect that makes it much more difficult to accurately fit potentials for six-atom systems reacting into six different channels than is the case when one has a smaller number of atoms with only one reaction channel present.

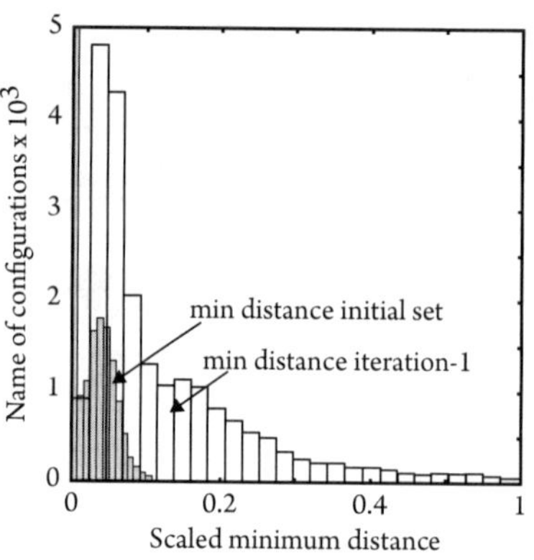

Figure 5-19 Illustration of the first cycle of the iterative novelty-sampling procedure. The shaded distribution is the modified minimum distance distribution of the 14,854 points obtained from the trajectories on the empirical potential surface. The unshaded histogram shows the distribution of minimum modified distances for the 19,821 new configurations from those in the shaded distribution. The failure of the two distributions to cover the same region of scaled minimum distances shows that there is not yet a sufficient sampling of configuration space to adequately represent the vinyl bromide potential surface[126] (reprinted with permission from American Institute of Physics).

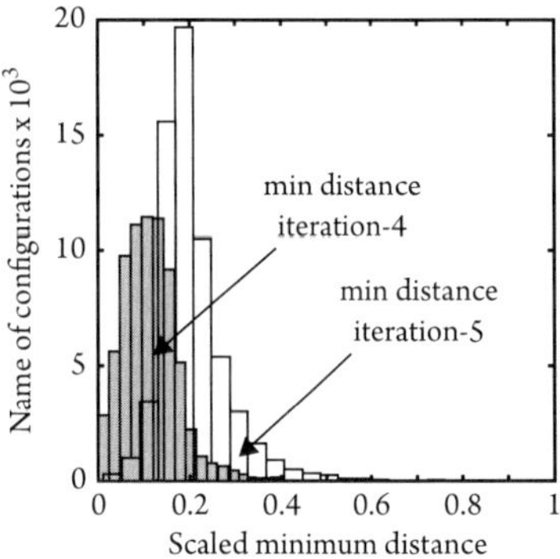

Figure 5-20 Illustration of the final cycle of the iterative novelty-sampling procedure. The shaded distribution is the modified minimum distance distribution of the 71,969 points obtained in the fourth cycle of the iterative procedure. The unshaded histogram shows the distribution of minimum modified distances for the new configurations from those in the shaded distribution. The new configurations are obtained in the fifth cycle of novelty sampling using a 15-140-1 NN fitted to the 71,969 points in the shaded distribution. As can be seen, the two distributions now span essentially the same region of scaled minimum distances. This is the novelty-sampling criterion for convergence[126] (reprinted with permission from American Institute of Physics).

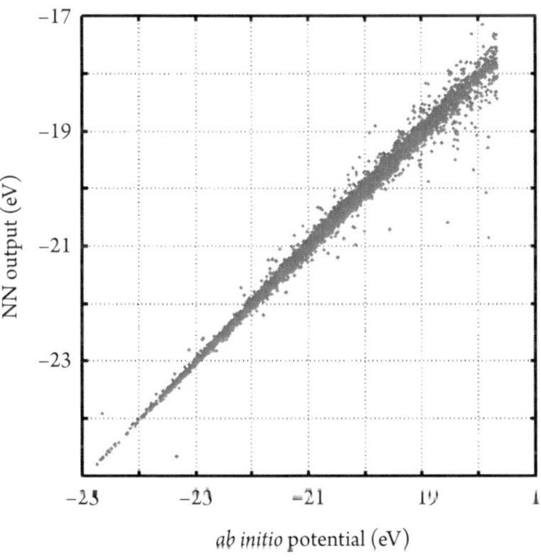

Figure 5-21 Comparison of the output from the 15-140-1 NN for the 7,196 configurations in the testing set with *ab initio* energies computed using MP4(SDQ) methods and the basis set described in the text. If the fit were perfect, all points would lie on a 45° line. The average absolute deviation is 0.065 eV[126] (reprinted with permission from American Institute of Physics).

Figure 5-22 shows the error distribution for all 71,969 configurations. It is nearly symmetric and sharply peaked at zero. Using this distribution, the average absolute error of the NN fit to the *ab initio* database was found to be 0.065 eV for the testing set.

The dissociation dynamics of vinyl bromide at an internal excitation energy of 6.44 eV were investigated[126] by the computation of 1,000 trajectories with the initial conditions being selected using projection methods.[148] Since feedforward neural networks may be easily differentiated analytically, the computation of the force field from the

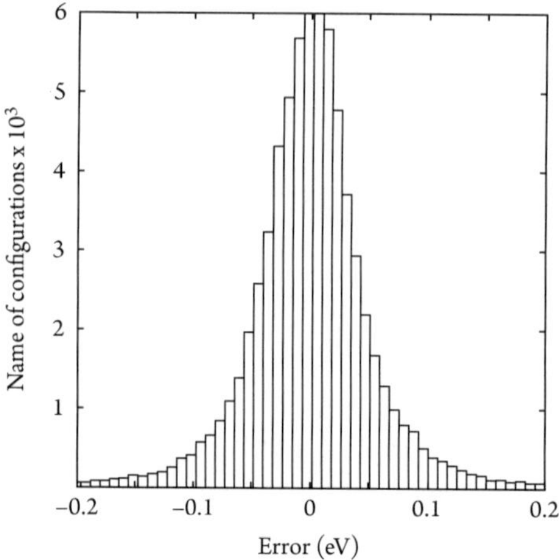

Figure 5-22 Histogram showing the distribution of the deviations of the predicted energies obtained from the 15-140-1 NN and those resulting from electronic structure calculations at MP4(SDQ) level using the basis set described in the text. The results for all 71,969 configurations are included in the histogram. The distribution average for the testing set is 0.065 eV[126] (reprinted with permission from American Institute of Physics).

NN fit to the *ab initio* database is straightforward. A velocity verlet algorithm with an integration step size of 1.018×10^{-5} ps was employed to integrate the 36 coupled, first-order differential equations describing the molecular motion. The required time per integration step is 9.8 ms. Final products of the dissociation reactions were determined using distance criteria whenever the C(2)-Br distance reached 4.0 Å or whenever a C-H bond length reached 5.0 Å.

Figure 5-23 shows a typical computed total decay curve for vinyl bromide dissociation at 6.44 eV, $k_T(6.44 \text{ eV})$. The good linearity of the result indicates that reactions into all open decomposition channels follow first-order rate laws and that no angular momentum is present. The slope of the least-squares line to the data yields a total dissociation rate coefficient of 0.91 ps^{-1}. This rate is a factor of five less than that obtained using the empirical potential developed by Rahaman and Raff.[16] The average lifetime is the reciprocal of $k_T(6.44 \text{ eV})$ or 1.099 ps. Consequently, with an integration stepsize of 1.018×10^{-5} ps, the average computation time required per trajectory using the (15-140-1) NN potential surface is 17.6 min on a computer with a 1.66 GHz clock speed.

Since all reactions appear to follow a first-order rate law, the decomposition rate coefficient for channel i at internal energy E, $k_i(E)$, may be obtained from

$$k_i(E) = \frac{k_T(E)}{\left[1 + \sum_j r_{ji}\right]}, \tag{5-10}$$

where the summation runs over all reaction channels save channel i, and r_{ji} is the branching ratio between reaction into channels j and i. This ratio can be estimated from the trajectory results as being equal to the ratio of the number of trajectories that react into channels j and i, respectively. This ratio is, in turn, equal to the ratio of the reaction probabilities into channels j and i. The computed branching ratios at 6.44 eV internal excitation energy were the following: three-center HBr formation

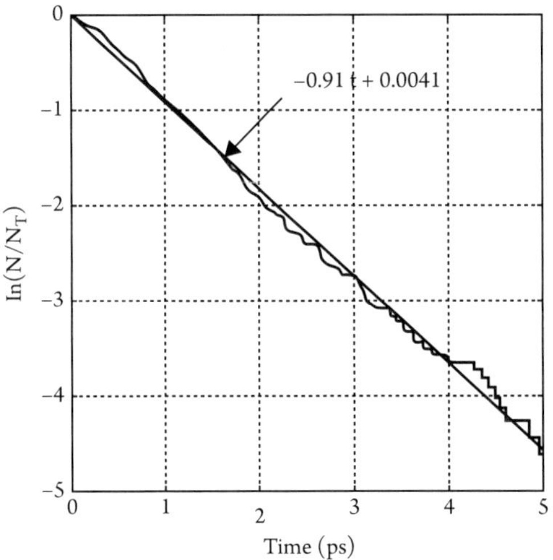

Figure 5-23 Computed decay curve for vinyl bromide at 6.44 internal excitation randomly distributed over all 12 vibrational degrees of freedom. The line is a least-squares fit to the data. Its slope yields a total decay rate coefficient of 0.91 ps^{-1} [126] (reprinted with permission from American Institute of Physics).

(0.48), three-center H_2 formation (0.28), bromine-atom dissociation (0.05), and the total hydrogen-atom dissociation (0.20).

The computed branching ratios show that at an internal excitation energy of 6.44 eV, the NN fit to the MP4(SDQ) database predicts that three-center HBr dissociation is the major decomposition channel. At 6.44 eV, the ordering of decreasing reaction probabilities is HBr > H_2 > H > Br. It is interesting to note that bromine atom dissociation has the lowest potential barrier, but it is not close to being the preferred dissociation pathway. These results demonstrate that the practice of using energy barriers to predict reaction mechanism can lead to erroneous conclusions when the energy present is sufficient to open several reaction pathways.

Liu et al.[208] have carried out photolysis experiments on vinyl bromide at 193 nm (6.44 eV). The IR emission from vibrationally excited HBr was measured and the pristine vibrational-state distribution determined. They found that their results are essentially Boltzmann with an effective temperature of $T_{vib}^{expt.} = 6999$ K. Malshe et al.[126] repeated these calculations using the *ab initio* NN potential surface. A total of 320 trajectories with 6.44 eV of vibrational excitation that dissociated via three-center elimination of HBr were examined to obtain the HBr vibrational energy distribution. The results are shown as the points in Figure 5-24. The curve is a least-squares fit of a Boltzmann distribution to these data. As can be seen, the fit is very good. The predicted HBr vibrational temperature is 6928 K, which is within 71 K of the experimental result reported by Lui et al.[208]

5.7. NON-ADIABATIC REACTIONS: $SIO_2 \rightarrow SIO + O$ AND $SIO_2 \rightarrow SI + O_2$

Non-adiabatic chemical reactions that involve more than a single PES present the investigator with an array of problems not encountered when the reactions of interest are adiabatic, i.e., when the reactions involve only a single PES. Two of the

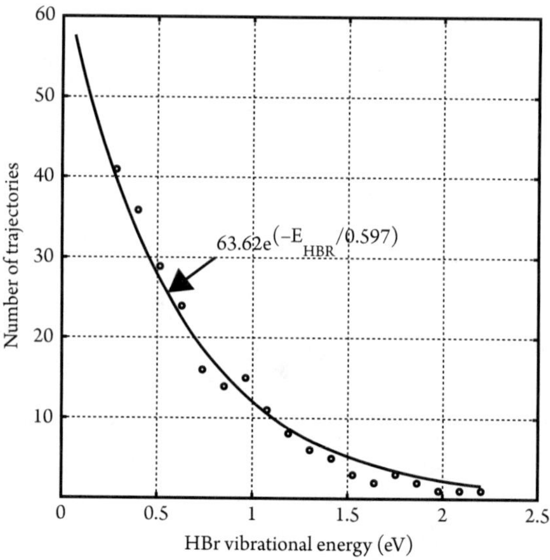

Figure 5-24 Distribution of vibrational energy in HBr subsequent to three-center dissociation. The points are the results obtained from 320 trajectories. The curve is a least-squares fit of a Boltzmann distribution. This fit corresponds to an HBr vibrational temperature of 6,928 K, which is in very good accord with the experimental results reported by Liu et al.[162] of 6,999 K[126] (reprinted with permission from American Institute of Physics).

principal difficulties are (1) the *ab initio* computation of sufficiently large databases to properly characterize all the potential surfaces involved in the non-adiabatic reactions and then obtaining appropriate analytic fits to these databases and (2) developing methods to treat the transitions between the diabatic surfaces. Tully[211] has suggested "surface hopping" methods to treat the second problem. Agrawal et al.[171] have addressed the first problem using NN methods to obtain the diabatic surfaces and analytic representations for a mixture of these surfaces. The test system for these procedures was the gas-phase dissociation of SiO_2 into either $Si + O_2$ or $SiO + O$. In this section, this application of NNs to reaction dynamics is reviewed.

It is well known that many chemical reactions for various systems proceed in such a manner that the system changes its spin state during and after the reaction. For example, for the dissociation,

$$SiO_2 \rightarrow SiO + O, \quad \text{Reaction (1)}$$

the system is initially in the singlet state but it is in the triplet state after the dissociation.

Similarly, the dissociation given by the following reaction

$$SiO_2 \rightarrow Si + O_2, \quad \text{Reaction (2)}$$

corresponds to an example in which the system is initially in the singlet state whereas the products are in the pentet state. Consequently, as SiO_2 undergoes gas-phase unimolecular decomposition, it may be doing so on either the singlet, triplet, or pentet PES. The approach adopted by Agrawal et al.[171] proceeds by *ab initio* computation of adequate databases for each of these surfaces and then using NNs to obtain analytic fits to each database and to a combination of them.

For the SiO_2 system, where only a three-dimensional hyperspace corresponding to three coordinates R_1, R_2, and R_3 is needed, one can attempt to investigate the reactions by sampling configuration space using chemical intuition and a narrow grid of points within an appropriate range of R_1, R_2, and R_3, where R_1, R_2, and R_3 refer to the Si-O$^{(1)}$, Si-O$^{(2)}$, and O$^{(1)}$-O$^{(2)}$ distances, respectively. Here, superscripts (1) and (2) have been used to distinguish the two oxygen atoms.

Ab initio calculations were performed using GAUSSIAN-03[141b] at B3LYP/6-31G* level at 3,753 configurations. For each configuration, three Gaussian calculations were executed by varying the spin multiplicity (singlet, triplet, and pentet). The 3,753 configurations were obtained from a non-uniform grid in R_1-R_2-θ space, where θ is the angle between R_1 and R_2 vectors. The grid spacing in the R_1 or R_2 dimension is varied from 0.01 to 1.0 Å. The minimum spacing value is used near the equilibrium value of the Si-O bond length and the maximum value is used for large values of R_1 or R_2. The bond distances are varied from 1 to 10 Å, and θ is varied from 0 to 180°. In the θ-dimension, the spacing is varied from 3 to 10°.

To obtain the final singlet, triplet, and pentet databases, all points that have potential higher than V_o, where V_o corresponds to the potential energy of the Si(Triplet) + O(Triplet) + O(Triplet) system, were discarded. When R_1 and R_2 are large, configurations having $R_3 > 2.5$ Å have been ignored. In addition, configurations that are very different from those associated with reactions (1) and (2) were ignored. For example, configurations corresponding to Si + O + O were not included in the database.

For convenience, all results were scaled by taking V_o as the zero. Finally, symmetry considerations were employed to increase the size of the database. That is, for a configuration with $R_1 = a$, $R_2 = b$, and $R_3 = c$, if $V = V_{abc}$, we added a point having $V = V_{abc}$ for the configuration with $R_1 = b$, $R_2 = a$, and $R_3 = c$. This addition raised the number of points in each database to 6,673. This procedure has been discussed in Chapter 5, Section 5.2.1. Agrawal et al.[171] obtained a fourth database by comparing singlet, triplet, and pentet energies at each configuration and selecting the minimum of these three.

To obtain analytic representations for the singlet, triplet, and pentet surfaces as well as for the database comprising the minimum of these three energies, (3-40-1) NNs were employed. During training, the database of 6,673 energies was partitioned into training and testing sets containing 5,500 and 1,173 data points, respectively. In each case, the fitting to the training set was accomplished iteratively using 3,000 cycles with the Levenberg-Marquardt algorithm.[117,133] The resulting NN surfaces for the singlet, triplet, pentet, and combined, minimum-energy databases are denoted as NN(S), NN(T), NN(P), and NN(STP), respectively. The quality of fits for the NN surfaces is measured by the rms deviation of the NN energies from the *ab initio* energies of the testing sets. The rms errors for NN(STP) and NN(S) are both 0.078 eV.

This fitting error is larger than that found in other applications of NNs to reaction dynamics that employ either NS,[136] gradient fitting,[124] or the CFDA method.[121,122] The primary reason for this difference is that when a grid method is employed for sampling configuration space, the location of the grid nodes is arbitrary and essentially unrelated to the gradients of the PES. Consequently, the configuration space sampling is not as good as that achieved using NS,[136] gradient fitting,[124] or CFDA methods.[121,122]

The SiO_2 dissociation dynamics were investigated using MD methods.[7] A Runge-Kutta method was used to integrate Hamilton's equations of motion.[7] An integration time step of 0.05 t.u. was used (1 t.u. = 1.018×10^{-14} second). With this time step size, total energy was conserved to an accuracy of better than 10^{-3} eV. To study dissociation of SiO_2 at an internal energy E, a set of 500 trajectories was computed. The initial configurations for the trajectories were obtained by randomly distributing this energy E into three normal modes of vibration. The initial position coordinates were taken to be the equilibrium configuration of SiO_2. The initial momenta were partitioned to the three atoms according to the method based on normal mode vectors.[148,82] The use of this method ensured that the initial states for the study of the decomposition reactions have a microcanonical distribution of the internal energy and contain no linear or angular momentum. Trajectories were allowed to run for 40,000 time steps or until dissociation occurred.

The Si-O bond lengths in equilibrium SiO_2 and SiO obtained on the NN(STP) PES are given in Table 5-13 where they are compared to other literature values.

The dissociation to Si + O_2 was studied in two steps. In step I, for a fixed value of θ, R_1 was varied, keeping R_2 equal to R_1, to obtain the minimum energy configuration. The value of minimum energy so obtained as a function of θ is as shown in Figure 5-25. As seen in the figure, there is a shallow minimum at $\theta = 55°$. R_1, R_2, and R_3 values at this minimum are 1.67, 1.67, and 1.54 Å, respectively. Agrawal et al.[171] inferred that although SiO_2 is most stable in the linear geometry, it also has a triangular metastable state which is 2.2 eV higher than the most stable linear state. The barrier height for the transition out of this metastable state is about 0.6 eV on the NN(STP) PES.

In step II, the minimum energy path on the NN(STP) surface was monitored for the transition from the triangular SiO_2 state to the products Si + O_2. To do so, in the non-linear geometry for a given value of R_1 ($= R_2$), R_3 was varied to obtain a

Table 5-13 Comparison of the properties of the NN(STP) PES with previous computations.[171] (Reprinted with permission from American Institute of Physics.)

Property	NN(STP)[a]	Other Work	Method	Reference
Si-O Bond Length	1.52 Å	1.54 Å	DFT(a)	212
Eq. Linear SiO_2		1.53 Å	DFT(a)	213
Si-O Bond Length in SiO	1.52 Å			

(a) Ref. 173.

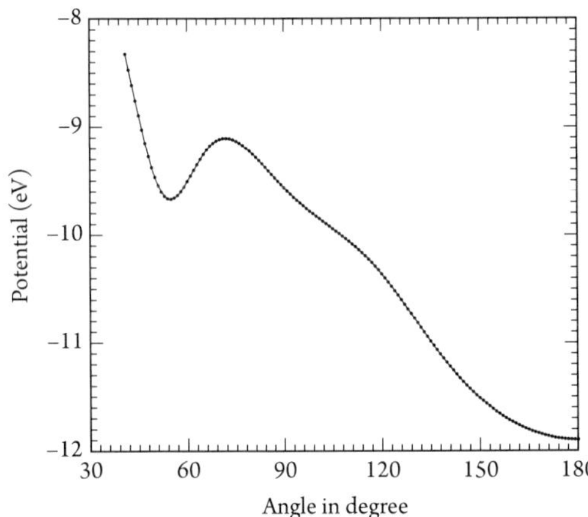

Figure 5-25 The potential energy of SiO_2 as a function of q corresponding to the minimum reaction path on the NN(STP) surface as described in the text for the reaction: SiO_2(linear) + 2.2 eV = SiO_2 (triangular) [171] (reprinted with permission from American Institute of Physics).

minimum in the energy versus R_1 curve. The minimum energy so obtained as a function of R_1 is as shown in Figure 5-26. From this curve, we note that the minima at $R_1 - R_2 - 1.67$ Å corresponds to the metastable state obtained in step I. The energy difference between this minimum and the state with large R_1 ($= R_2$) given by Figure 5-26 is equal to 4.2 eV. It was found that the value of R_3 corresponding to such a large R_1 is 1.22 Å. As expected, this value of R_3 is in agreement with the experimental value of 1.207 Å for the equilibrium bond length of O_2.[214]

Some of the results thus far can be expressed succinctly as follows:

$$SiO_2 + E_1 = SiO + O,$$

$$SiO_2(\text{linear}) + E_2 = SiO_2 (\text{triangular}),$$

$$SiO_2 (\text{triangular}) + E_3 = Si + O_2,$$

where $E_1 = 4.1$, $E_2 = 2.2$, and $E_3 - 4.2$ eV.

Combining the above equations, we obtain

$$SiO_2 (\text{linear}) + E_4 = Si + O_2,$$

where $E_4 = 6.4$ eV. The barrier heights and energies shown have been obtained without any consideration for the small corrections due to zero point energy.

The value of $E_1 = 4.1$ eV given by the NN(STP) PES is very close to 4.013 eV given directly by the *ab initio* data without the use of NN fitting. Lu et al.[215] also performed similar B3LYP/6-31G(d) ab initio calculations using the GAMESS[216] code and found the value of $E_1 = 3.953$ eV. Similarly, against $E_4 = 6.4$ eV given by the NN(STP) PES, the ab initio database yields $E_4 = 6.432$ eV. Lu et al.[215] obtained $E_4 = 6.310$ eV using the GAMESS software. These results suggest that the error associated with the NN fitting is comparable to the inaccuracy associated with the *ab initio* data. The differences in the *ab initio* results may be due to small differences in the numerical procedures in the GAMESS[216] and GAUSSIAN[141] codes.

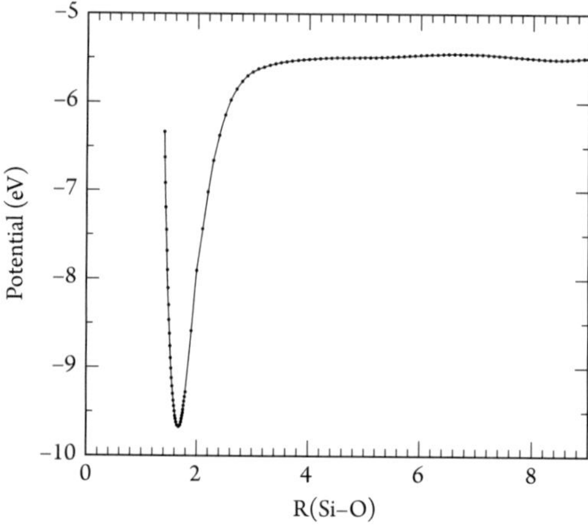

Figure 5-26 The potential energy of SiO_2 as a function of R_1 corresponding to the minimum reaction path on the NN(STP) surface as described in the text for the reaction SiO_2 (triangular) + 4.2 eV = Si + O_2[171] (reprinted with permission from American Institute of Physics).

As expected, the value of E_2 = 2.2 eV given by the present studies is also in good agreement with 2.108 eV computed by Lu et al.[215] using the GAMESS code at B3LYP/6-31G(d) level. The Si-O equilibrium bond length of 1.67 Å predicted by the NN(STP)-PES for SiO_2 in the triangular geometry is comparable with the Si-O bond lengths computed by Chu et al.[212] for various large clusters of oxides of silicon.

With respect to comparison with experiment, the values of E_4 given by the NN(STP) PES as well as by Lu et al.[215] differ from the experimental value[217] of 6.769 eV by ~5%. For the binding energy of SiO, we note a difference of ~5% from the experiment. We obtain a value of 7.828 eV for this quantity in good agreement with 7.738 eV obtained by Lu et al.[215] using the GAMESS code at B3LYP/6-31G(d) level. The reported experimental value[217] is 8.237 eV.

Trajectories were computed on both the NN(STP) and NN(S) surfaces to study the dissociation reaction to SiO + O products. The resulting decay plots all exhibit good linearity demonstrating that the dissociation process is first order. A typical trajectory running for 1,500 time steps takes about one second of CPU time when integrated on a machine with a clock speed of 1.4 GHz.

The dissociation rate coefficients, k, are obtained directly from the slopes of the decay plots. The rate coefficients for a few values of E in the range 4.3 to 5.0 eV were computed with the NN(STP) surface. These coefficients lie in the range from 0.304 ps^{-1} to 4.274 ps^{-1}. If the reaction is assumed to occur adiabatically on the singlet NN(S) surface, the corresponding rate coefficients are much smaller due to the reaction endothermicity being 2.7 eV larger for the process. All results are given in Table 5-14.

To investigate the dissociation of SiO_2 into Si + O_2, the vibrational energy must be in excess of 6.4 eV (see Equation (5-10)). Therefore, a total vibrational energy E as

Table 5-14 Reaction rate coefficients (k) for the reaction SiO2 = SiO + O.[171] (Reprinted with permission from American Institute of Physics.)

Energy (eV)	k (ps⁻¹) NN(S)	k (ps⁻¹) NN(STP)
4.3	-	0.304
4.4	-	0.548
4.5	-	0.867
4.6	-	1.123
4.8	-	2.266
5.0	-	4.274
7.0	0.0314	-
7.2	0.225	-
7.4	0.467	-
7.5	0.618	-
8.0	1.456	-
8.5	2.397	-

high as 8.0 eV was inserted randomly in all modes and more than 2,000 trajectories were computed but all trajectories were found to lead to dissociation to SiO + O. The same total energy was also placed in a single bending mode but no dissociations leading to Si + O_2 were observed. Agrawal et al.[171] suggested that the reason for this result may be that the SiO + O channel has a very low barrier height relative to that for the Si + O_2 channel. Therefore, the dissociation into SiO + O occurs with very high probability, precluding the formation of Si + O_2 products.

To test the ability of the NN(STP) PES to properly represent the reaction channel leading to Si + O_2 products, the reaction Si + O_2 → SiO_2 was investigated. The results showed that although SiO_2 is formed in the collision of Si with O_2, in all cases SiO_2 immediately dissociates into SiO + O.

The investigations of the SiO_2 dynamics reported by Agrawal et al.[171] represent two limiting cases. If dissociation to SiO + O occurs adiabatically on the singlet surface represented by NN(S), it will be a high-energy process with a threshold slightly below 6.8 eV when zero-point energy effects are considered. If, however, the dissociation follows the lowest-energy pathway leading from SiO_2 to SiO + O, dissociation can occur at significantly lower internal energies with a threshold around 4.0 eV. In addition to these two limiting cases, there is a very large number of other possible pathways that involve transitions between the singlet and triplet surfaces at points other than the crossing points. Such pathways might be investigated using a surface hopping procedure such as that suggested by Tully.[211]

A simple thermodynamic argument suggests that the minimum pathway mechanism represented by the NN(STP) surface would be favored. However, such a view ignores the fact that the system must be able to undergo a change in spin state in the time available when the singlet-triplet seam is reached. If the spin-orbit and non-adiabatic coupling terms are not sufficiently strong to induce such a multiplicity change, the dissociation process might pass through the seam without a change of spin state and continue on the adiabatic singlet surface. In the absence of a careful computation of these coupling terms in a surface hopping investigation,[211] questions related to the SiO_2 dissociation mechanism cannot be definitively answered.

One factor that favors a near minimum-energy pathway mechanism is the location of the singlet-triplet crossing point along the reaction coordinate for dissociation to SiO + O. This is seen in Figure 5-27, which shows the computed *ab initio* energies for the singlet, triplet, and pentet states along the dissociation coordinate. The NN fits, NN(S), NN(T), and NN(P), are shown as points. The minimum-energy path on the NN(STP) surface is the solid curve. The crossing point between the NN(T) and NN(S) surfaces occurs at an energy 3.5 eV above the SiO_2 minimum. Therefore, dissociating trajectories with a total internal energy of 4.3 eV will have only 0.8 eV of internal vibrational energy remaining when the seam is reached. This energy will be distributed throughout three vibrational degrees of freedom. Consequently, we would expect the energy remaining in the reaction coordinate to be about 0.3 eV. The relatively heavy silicon and oxygen atoms will not be moving very fast at such energies. Under these conditions, we might expect the system to remain in the region of the crossing point for a sufficient time to permit spin-orbit and other adiabatic coupling terms to induce a spin transition to the triplet state.

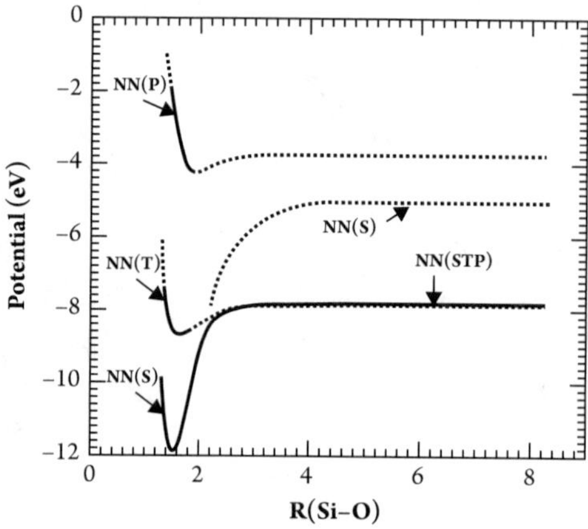

Figure 5-27 Variation of the four potentials along the $SiO_2 \rightarrow$ SiO + O reaction coordinate. The results for the singlet, triplet, pentet, and minimum-energy reaction pathways are labeled in the figure as NN(S), NN(T), NN(P), and NN(STP), respectively. The crossing point of the NN(S) and NN(STP) potentials is seen at a Si-O distance of 2.1 Å. The energy at the crossing point relative to that for equilibrium SiO_2 is 3.5 eV[171] (reprinted with permission from American Institute of Physics).

5.8 GENERALIZED NN REPRESENTATION OF HIGH-DIMENSIONAL POTENTIAL-ENERGY SURFACES

Neural network methods are ideally suited for fitting large databases of *ab initio* electronic structure data or databases obtained by any other procedure provided the number of atoms involved in the system of interest is relatively small and the chemical reactions of interest are not large in number or extremely complex. When either of these become large, several problems arise. The most demanding of these is the large increase in the volume of configuration space that must be adequately sampled to permit the NN to properly represent the system in all important regions of configuration space. The unimolecular decomposition reactions of the six-atom, vinyl bromide system serve as an example. This system involves six distinct reaction channels (12 if back reactions are counted as separate channels). As a result, about 70,000 points were needed to adequately sample the configuration space (see Section 5.6). If the number of atoms present approaches or exceeds 10 and the reactions become even more complex, it is likely that 10^5 to 10^6 points will be required.

In addition to the computational demands of conducting the electronic structure calculations required to obtain such large databases, the NN itself becomes increasingly difficult to train and employ in MD calculations. With N atoms, the input vector must contain at least $3N$-6 elements. If a two-layer NN is employed with M neurons in the hidden layer and one in the output layer, the number of weight and bias parameters required will be $(3N - 4) M +1$.

In view of this, consider the situation if we attempt to represent the PES of a 20-atom system using an NN. In this case, the input vector will need to contain a minimum of 54 elements. Since the NN Malshe et al.[126] employed to fit the vinyl bromide data base contained 140 neurons in the hidden layer, we might well expect a 20-atom system to require something on the order of 300 neurons. Such an NN will contain 16,800 weight and bias parameters. The database will probably need to contain something like 250,000 points. Such a large database does not present great difficulties because the Jacobian required by the usual Levenberg-Marquardt training algorithm[117,127] can be computed in pieces. Alternatively, the "data blocking" procedure introduced by Manzhos and Carrington[50] can be used. In this method, sequential fitting of the NN is executed using only a subset of the data to fit the PES. Once good fits are obtained to the subset, the process is repeated by cycling through the entire data set so that at the conclusion of the iterative procedure, all of the database has been used to find the best parameter values. Another option with large databases is incremental training in which the weight and bias parameters are updated as each input is presented to the network. One of the fastest incremental algorithms is the extended Kalman filter.[218]

The large number of weight and bias parameters presents a different problem. The Levenberg-Marquardt training algorithm requires the approximate Hessian to be inverted, which is an operation that cannot be done in parts. This becomes a problem when the size of the Hessian is large. In this case, other training functions can be used such as conjugate gradient algorithms. The scaled conjugate gradient algorithm of Moller[219] is a good choice for very large databases. This method needs to store only the gradient, which is much smaller than the Hessian.

The problems associated with treating systems containing large numbers of atoms, such as workpieces undergoing machining applications or bulk crystals involved in phase transitions, have recently been effectively addressed by Behler and Parrinello.[173] Their method combines the generalized fitting ability of NNs with DFT-based calculations. These methods have *ab initio* accuracy in a framework that incorporates concepts originally developed for empirical potentials to produce a procedure that yields many-body potentials for systems of arbitrary size that automatically incorporates the required symmetry for the exchange of identical atoms. The details of the Behler-Parrinello (BP) method are described in this section.

The principal idea underlying the BP method is write the total potential V_T as the sum of individual atomic potentials,

$$V_T = \sum_i^N V_i,\qquad (5\text{-}11)$$

where V_i represents the contribution of atom i to the total potential. Equation (5-11) is often employed in developing empirical potentials. If Equation (5-11) is accurate, then it is clear that PESs for systems of arbitrary size can be obtained by simply increasing N, the upper limit on the summation. The difficulty associated with using such a simple summation to obtain the total potential is that V_i depends not only upon the chemical identity of atom i but also upon its chemical environment. Since this environment continuously changes during a chemical process, this is a difficult problem to handle. The power of the BP method lies in the NN methods Behler and Parrinello[173] have developed to handle this problem and simultaneously incorporate exchange symmetry.

The BP procedure begins by defining an energetically relevant local environment for an atom as being all atoms about atom i that are within some chosen cutoff radius R_C. If the set of atoms within the cutoff radius is denoted by α, the input to the network is the Cartesian coordinates of these atoms. The set Cartesian coordinates of atom k in set α is represented by $\{R_k^\alpha\}$. In order to incorporate the energetically relevant environment of atom i, the coordinates of all atoms in set α are transformed into a set of symmetry function values denoted by $\{G_i^\mu\}$ that are then used as input for the NN. Figure 5-28 shows the structure of the BP network architecture for the simple case of a system comprising three identical atoms. The incorporation of the coordinates of all atoms in set α into the set of symmetry functions is indicated in Figure 5-28 by the arrows connecting the $\left\{ R_i^\alpha \, (i=1,2,3) \right\}$ to $\{G_i^\mu\}$.

The fundamental concept being employed by Behler and Parrinello[173] is similar in spirit to Manzhos and Carrington's[186] use of redundant coordinates as input elements to the mode terms in a high-dimensional model representation (HDMR). Manzhos and Carrington found that if linear combinations of the interparticle coordinates were employed as input elements for their NNs, the rate of convergence of the HDMR expansion was increased significantly. Section 6.2 discusses this work

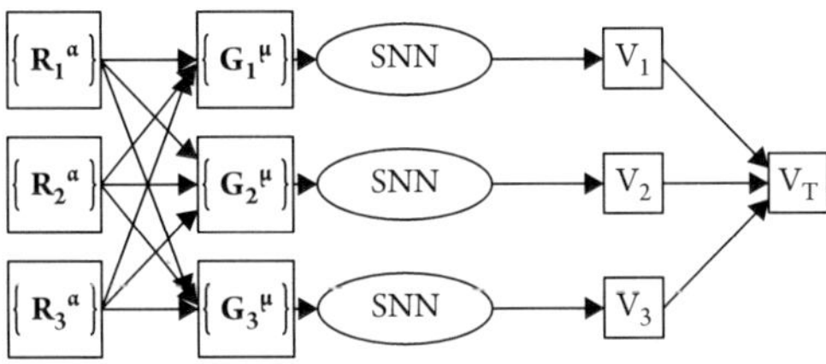

Figure 5-28 Specific case of a BP network applied to the case of three identical atoms. The set of Cartesian coordinates for the atoms within the cutoff radius is denoted by $\{R_i^\alpha\}$ The $\{R_i^\alpha\}$ (i = 1, 2, 3) are transformed to a set of symmetry function values, $\{G_i^\mu\}$ (i = 1, 2, 3), each of which is dependent upon the positions of all atoms in set a as indicated by the arrows connecting the $\{R_i^\alpha\}$ to the $\{G_i^\mu\}$. The notation SSN denotes the "subnet" for each identical atom. Its output it the potential energy of atom i, V_i. The V_i are then summed as denoted by Eq. (5-11) to yield the total system PES. The notation is that employed in Reference 173.

in much greater detail. Similarly, in the BP procedure, the use of the coordinates for all atoms within a given cutoff radius of atom *i* permits the NN to accurately assess the energetic effects of the environment on the energy of atom *i*.

The individual symmetry functions in set µ that have been suggested by Behler and Parrinello[173] are easily computed since they are defined in terms of simple summations of Gaussians and cosine terms. Specifically, a given symmetry function, *v*, in the µ set about atom *i*, G_i^v is written as the sum of a radial and an angular term,

$$G_i^v = G_i^v(r) + G_i^v(\theta),\qquad(5\text{-}12)$$

where

$$G_i^v(r) = \sum_{j\neq i}^{all}\exp\left[-\eta(v)\left(R_{ij} - R_{S(v)}\right)^2\right]f_C\left(R_{ij}\right).\qquad(5\text{-}13)$$

In Equation (5-13), R_{ij} is the interparticle distance between atoms *i* and *j*, which is easily computed using the Cartesian coordinates, $f_C(R_{ij})$ is a cutoff function used to define the energetically relevant local environment of atom *i*, and $\eta(v)$ and $R_{S(v)}$ are adjustable parameters for the vth symmetry function in set µ.

The angular term in the symmetry functions is defined to be

$$G_i^v(\theta) = 2^{1-\zeta(V)}\sum_{j\,k\neq i}^{all}\left[1+\lambda(v)\cos\theta_{ijk}\right]^{\zeta(v)}\exp\left[-\eta(v)\left(R_{ij}^2 + R_{ik}^2 + R_{jk}^2\right)\right]$$

$$\qquad(5\text{-}14)$$

$$\times\ f_C\left(R_{ij}\right)\ f_C\left(R_{ik}\right)\ f_C\left(R_{jk}\right).$$

In Equation (5-14), θ_{ijk} is the angle formed by vectors $\mathbf{R}_{ij}$ and $\mathbf{R}_{ik}$. $\lambda(v)$ (= +1 or -1) and $\zeta(v)$ are adjustable fitting parameters for the vth symmetry function in set µ. The multiplication by the three cutoff functions and by the Gaussian ensures a smooth decay to zero in the case of large interatomic separations. The total number of such symmetry functions, each with different parameter values, is arbitrary and may be adjusted to fit the complexity of the system under investigation.

The cutoff functions are similar in form and spirit to analogous cutoff functions employed by Tersoff.[13] The one used by Behler and Parrinello[173] is

$$f_C(R_{ij}) = 0.5\left[\cos\left(\frac{\pi R_{ij}}{R_C}\right) + 1\right] \quad \text{for } R_{ij} \le R_C = 0 \quad \text{for } R_{ij} > R_C, \quad (5\text{-}15)$$

where R_C is the cutoff distance.

The essential feature of the BP method is the use of NNs to fit Equation (5-11) to the results of *ab initio* electronic structure calculations, which will most often be performed at the DFT level of accuracy owing to the large number of atoms present. These NNs are indicated in Figure 5-28 by notation SNN when all atoms are identical. Behler and Parrinello[173] term these NNs as "subnets." In their initial application of the procedure, Behler and Parrinello[173] employed three-layer NNs with hyperbolic tangent and linear transfer functions in the hidden and output layers, respectively. In order to incorporate exchange symmetry between identical atoms, the parameters of the symmetry functions and the weight and bias parameters of SNN are required to be the same for identical atoms. The forms chosen for the symmetry functions also ensure that they will be invariant with respect to the location of the center of mass of the system and with respect to rotation of the entire system about its center of mass. Consequently, if the system under investigation contains only one type of atom, only one set of symmetry functions and one SNN will be required.

Behler and Parrinello[173] applied the BP procedure to the calculation of the PES for bulk silicon using DFT in the local density approximation. In this initial application, bulk silicon was represented by an ensemble of 64 silicon atoms. The details of the electronic structure calculations are given in References 173 and 220.

The NN employed in the investigation is schematically represented by Figure 5-29. Each of the hidden layers of the three-layer, feedforward NN contained about 40 neurons. A cutoff radius of 6 Å was employed in the calculations.

The set of symmetry functions, $\{G_i^\mu\}$, comprised 48 different members, that is, ν in Equations (5-12) through (5-14) spanned the range 1 to 48. With these choices, the NN contained about 3,833 parameters.

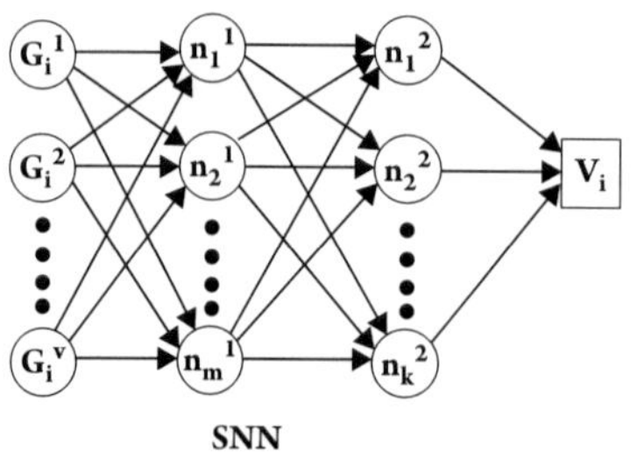

Figure 5-29 Structure of the subnet, SNN. In general, the subnet takes its input from the symmetry function set $\{G_i^\mu\}$, contains m and k neurons in the first and second hidden layers, respectively, and produces the potential energy associated with atom i.

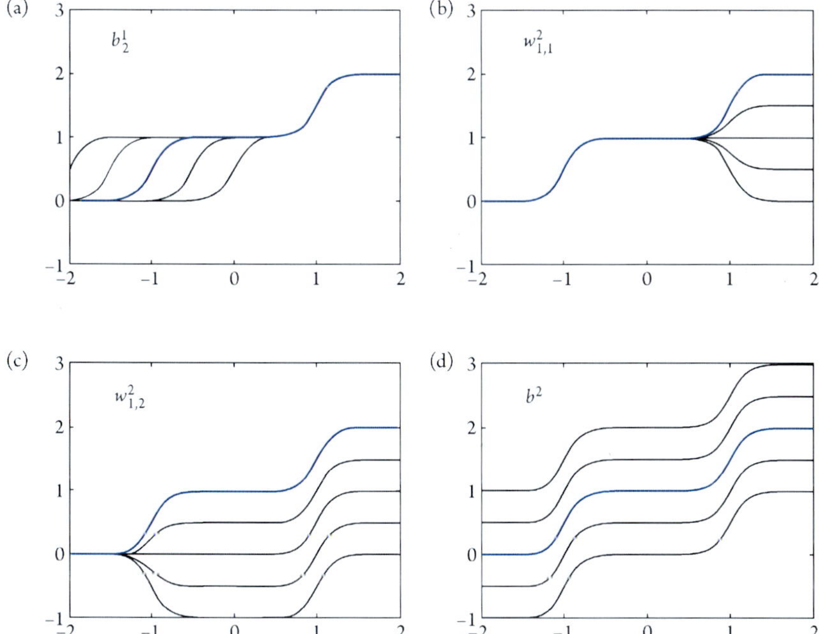

Figure 3-11 Effect of parameter changes on network response (a^2 vs. p). [105]

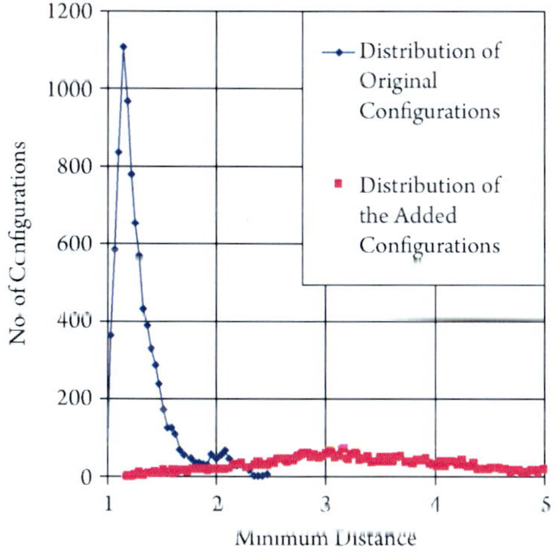

Figure 5-10 Variation of the number of configurations verses minimum distance (first iteration) [125] (reprinted with permission from American Institute of Physics).

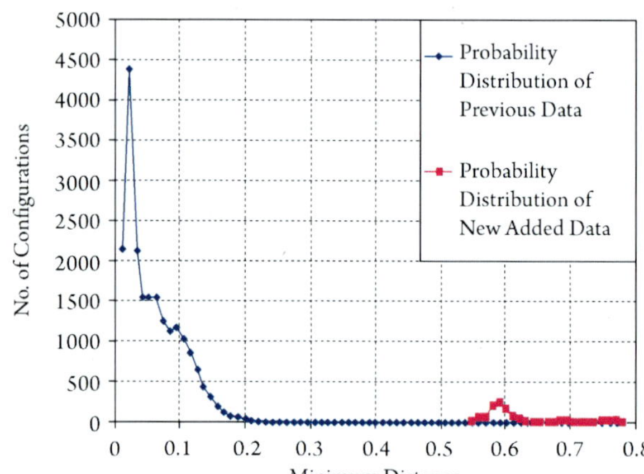

Figure 5-11 Variation of the number of configurations verses minimum distance (final iteration)[125] (reprinted with permission from American Institute of Physics).

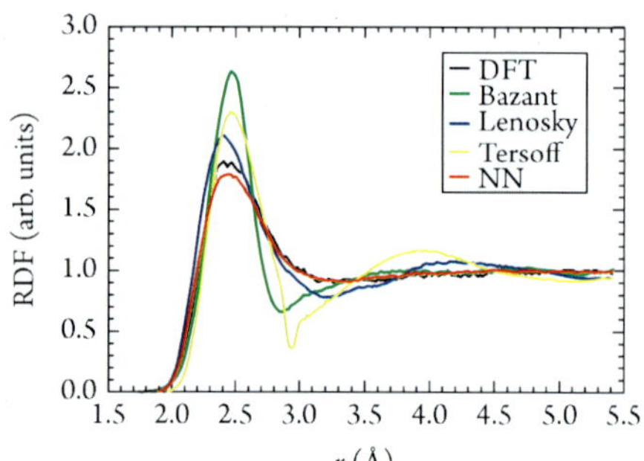

Figure 5-30 Radial distribution function (RDF) of a silicon melt at 3000 K as obtained using a cubic 64-atom cell with a = 20.526 bohr. The curves shown were obtained from the Bazant,[227,228] the Lenosky,[229] the Tersoff,[13] the NN potential, and from DFT calculations[226] (reproduced with permission from the American Physical Society; see ref. 173).

The configuration sampling to obtain the silicon structures on which the DFT calculations were executed was initiated by a selection of crystal structures that included high-pressure phases[221] and MD simulations at different pressures and temperatures. DFT calculations were then conducted to obtain the energies of these structures. The ensemble of energies thus obtained provided the initial database for the system. A temporary NN was then fitted to this database and MD, Monte Carlo,[222,223] and metadynamics[224,225] calculations performed to obtain additional structures. The DFT energies of these additional configurations were then computed, and the root mean square error (RMSE) between these DFT energies and those predicted by the temporary NN obtained. If this RMSE is larger than the previous error of the NN fit, the new DFT energies are added to the database and a new temporary NN is determined. This procedure is repeated iteratively until the NN converges to the database. This overall procedure is essentially the same as those described in Section 4.2.

Behler and Parrinello[17] found that about 9,000 DFT energies were required to obtain sufficient convergence of the NN energies to the DFT results. The SNN subnet was trained using 8,200 points with the remaining 800 points be withheld and used as an independent testing set. RMSE for the training set was found to be about 4-5 meV per atom. Thus, with 64 silicon atoms present, the total RMSE for system was between 0.256 eV to 0.320 eV. The authors do not report the total range of DFT energies present in the database.

Behler and Parrinello tested the resulting PES in various ways. First, they computed the energy versus volume curves for different crystal structures of silicon.[221] It was found that the NN PES accurately reproduces the curves and the transition pressures predicted by DFT. The ability of the NN PES to properly describe disordered structures was tested by computation of the radial distribution function (RDF) of a silicon melt at 3,000 K. MD simulations on the NN PES were run for 20 ps while those using DFT were continued for 8 ps.[226] It was found that the RDF obtained from the NN PES is very close to the DFT result. The comparison is shown in Figure 5-30 where the NN and DFT results are shown as the red and black curves, respectively.

The authors[173] suggested that the origin of the small differences between the NN and DFT results is due to the lack of configuration space sampling on the Γ point used to sample the Brillouin zone while the NN was trained to reproduce the energetics of a converged k-point mesh.

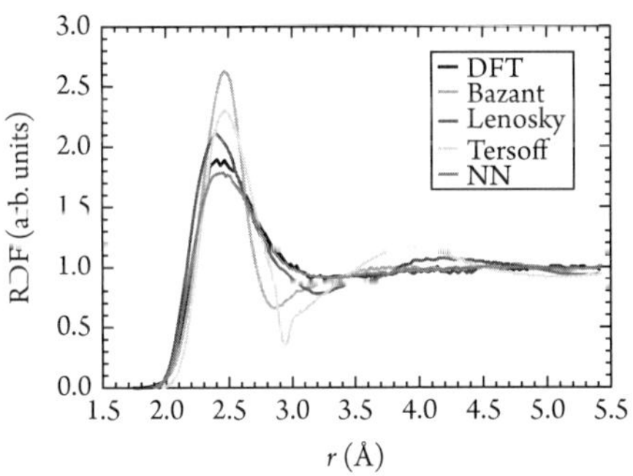

Figure 5-30 Radial distribution function (RDF) of a silicon melt at 3000 K as obtained using a cubic 64-atom cell with a = 20.526 bohr. The curves shown were obtained from the Bazant,[227,228] the Lenosky,[229] the Tersoff,[13] the NN potential, and from DFT calculations[226] (reproduced with permission from the American Physical Society; see ref. 173).

The accuracy of the NN energies was also checked by running a metadynamics simulation[225] using the NN PES and comparing the energies of the resulting structures at the start and the end of 50 intervals of 2 ps each with the computed DFT energies. The results[173] showed that the difference $E_{\mathrm{DFT}} - E_{\mathrm{NN}}$ varied from about -0.45 eV to $+0.80$ eV for the 64-atom system. Figure 5-31 shows a plot of the differences at start and end of each metastep of 2 ps duration.

For the 64-atom silicon system used to test the BP method, Behler and Parrinello[173] found that the dynamics simulations performed using the NN PES are about five orders of magnitude faster than DFT calculations. They also note that the computational requirements of NNs scale linearly with system size.

As usual, the accuracy of the NN method is limited by the accuracy of the electronic structure calculations that are used to train the NN. As discussed in Chapter 3, NNs cannot be expected to extrapolate accurately. Therefore, they are reliable only in those regions of configuration space that have been adequately sampled during the training process.

The drawbacks of the BP method include those always associated with the use of cutoff functions of the type represented by Equation (5-15). Such functions produce continuous function and first derivative values for the PES, but the second derivatives of the PES have discontinuities at the point $R_{ij} = R_C$. This is easily shown by the fact that

$$\frac{d^2 f_C\left(R_{ij}\right)}{dR_{ij}^{\,2}} = -\frac{\pi^2}{2R_C^{\,2}} \cos\left(\frac{\pi R_{ij}}{R_C}\right) \quad \text{for } R_{ij} \leq R_C$$

so that when $R_{ij} = R_C$, we have

$$\frac{d^2 f_C\left(R_{ij}\right)}{dR_{ij}^{\,2}} = \frac{\pi^2}{2R_C^{\,2}}.$$

However, when $R_{ij} > R_C$, the second derivatives are always equal to zero over the range. This feature produces discontinuities in results for any bulk quantity that depends upon the second derivatives, such as stress-strain curves.

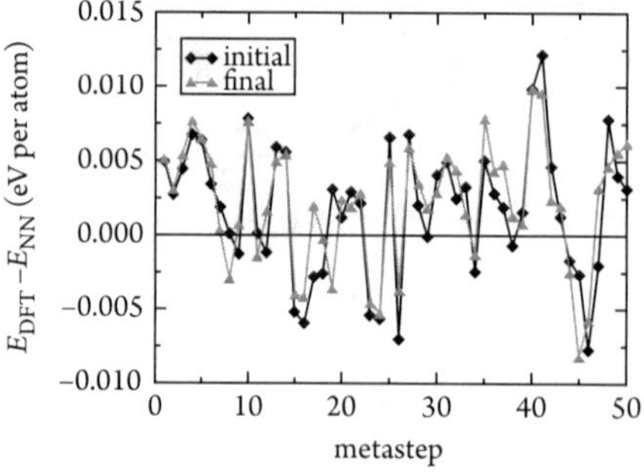

Figure 5-31 Difference between the energies predicted by the NN and those obtained from DFT calculations for the initial and final structures in each step of a metadynamics simulation of bulk silicon. Each metastep involves an MD simulations of 2 ps starting from the b-tin structure with a pressure of 15 GPa at 300 K (reproduced with permission from the American Physical Society).

There are no formal difficulties treating systems with multiple types of atoms present, but the symmetry functions must be modified to include different terms for the interactions of all possible different two- and three-body combinations of the different types of atoms present.

At present, the BP method is probably the most powerful technique for executing dynamics or Monte Carlo studies of large systems using PESs obtained from *ab initio* electronic structure calculations.

6

POTENTIAL-ENERGY SURFACES
USING EXPANSION METHODS
AND NEURAL NETWORKS

6.1. INTRODUCTION AND OVERVIEW
OF EXPANSION METHODS

Expansion methods have been employed for some time to represent the potential-energy surface for molecular systems. The basic concept involved with any expansion method is to write the PES expression for an N-atom system that requires the specification of 3N-6 internal coordinates as a sum of terms each of which involves fewer than N atoms and/or fewer than (3N-6) coordinate variables.

Two approaches to the implementation of this concept have been suggested. In the first approach, the focus of attention is the number of internal coordinates upon which each term in the expansion depends. That is, the potential V is written in the form

$$V = \sum_{i}^{3N-6} f_1(i) + \sum_{\substack{i>j}}^{3N-6} f_2(i,j) + \sum_{\substack{i>j \\ k>i,j}}^{3N-6} f_3(i,j,k) + \cdots + \sum_{\substack{i>j \\ k>i,j \\ \cdots \\ m>i,j,k,\ldots l}}^{3N-6} f_n(i,j,k,l,\ldots,m)$$

$$(6\text{-}1)$$

where the i, j, k, ... m subscripts denote one of the 3N-6 internal coordinates required to specify the configuration of the N-atom molecular system, and f_n ($n = 1$, 2, 3, ...) are appropriate expansion functions each dependent upon n of the 3N-6 internal coordinates. When this approach is taken, the expansion is termed a "high-dimensional model representation" (HDMR) and the f_n are called n-dimensional component functions or n-d component functions. The basic idea is to truncate the expansion when the largest value of n appearing in the expansion is less than $(3N\text{-}6)$. If the n-d component functions in the resulting truncated expansion can be adjusted so that an accurate fit of the potential is obtained, a $(3N\text{-}6)$ dimensional potential can be represented by functions whose dimensionality is less than this value.

The second approach appears to be the same, but is actually much different. In this approach, the potential for the N-atom molecule is written in the form

$$V = \sum_{\substack{i>j}}^{N} f_2(i,j) + \sum_{\substack{i>j \\ k>i,j}}^{N} f_3(i,j,k) + \sum_{\substack{i>j \\ k>i,j \\ l>i,j,k}}^{N} f_4(i,j,k,l) + \cdots + \sum_{\substack{i>j \\ k>i,j \\ l>i,j,k, \\ \cdots \\ m>i,j,k,\ldots,l}}^{N} f_n(i,j,k,l,\ldots,m)$$

$$(6\text{-}2)$$

where the i, j, k, ... m subscripts now denote one of the N atoms in the molecule, and f_n $(n = 2, 3, 4, \ldots)$ are appropriate expansion functions each dependent upon n of the N atoms in the molecule. That is, the summations run over the N atoms present rather than over the $(3N\text{-}6)$ internal coordinates of the system. When this approach is taken, the expansion is termed a many-body expansion and the f_n are called n-body terms. Again, the basic idea is to truncate the expansion when the largest value of n appearing in the expansion is less than N.

Rabitz and co-workers[54-70] have utilized an HDMR to represent a number of different systems. They determine the individual component functions by minimization of the weighted sum square errors between the target values and those predicted by the component function integrated over the entire space of the database. This evaluation generally requires the evaluation of many high-dimensional integrals.[59,68]

In its most general form, all possible m-d component functions are included in the summation over the m-dimensional components of an HDMR expansion. If the system contains N atoms, the total number of internal coordinates required will be $(3N\text{-}6)$ or more if a redundant set of variables is employed. Therefore, the number of terms in the summation over the m-d components is at least the number of combinations of $(3N\text{-}6)$ objects taken m at a time, $^{3N\text{-}6}C_m$, where $^{3N\text{-}6}C_m = \dfrac{(3N-6)!}{m!(3N-6-m)!}$.

If N is large, the computational demands of the procedure will be extremely large. For example, for a six-atom system using the minimal set of 12 internal coordinates, the summation over the 4-d components will contain 495 terms. If the redundant set of 15 interparticle distances is used, the number becomes 1,365. This difficulty is referred to as the combinatorial problem.

Depending on the complexity of a problem, one can decide on the number of component functions or many-body terms needed to characterize the system accurately. For example, a two-body potential or a 1-d component function would be sufficient to represent the interaction between two hydrogen atoms in an H_2 molecule. However, a pair potential or a 1-d component function will inadequately represent a more complex structure such as diamond. The selection of the number of many-body terms or component functions required to represent a potential accurately is completely dependent on the application. The potential advantage of using a many-body expansion for an N-body system lies in the fact that in many cases, the expansion will converge after M terms, where $M < N$. When this is the case, the necessity of using an N-body expression for the potential surface is avoided. The same is true for an HDMR expansion.

The concept of using a many-body expansion is not new. They have been frequently employed to develop empirical potential surfaces. For example, the Tersoff potential[13] has been frequently employed to represent the potential force field for covalently bonded, non-metallic systems. This potential truncates the expansion given by Equation 6-2 after the three-body terms. The Brenner potentials[15] for hydrocarbon systems are also many-body expansions that include only the two- and three-body terms. Bolding and Anderson[14] have used such an expansion for Si_nH_m systems. Their empirical potential includes one four-body term. All of these empirical potentials employ parameterized functional forms for the various terms in the expansion. In general, the parameters contained in the functional forms for the various terms are obtained by fitting the expansion to some database of experimental data, configuration energies obtained from electronic structure calculations, or both. Since the functional forms used in the expansion are arbitrary, they automatically limit the fitting accuracy of the expansion.

Many-body expansions have previously been exploited by Murrell and co-workers[52,53] who used an expansion of functions dependent upon interatomic coordinates. This approach was modified by Carter, Bowman, and co-workers[230-232] by making the terms in the many-body expansion dependent upon normal coordinates.

The principal limitation on the accuracy of expansion methods is the unknown nature of the expansion functions, that is, the f_n in Equations (6-1) and (6-2). In most applications of expansion methods, these functions are chosen arbitrarily. Usually, an attempt is made to choose their functional forms based on chemical and physical intuition. These functional forms are parameterized, and the parameters then adjusted to some experimental and/or *ab initio* database by a suitable non-linear method. It is the assumption of specific functional forms for the expansion functions that has limited the accuracy and hence the use of expansion methods.

The substitution of NNs for the expansion functions greatly empowers the overall method. Since two-layer feedforward NNs with sigmoid transfer functions in the hidden layer and linear functions in the output layer have been shown by Hornik et al.[108] to be universal approximators for analytic functions, this substitution effectively eliminates the problems associated with choosing the expansion functions.

This method can be used with both HDMR and many-body expansions. Examples of each of these approaches are given in Sections 6.2 and 6.3 of this chapter.

6.2. HIGH-DIMENSIONAL MODEL REPRESENTATION (HDMR) AND NNS

Manzhos and Carrington[49,50,186] have combined HDMR expansions with NNs using linear combinations of internal coordinates to obtain an accurate and robust means of building multi-dimensional potentials. The NNs are employed to represent the component functions present in the HDMR expansion. The fitting error of the expansion of Equation (6-1) to the target energies is minimized sequentially mode term by mode term by adjustment of the weights and biases of the NNs.

Manzhos and Carrington[49] have noted that NNs are good for fitting component functions because of the generality and robustness of the method. The procedure works for any choice of coordinates, and the same fitting method is applicable for all component functions and for any potential. In addition, the results can be improved by simply increasing the size of the NNs. In contrast to expanding the component functions in a basis set such as polynomials,[54-70] there are no problems with oscillations or nonorthogonality, and the sampling method utilized to obtain the database for fitting the expansion can be adjusted to reflect the relative importance of different regions of the potential.

The NN fitting is executed with the Levenberg-Marquardt algorithm with Bayesian regularization.[117,120,135] Manzhos and Carrington[49] discuss two procedures to determine the mode terms. In the first procedure, the mode terms are obtained sequentially. That is, one first executes a fit of the 1-d terms under the first summation of Equation (6-1). Subsequently, the 2-d mode terms under the second summation are fitted to the difference between the target potentials and the 1-d mode terms. This procedure is continued until the desired accuracy of the total fit is obtained. In the second procedure, one decides the maximum order of the mode terms to be included and "lumps" all the mode terms of this and lower orders together and fits only the mode terms of the highest order. Clearly, the second method requires an a priori knowledge of the maximum order of HDMR needed. If this information is not available, the first procedure is employed.

In the HDMR-NN approach, there is one NN per component function. As discussed previously, the number of required component functions and, therefore, the number of fitting parameters increases combinatorially with the dimensionality of the problem. This difficulty makes the use of an HDMR expansion very costly if the number of coordinates is greater than about six. Manzhos and Carrington[49] have developed a partial solution to the combinatorial problem by grouping parameters. That is, the parameters of the expansion are grouped into K sets. The optimization then proceeds by fixing the values of K-1 of the sets and the error is minimized with respect to the parameters in the remaining set. When completed, the procedure is repeated with a different set being varied to obtain the minimum error. By cycling through this procedure iteratively, one arrives at a final optimization that is close to what would have been obtained by a more costly, global optimization procedure. If the parameters are uncorrelated, the final result will be identical to that achieved by global optimization. However, Manzhos and Carrington[49] report that even in the case of strong correlation, it is possible to obtain a good fit.

In practice, the minimization during each cycle of the grouping process is not carried out to convergence. Only a partial minimization is executed. The result is a reduction in the computational time required without a significant reduction in the accuracy of the final fit.

In their initial test, Manzhos and Carrington[49] applied the HDMR-NN method to the fitting of the Kuhn potential[51] surface for hydrogen peroxide, H_2O_2. The description of the database used for the test and the methods employed to obtain it have already been described in Chapter 5, Section 5.2.2. The quality of the HDMR-NN fit was assessed by computing the root mean square error (RMSE) for a set of test points. For this system, one NN was employed for the one-mode term. The two-mode, three-mode, … , and M-mode terms were lumped together into another NN. Typical results are given in Table 6-1.

In general, Manzhos and Carrington[49] found that for a fixed value of M, the RMSE decreases as the number of neurons used in the NN for the component functions is increased up to a point. When the number of neurons is sufficiently large, the error attains an approximate constant value. Further increases will have little effect. If overfitting occurs, increasing the number of neurons can cause the RMSE on the testing set to increase. They also found that it is the number of points in the database that determines the optimum value of M in the HDMR-NN expansion.

As discussed in Chapter 5, Section 5.2.2, the H_2O_2 potential sampling was only over the configuration space important to describe the molecular vibrational levels of the molecule. For their H_2O_2 investigation using the HDMR-NN method, Manzhos and Carrington included in the database energies up to 15,000 cm^{-1}. Since this is less than the energies that must be included in the database when chemical reactions are occurring, the RMSE errors obtained by Manzhos and Carrington[49] using an HDMR-NN expansion are very low. They are not, however, as low as the results obtained by these investigators when using their dual NN approach.[47] One

Table 6-1 H_2O_2 fit quality with different orders of HDMR-NN.[49] (Reprinted with permission from American Institute of Physics and the authors.)

Order M	No. Neurons per Component Function	No. of Component Functions	RSME (cm^{-1}) Testing Set
2	20	15	156
3	30	20	23.4
4	40	15	8.9
5	60	6	7.7
6	90	1	7.4

of the major reasons for this difference is the fact that the database employed for the dual NN fit included primarily points with energies up to 10,000 cm^{-1}. The volume of configuration space is, therefore, smaller than that present in the database used for the HDMR-NN investigation. This difference produces a more accurate fit in the dual NN study.

Table 6-2 shows the errors in the computed H_2O_2 vibrational energy levels relative to those obtained using the analytic Kuhn et al.[51] potential.

In a much more demanding test of the HDMR-NN method, Manzhos and Carrington[50] applied the method, modified by incorporation of optimized redundant coordinates as described later, to fitting the database obtained by Malshe et al.[126] for the unimolecular decomposition reactions of vinyl bromide (VB). These reactions and the methods employed to obtain the database have been previously described in Chapter 5, Section 5.6.

The specification of the VB configurations requires a minimum of 12 internal coordinates. However, as discussed in Chapter 3, a redundant set of coordinates can be employed. Malshe et al.[126] employed the 15 interparticle distances to obtain a (15-140-1) NN fit to the VB database. This fit resulted in a mean absolute fitting error of 0.065 eV or about 524 cm^{-1}.

When using an HDMR expansion method to fit the database, the combinatorial problem makes it much more convenient to employ the minimum set of 12 coordinates than the redundant set of 15. Manzhos and Carrington[50] have investigated both choices using an HDMR-NN method.

A straightforward application of Equation (6-1) within the framework of the HDMR-NN method to VB is essentially computationally intractable. If the problem is treated in the 15-dimensional space of the redundant interparticle distances and the expansion is truncated after the 6th-order terms, there will be 5005 6-d (6-dimensional) terms, 3003 5-d terms, 1365 4-d terms, 455 3-d terms, and 105 2-d terms present. Each of these will have to be represented by an NN. The magnitude of the combinatorial problem is obvious. To bring the problem within range of available computational resources, Manzhos and Carrington[50] developed some simplifying procedures in addition to the ones already described.

Table 6-2 Mean absolute errors of the first 48 vibrational levels of A_g symmetry of H_2O_2 on top of the zero-point energy with different orders of HDMR-NN.[49] (Reprinted with permission from American Institute of Physics and the authors.)

Order M	No. Neurons per Component Function	No. of Component Functions	Average Level Error (cm^{-1})
2	20	15	10.35
3	30	20	1.14
4	40	15	0.44
5	60	6	0.37
6	90	1	0.35

One of more important of these procedures is the use of linear combinations of the original 15 interparticle distances as the elements in the input vector to the NNs. Since these combinations are generally redundant, Manzhos and Carrington[186] have termed this redundant coordinate (RC) procedure an RC-HDMR-NN method. The RC procedure has the effect of reducing the number of component functions needed for each mode term in the HDMR expansion.

A qualitative understanding of the RC method may be gained as follows: Consider the 6-d mode terms for VB being treated in the 15-d space of the interparticle distances as the original coordinates. The 6-d mode terms of the HDMR will contain 5,005 terms as that is the number of combinations of 15 distances taken six at a time. In the HDMR-NN method, each of these terms is represented by an NN. A given NN will have a (6×1) input vector with the six elements being six of the 15 interparticle distances. The sum of all 5,005 NNs will provide the HDMR expansion with the contribution of all combinations of six coordinates. If instead of using a single distance as each of the six input elements, linear combinations of all 15 coordinates are employed as the elements of the input vector, each NN will contain contributions of all combinations of six coordinates. Therefore, convergence can be obtained with a smaller number of component functions.

In practice, Manzhos and Carrington[50,186] optimize the linear combinations being employed during the process of fitting the RC-HDMR-NN expansion to the VB database. The new set of coordinates is redundant because no component functions share the same coordinates. The result is that the number of component functions needed for a given mode term becomes a convergence parameter of the method rather than an absolute number determined by the dimensionality of the fit and the order of the mode term.

The process of parameter grouping developed to treat the H_2O_2 system is also employed for VB. Manzhos and Carrington[50] do sequential fitting of the NNs using only a subset of the data to fit the component functions. Once good fits are obtained to the subset, the process is repeated by cycling through the entire data set so that at the conclusion of the iterative procedure, all of the database has been used to find the best parameter values. This process was termed "data blocking." The concept of "lumping" mode terms was also employed. For the VB case, it was determined that the HDMR-NN expansion had to include the 6-d mode terms. Therefore, using lumping concepts, the lower-order mode terms were omitted or "lumped" into the 6-d terms. The actual fitting was accomplished using the Levenberg-Marquardt algorithm.[117,135]

In order to facilitate quantum scattering calculations, exponential transfer functions were employed in all NNs rather than the usual sigmoid functions. Because an exponential of a linear combination of the original coordinates can be written as a product of 1-d functions of the original coordinates, exponential neurons result in a potential that has sum-of-products form. Previous investigations by Manzhos and Carrington[48] have shown that although NN fitting error using exponential transfer functions is greater than that obtained with sigmoid functions, the difference is not large.

Using these methods, Manzhos and Carrington[50] have refitted the VB data base of 71,969 points obtained by Malshe et al.[126] This database was divided into a set of 64,773 for NN training and 7,196 for testing. Using the RC-HDMR-NN method with lumping, parameter grouping, and data blocking, they found that their best fit was obtained with 15 6-d component functions. The test set error for this expansion was 0.095 eV or 769 cm^{-1}. Figure 6-1 shows a histogram giving the dependence of the testing set error in cm^{-1} as a function of the number of component functions employed in the RC-HDMR-NN expansion when lumping is employed with 2-d, 3-d, 4-d, and 6-d mode terms incorporated.

The PES for the dissociation of VB into six open reaction channels poses an extremely difficult problem because of the volume of configuration space involved and the very wide range of energies present. The best fit that Malshe et al.[126] were able to obtain using a single (15-140-1) NN had a mean absolute error of 0.065 eV or 524 cm^{-1}. Since the root mean square error is usually about 50% larger than the mean absolute error, the RMSE error of 769 cm^{-1} obtained by Manzhos and Carrington[50] using a RC-HDMR-NN method is very similar in accuracy to the result reported by Malshe et al.[126] The fact that Manzhos and Carrington[50] were able to achieve a fit of very similar quality using only 15 6-d functions attests to the power of the RC-HDMR-NN procedure. Manzhos et al.[233] have recently applied this method to obtain a lower-dimensional PES for the interaction of N$_2$O with a Cu(100) surface.

6.3. MANY-BODY EXPANSIONS, MOIETY ENERGY APPROXIMATIONS, AND NNs

Malshe et al.[71] have utilized the many-body expansion represented by Equation (6-2) in which the summations run over the two-, three-, ... , M-body atomic

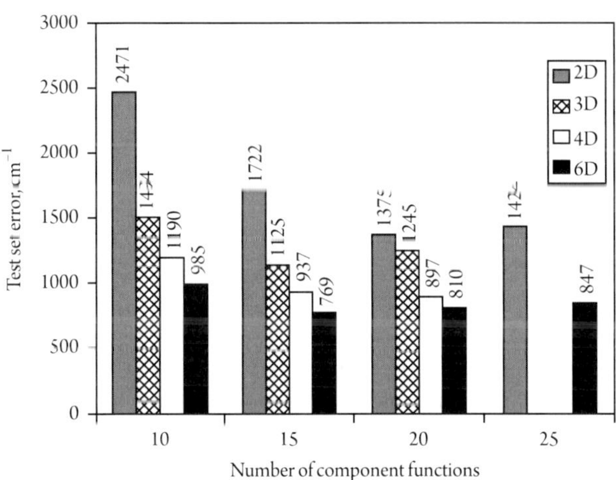

Figure 6-1 Test rmse for RC-HDMR-NN with different numbers of component functions using 2-d, 3-d, 4-d, and 6-d mode terms with lumping[50] (reprinted with permission from American Institute of Physics; see ref. 173).

clusters rather than over the two-, three-, ..., m-dimensional component functions. NNs are employed for each of the M-body terms in the expansion. The bottleneck presented by the combinatorial problem is significantly reduced by summing over atomic clusters rather than over components of different dimensions. This problem is reduced still further by an extension of the concept of "bond energy" to "moiety energy" (ME). The prime advantages of using NNs are that it avoids the assumption of a specific functional form and permits the use of Jacobian methods, such as the Levenberg-Marquardt (LM) procedure,[117,135] for automatic adjustment of weights and biases of the NNs. Since Hornik et al.[108] have shown that two-layer NNs with sigmoid transfer functions in the first hidden layer and linear functions in the output layer are universal approximators for analytic functions, such forms can take any shape. Therefore, employing NNs for each of the M-body terms provides the flexibility needed to accurately fit a complex potential surface. The use of the ME concept reduces the number of NNs required to a very small number.

In a system of N atoms, there are $^{N}C_2$ pairs, $^{N}C_3$ 3-body, ... $^{N}C_k$ k-body terms, where

$$^{N}C_k = \frac{N!}{(N-k)!\,k!}.$$ (6-3)

Using Equation (6-2), the total potential is written as a sum of these k-body potentials for $2 \le k \le M$. Each of these terms is represented by a two-layer NN with sigmoid transfer functions in the hidden layer and a linear transfer function for the output layer. It should be noted that Equation (6-2) significantly reduces the combinatorial problem of an HDMR. For example, consider a six-atom system in which the redundant set of 15 interparticle distances is used as the configuration variables. The 2-d + 3-d + 4-d summations of an HDMR will contain 1,925 component functions. The many-body expansion represented by Equation (6-2) truncated after the four-body summations will contain 50 terms.

To reduce the number of NNs required still further, the approximation of "bond energy" is now extended to include any set of k atoms in which the same number of each type of atom is present with similar bonding characteristics. Such sets are termed "equivalent sets." It is assumed that all equivalent sets may be represented with a single NN. If there exist s non-equivalent k-body sets in the N-atom system, then s NNs will be required in the summation over the k-body terms. For example, in the A_2B_2 system, there exist three non-equivalent two-body sets, A-A, B-B, and A-B. Therefore, the two-body summation in Equation (6-1) would contain three NNs. This system has two non-equivalent three-body sets, A-A-B and A-B-B, so that the three-body summation would contain two NNs. In the simplest case in which all N atoms are identical, a single NN will suffice for each of the k-body summations. It is obvious that the use of this ME approximation significantly reduces the number of NNs required by Equation (6-2).

The input vector for the NNs contained in the k-body terms can be any set of coordinates that is sufficient to specify the internal configuration of the k-body system. In general, for k bodies, a minimum of $3k$-6 coordinates must be specified. However, if desired, the system may be overspecified. A common choice is the set of all interatomic distances or inverse distances between the k atoms. Such a set will contain $k(k$-$1)/2$ elements. The input vector for the NNs can be a linear combination of internal coordinates. Carter, Bowman, and co-workers[230-232] used this concept in their work when they employed normal coordinates as the variables in the various terms of Equation (6-1). More recently, Manzhos and Carrington[50,186] have effectively employed linear combinations of interatomic distances as the input vectors for the NNs used in their HDMR treatment of the potential surface for the vinyl bromide system. They find that the use of such optimized, redundant coordinates significantly reduces the number of component functions needed for each k-d summation. In the present investigation, the set of all interatomic distances between the k atoms is employed.

In general, each NN will contain p elements in the input vector with $p \geq 3k$-6, n neurons in the hidden layer, and a single neuron for the output layer, which will be the predicted contribution to the potential for that particular k-body term. Such NNs will each contain $(p+2)n + 1$ parameters that are the elements of the weight and bias matrices of the network.[105] In the present method, the elements of the weight and bias matrices are adjusted to fit the potentials in the database using the LM[117,135] iterative procedure.

The iterative procedure for parameter adjustment is illustrated in Figure 6-2 for the case where Equation (6-2) is truncated after the three-body terms for a system containing three types of atoms, A, B, and C. In Loop 1 of this illustration, the input vectors comprising the interatomic distances for two-body and three-body terms for configuration n are input into the two-layer NN appropriate for the specific two-body or three-body interaction involved. The outputs from these NNs are then summed and compared to the target potential for configuration n. The difference in these values is the error e_n for configuration n. If there are Q configurations in the database, Loop 1 is repeated until all errors, e_1 through e_Q, are computed.

After computation of the fitting errors for all Q configurations in the database, control is transferred to Loop 2 of Figure 6-2. In this loop, the Jacobian matrix, **J**, required by the LM algorithm is computed. This is followed by the execution of one-step of the parameter adjustment. Subsequently, the updated weight and bias matrices, **W** and **b**, are returned to Loop 1 for the execution of the next cycle of iterative parameter adjustment.

In general, the Jacobian matrix, **J**, will contain Q rows and $\{s_2[3S_2 + 1] + s_3[5S_3 + 1] \ldots + s_M[(3M$-$4)S_M +1]\}$ columns, where S_t is the number of neurons in the first hidden layer of the NN for the t-body term, and s_t is the number of non-equivalent t-body terms. The elements of **J** all have the form $[\partial e_n/\partial w_{ij}^v]$ or $[\partial e_n/\partial b_i^v]$, where e_n represents the fitting error of the potential for the nth data point, w_{ij}^v is the NN weight connecting the jth input to the ith neuron of the vth layer of network, and b_i^v is the bias for the ith neuron of the vth layer of the network. These derivatives can be computed using the Jacobian backpropagation algorithm.[105] It is important to note that this overall procedure adjusts the weights and biases of all NNs

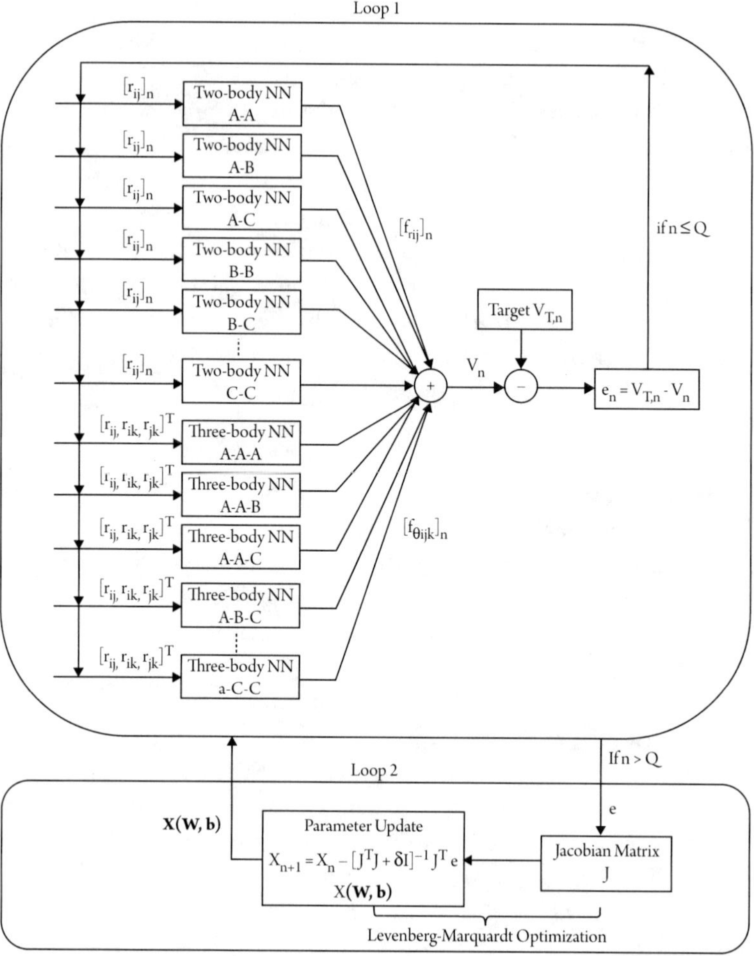

Figure 6-2 Flowchart for the general many-body expansion/NN fitting algorithm[71] (reprinted with permission from American Institute of Physics).

simultaneously, thereby properly treating their coupled nature of the terms in the expansion of Equation (6-2).

Malshe et al.[71] have applied this many-body method to several systems of increasing complexity. As a first example, the many-body/NN/ME method was used to fit the energies of Si_5 clusters observed in an MD simulation of silicon machining with a single-point diamond cutting tool.[140] The convergence properties of the expansion were illustrated by first including only two-body and three-body terms. The sufficiency of the interactions terms considered is reflected by the testing set error of the expansion to the database. If the error is too high, higher-order, many-body terms are needed.

There are 10 two-body and 10 three-body terms in each Si_5 cluster. Since all atoms are identical, all two- and three-atom moieties are equivalent, and only a single NN is required for the two-body interactions and a second NN for the three-body terms. For notational purposes, let $f(r_{ij})$ and $\theta(r_{ij}, r_{ik}, r_{jk})$ represent the two- and three-body terms, respectively, where r_{ij} is the interatomic i-j distance with analogous definitions for r_{ik} and r_{jk}. Figure 6-3 shows the flow diagram for the general case modified for the specific case of an Si_5 cluster system. Only two NNs are required. The input **p** vector for the two-body term is a single distance while that for the three-body NN is a (3×1) column vector whose elements are r_{ij}, r_{ik}, and r_{jk}.

In this application, a (1-25-1) NN was used for the two-body term and a (3-25-1) NN for the three-body term. However, this choice is arbitrary. If required, larger networks can be easily employed. In both networks, log sigmoid-linear transfer functions were used.

The Si_5 database comprised 10,202 five-atom silicon configurations. This database was partitioned into training (6,120 points), validation (1,530 points), and testing (2,552 points) sets that were employed in the usual manner to fit Equation (6-2), to avoid overfitting using early stopping, and to test the accuracy of the final fit, respectively.[105] The target values (*ab initio* energies) were produced by using a density functional theory (DFT) method with a 6-31G** basis set and the B3LYP procedure for incorporating correlation energy. The total range of energies spanned by the Si_5 clusters is 0.8138 eV. The 10,202 Si_5 configurations were obtained from MD simulations of nanometric cutting of a silicon workpiece using a single-point

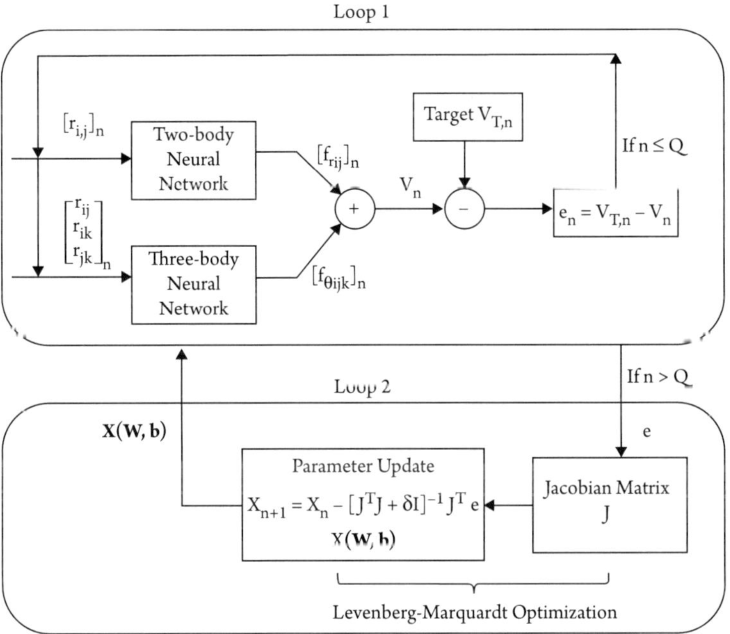

Figure 6-3 Flowchart for the many-body expansion/NN fitting algorithm for the specific case of Si_5 clusters with the expansion of Eq. (6-2) truncated after the three-body terms[71] (reprinted with permission from American Institute of Physics).

cutting tool and sampling configurations within a 3 Å cutoff radius in front of and immediately below the cutting tool.[140] With these choices, the Jacobian required by the training algorithm contains 6,120 rows and 202 columns.

A root-mean-squared (RMS) testing set error of 0.0008143 eV was obtained from the fitting with a mean absolute testing set error of 0.0005693 eV. This accuracy, which corresponds to 4.6 cm^{-1} or 0.055 kJ mol^{-1}, is among the best thus far reported for a system of this complexity with such an extensive database. As a percentage of the total energy range spanned by the Si_5 clusters, the mean absolute testing set error represents a percentage error of 0.070%. Clearly, there is no need to include four-body or higher terms for this Si_5 system.

The general method can be easily extended to a multi-cluster system such as Si_n ($n = 3, 4, 5$) clusters. The procedure for training the network described above for the Si_5 system is repeated, but the database now includes energies of silicon clusters containing three to five atoms that are observed in the MD cutting simulations.[140] This database comprises 300 Si_3, 260 Si_4, and 500 Si_5 configurations yielding a total of 1,060 configurations. Since the number of Si_3 and Si_4 clusters observed in the cutting experiments is much less than the number of Si_5 configurations, 500 Si_5 configurations were randomly selected so that the total number of configurations for each type of cluster is roughly the same. If this is not done, the fitting will be dominated by the Si_5 data. An additional 597 configurations were employed as a testing set to evaluate the accuracy of the final fit.

With this training database, one iteration of the training cycle is illustrated in Figure 6-3; it requires about 10 minutes of CPU time on a PC with a clock speed of 2.6 GHz. A total of 860 iterations were required to obtain convergence of the weights and biases of the NNs.

The electronic structure calculations employed to obtain the energies of these configurations was the same as that used for the Si_5 clusters. For the training set, a root-mean-squared error of 0.0338 eV was obtained for the fitting with mean absolute error of 0.0192 eV. For the testing set, the RMS error is 0.0437 eV and the mean absolute error is 0.0249 eV. Examination of the results for each type of cluster shows that the testing set error is relatively higher for Si_4 clusters than for either Si_3 or Si_5. If it is judged that a testing set error with an RMS deviation of 0.0437 eV is too large, the results can be further improved by simply adding higher M-body terms.

As an illustration, the four-body terms in the expansion were added. The resulting iterative procedure to fit the potential was the same as that shown in Figure 6-3 except that an additional (6-25-1) NN was added in parallel to compute the four-body terms. With this choice and the database previously described, the Jacobian matrix required for the fitting contains 1,060 rows and 403 columns.

For the training set, the resulting RMS error is 0.0038 eV, and the mean absolute error is 0.0030 eV. For the testing set, the RMS error is 0.0076 eV and the mean absolute error is 0.0056 eV. Thus, the inclusion of four-body terms in the potential reduces the RMS error for the testing set by a factor of 5.8 and the mean absolute error by a factor of 4.4. The mean absolute error for the testing set corresponds

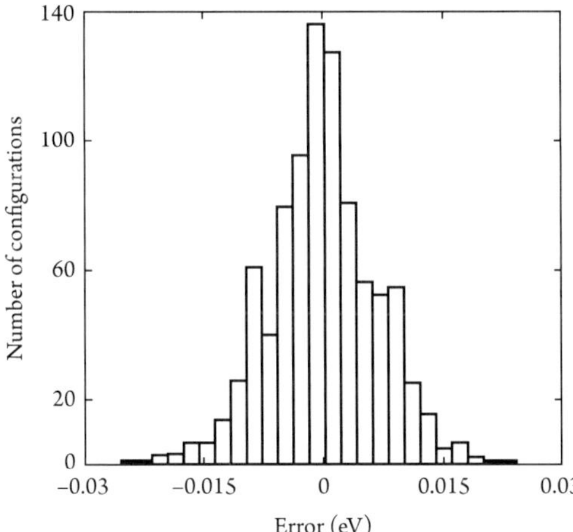

Figure 6-4 Distribution of fitting errors for Si_3, Si_4 and Si_5 clusters with four-body terms included in the many-body expansion. The mean absolute testing set error is 0.0056 eV. With the database covering an energy range of about 6 eV the error corresponds to a fitting accuracy of about 0.093%[71] (reprinted with permission from American Institute of Physics).

to 0.54 kJ mol[-1] or 45 cm[-1]. The distribution of testing set errors for Si_3, Si_4, and Si_5 clusters is shown in Figure 6-4. Since the Si_3, Si_4, and Si_5 database spans an energy range of about 6 eV, the percentage error represented by the mean absolute error is about 0.093%.

The silicon clusters occurring in the MD simulations of silicon machining[140] also include Si_6 and Si_7 clusters for a cutoff radius of 3.0 Å. The many-body expansion/NN/ME method can easily be extended to include such clusters. The number of Si_n clusters for the various values of n used in the training set for the NNs is given in Table 6-3. The *ab initio* electronic structure calculations for these cluster configurations were performed in the same manner as previously described for the Si_3, Si_4, and Si_5 clusters.

Malshe et al.[71] have examined the effect of network configuration upon the average testing set error by fitting the above database with three different NN configurations each having 403 to 468 total weight and bias parameters. In each case, overfitting is avoided by use of a validation set. Table 6-4 gives the NN specifications for each of the three NN configurations examined. The mean absolute errors for the testing set varied from a low of 0.0212 eV for Calculation #3 to 0.0353 eV for Calculation #1. Since the configuration space spanned by the Si_n ($n = 3, 4, \ldots, 7$) clusters is about 17 eV, these errors correspond to percentage errors of 0.12% and 0.21%, respectively.

Figure 6-5 shows a comparison of the fitted energies of the Si_n clusters with the computed *ab initio* energies when the NNs are those specified under Calculation #1 in Table 6-4. If the results were perfect, all points would fall on the 45° line in the figure. As can be seen, the results are very accurate for all clusters even when a single set of three NNs is used for the many-body expansion for all cluster sizes. As shown in Table 6-4, the mean absolute error of the testing set energies is 0.0353 eV, which corresponds to a mean absolute percentage testing set error of about 0.21%.

The most demanding test of the many-body/NN/ME method is the vinyl bromide (VB) system. The VB database,[126] the fitting results when using a single

Table 6-3 Number of configurations of each of the clusters in the training set.[71] (Reprinted with Permission from American Institute of Physics.)

Cluster Size	Number of Configurations
3	300
4	260
5	500
6	360
7	117

Table 6-4 NN specifications for the many-body expansions for Si_n ($n = 3, 4, \ldots 7$) clusters. Errors are given in eV. |error| denotes the mean absolute testing set error. The entry labeled % error is |error| x 100/range, where range is the total energy range spanned by the database employed in the fitting of the potential. The entry labeled # Parameters gives the total number of weight and bias parameters contained in the three NNs.[71] (Reprinted with permission from American Institute of Physics.)

	NN Specifications		
Term	Calculation #1	Calculation #2	Calculation #3
2-Body	(1-25-1)	(1-15-1)	(1-10-1)
3-Body	(3-25-1)	(3-20-1)	(3-15-1)
4-Body	(6-25-1)	(6-40-1)	(6-45-1)
# Parameters	403	468	468
rms error	0.0540	0.0419	0.0360
mean |error|	0.0353	0.0276	0.0212
% error	0.21 %	0.16 %	0.12 %

(15-140-1) NN, and the results obtained with the RC-HDMR-NN method[50] have been previously described in this chapter.

There are five non-equivalent two-body sets, six non-equivalent three-body sets, and five non-equivalent four-body sets in vinyl bromide. Therefore, the general method illustrated in Figure 6-2 requires a corresponding number of NNs be employed in the two-, three-, and four-body summations of the expansion for a total of 16. Without the use of ME approximation, this number would be 50. Using novelty sampling methods,[136] Malshe et al.[126] selected the energies for 45,427 configurations as the training set. The remaining 25,542 points in the database were divided equally and randomly between the validation and testing sets.

The NN architecture for the each non-equivalent set of the t-term ($t = 2, 3, 4$) was the same. Several different choices for the network architectures for the

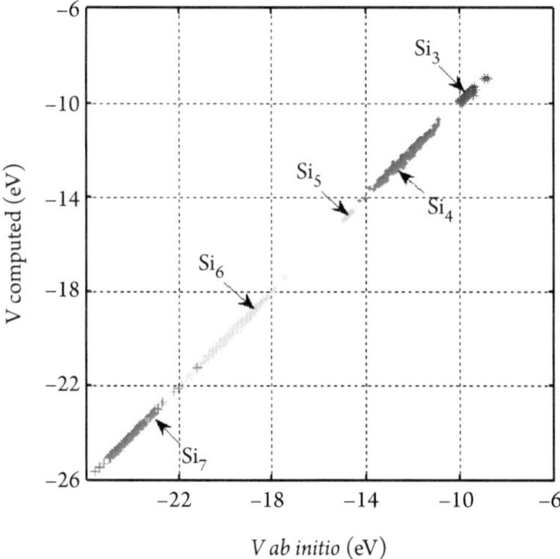

Figure 6-5 Comparison of the fitted energies obtained using Eq. (6-2) with the NNs specified under Calculation #1 in Table 6-4 with the *ab initio* DFT energies for Si_n (n = 3, 4, ... 7) clusters. If the fitting were perfect, all points would fall on the 45° line in the figure. The mean absolute testing set error is 0.0353 eV, which corresponds to a mean percent testing set error of 0.21%[71] (reprinted with permission from American Institute of Physics).

two-, three-, and four-body NNs were examined. The architectures, number of parameters, and fitting results for two of these choices are given in Table 6-5.

The best result obtained by Malshe et al.[71] was a mean absolute error of 0.0782 eV (631 cm^{-1}) for the testing set using the NN architecture for Fit #2 in Table 6-5. Thus, the many-body expansion of Equation (6-2) has a testing set error about 20.3% larger than the global fit obtained by Malshe et al.[126] using a (15-140-1) NN.

On the basis of their investigations, Malshe et al.[71] conclude that the advantages of the many-body/NN/ME method are the following:

1. The use of the many-body expansion described by Equation (6-2) with NNs representing the various k-body terms (k = 2, 3, ... , M) provides a well-defined, practical method for constructing analytic surfaces to databases obtained from *ab initio* electronic structure calculations from a combination of terms each of lower dimensionality than the full N-body system under investigation. Since the summations of Equation (6-2) run over atoms rather than coordinates, the combinatorial bottleneck present in an HDMR expansion is significantly reduced in a well-defined manner without empirical testing or adjustment.
2. The use of the ME approximation reduces the combinatorial problem still further in a straightforward, well-defined manner. In the most favorable case of clusters of the same type of atom, only a single NN is required for each summation in the expansion. In other cases, such as in the case of vinyl bromide, the ME approximation results in a significant simplification.

Table 6-5 Many-body expansion/NN/ME results for two different network architectures for fitting the vinyl bromide database. Testing set errors are given in eV.[71] (Reprinted with permission from American Institute of Physics.)

Fit #	Network Architecture			# Parameters	Mean Absolute Error(eV)	
	2-Body	3-Body	4-Body		Training	Testing
1	(1-25-1)	(3-50-1)	(6-65-1)	4491	0.0735	0.0808
2	(1-25-1)	(3-50-1)	(6-75-1)	4891	0.0708	0.0782

3. Any iterative Jacobian method can be employed to affect the fitting of the weights and biases of the NNs. The method described here adjusts all weights and biases simultaneously so that the coupled nature of these parameters is properly treated in the fitting. However, when this method is employed, the standard MATLAB NN toolbox cannot be used.

4. Although we have employed sigmoid transfer functions for the hidden layer of the NNs used in our investigations, any transfer function can be employed. Specifically, exponential transfer functions may be preferred because of the computational advantage they provide in quantum mechanical scattering calculations as Manzhos and Carrington have pointed out.[48] The method also lends itself easily to the use of any type of input coordinates including redundant, linear combinations (RC) that Manzhos and Carrington have used so effectively in fitting the vinyl bromide database.[50,186]

5. The fitting accuracy provided by the methods described here are excellent for silicon clusters containing up to seven atoms. For the much more complex and difficult vinyl bromide system, the fitting accuracy obtained using 5 one-dimensional NNs, 6 three-dimensional NNs, and 5 six-dimensional NNs with interparticle distances as input elements is about 20% less than the 15-dimensional, global fit obtained by Malshe et al.[126] using a (15-140-1) NN. This is apparently the price of the added simplicity of the many-body expansion, the ME approximation, and the simple input elements.

7

GENETIC ALGORITHM (GA) AND INTERNAL ENERGY TRANSFER CALCULATIONS USING NEURAL NETWORK (NN) METHODS

7.1. GENETIC ALGORITHM (GA) CALCULATIONS USING NN METHODS

7.1.1. Introduction

Genetic algorithms (GA), like NNs, can be used to fit highly nonlinear functional forms, such as empirical interatomic potentials from a large ensemble of data. Briefly, a genetic algorithm uses a stochastic global search method that mimics the process of natural biological evolution.[234] GAs operate on a population of potential solutions applying the principle of survival of the fittest to generate progressively better approximations to a solution. A new set of approximations is generated in each iteration (also known as generation) of a GA through the process of selecting individuals from the solution space according to their fitness levels, and breeding them together using operators borrowed from natural genetics. This process leads to the evolution of populations of individuals that have a higher probability of being "fitter," i.e., better approximations of the specified potential values, than the individuals they were created from, just as in natural adaptation.

The most time-consuming part in implementing a GA is often the evaluation of the objective or the fitness function. The objective function $O[P]$ is expressed as sum squared error computed over a given large ensemble of data. Consequently, the time required for evaluating the objective function becomes an important factor. Since a GA is well suited for implementing on parallel computers, the time required for evaluating the objective function can be reduced significantly by parallel processing. A better approach would be to map out the objective function using several possible solutions concurrently or beforehand to improve computational efficiency of the GA prior to its execution, and using this information to implement the GA. This will obviate the need for cumbersome direct evaluation of the objective function. Neural networks may be best suited to map the functional relationship between the objective function and the various parameters of the specific functional form. This study presents an approach that combines the universal function approximation capability of multilayer neural networks to accelerate a GA for fitting atomic system potentials. The approach involves evaluating the objective function, which for the present application is the mean squared error (MSE) between the computed and model-estimated potential, and training a multilayer neural network with decision variables as input and the objective function as output. This trained neural network is then used as part of a GA for computing the objective function. This approach has been found to speed up a GA in general by eliminating redundant computation of the objective function and by precomputing offline relative to the GA. The results show that the set of parameters obtained using the approach described in this chapter yields closer fits for Tersoff potential energies (MSE $< 1.32 \times 10^{-6}$ eV^2) and *ab initio* potential energies (MSE < 0.0025 eV^2) for isolated 5-atom silicon clusters. Thus, in cases where the objective function is expressed as MSE over a large number of data points, an NN can be a suitable means for faster estimation of the objective function to accelerate GA. It is also apparent, based on the present investigation, that a Tersoff potential, albeit with different (GA-parameterized) coefficients, is adequate for representing the *ab initio* potentials of 5-atom Si clusters.

The organization of the remainder of this chapter is as follows: an overview of GA is presented in Section 7.1.2. This is followed by Section 7.1.3, where some of the previous methods that combine GA with NN are briefly reviewed. A brief background on the interatomic potential function used to demonstrate the GA-NN approach is presented in Section 7.1.4. The basic GA-NN approach is described in Section 7.1.5. This is followed by Section 7.1.6 on the application of GA-NN approach to PES function fitting; and concludes in Section 7.1.7 with a chapter summary.

7.1.2. Brief Overview of GA

An advantage of a GA is that it can be used to fit physically meaningful functional forms using the experimental data. GAs are among the most widely used tools for deriving optimal models for highly nonlinear, uncertain, and multimodal spaces.[234] Their wide acceptance is attributed to their ability to search the space for globally optimal solutions instead of getting stuck in local optima. They use the processes of biological selection and evolution to iteratively create new generations of solution (here, a vector of model parameters) [P] alternatives from the superior "strains" of previous generations. They also allow for designing adequate mathematical structures to capture multiple interrelated and/or competing phenomena. The models designed and parametrized through GAs bear physically meaningful structures. Therefore, they tend to be more robust for complex function fitting applications.

The building blocks of a GA are called genes. A sequence of genes constitutes a chromosome. Individual candidate solutions, represented in the present context by vectors, or alleles, of model parameter values $[P]$ are encoded as chromosomes. The most common encoding involves representing an allele as a single fixed-width binary string. The first step in the execution of a GA involves generating the set of initial chromosomes known as the parent chromosomes. A chromosome can simply be a fixed length binary string obtained from concatenating the binary values of a parameter vector P. New chromosomes (offspring) and hence a new set of parameters $[P]$ are created in every generation (i.e., iteration) by using a combination of operators, such as selection, crossover, mutation, etc. on the parent chromosomes.[234] The major GA operators are concisely described in the following paragraphs.

Roulette wheel selection. This operation follows Darwin's theory of evolution's survival of the fittest principle. The chromosomes having higher fitness $O[P]$ value bear the better chance for duplication in the next generation. In a GA, the number of offspring generated in each iteration is restricted by the operation rate. Operation rate is the ratio of the number of the offspring to the number of parents. Typically, parents with higher fitness (i.e., lower model fitting errors) will have a higher probability of being selected to generate the offspring.

Mutation. Mutation is aimed at producing new individuals by randomly selecting the single chromosomes and the mutation point. The sub-tree or a sub-array at the mutation point is replaced by a new sub-tree. The new sub-tree is randomly created.

Crossover. A crossover is the operation for producing new individuals. Here, two parental chromosomes are selected from the population and a crossover point is randomly selected from each parent. Two new individuals are created in the next generation by swapping the two sub-trees.

A GA terminates if either the set number of iterations is executed or if all the chromosomes converge to an optimal, almost invariant set. Thus, the essence of a GA is to create multiple generations of a population through the application of one or more genetic operators that manipulate the individuals so as to progressively and concurrently improve a fitness function of the individuals in a population to reach a globally near-optimal fit for such complicated nonlinear structures, such as PES functions.

7.1.3. Applications of NNs for GA Acceleration

Since GAs use physically meaningful functional forms and possess excellent extrapolation capabilities they have been used in a variety of applications. It may be noted that the literature on GAs is quite extensive covering a wide range of applications. Comprehensive reviews of GA applications are available in the literature (see, e.g., Tang et al.,[235] Bäck et al.[236]) and may be referred to for details. Although GAs can be applied to any function fitting process, repeated evaluations of the fitness (or the objective) function $O[P]$ often present a major bottleneck for GA-based fitting of complex functions. Surrogate modeling approaches, including the use of NNs, have been attempted to minimize the computational efforts involved in fitness (objective) function evaluations.[105,237–240] Such combinations of NNs with GAs have been attempted to address a variety of issues arising in aerospace applications from shape optimization to cost minimization. All these approaches use NNs and/or other modeling approaches as cheaper surrogates for costly simulations. The nonlinearity of the underlying real functional forms has necessitated acceleration using NNs and other surrogate models. In the present context, in addition to the complexity involved in function computation, the combinatorial nature of the objective function offers additional variables for NN acceleration. In fact, the application of GAs was almost intractable for parameterization without acceleration, as will become evident in Sections 7.1.3 and 7.1.4. Therefore, although GAs have been applied for the derivation of optimal atomic configurations,[42] the likely computational overhead involved in repetitive determination of the objective function has prevented GAs from being applied for parametrizing interatomic potentials. The use of NNs renders GAs attractive for such applications where the repetitive computation of the objective function is a major bottleneck.

7.1.4. Interatomic Potential Functions for Si

The Tersoff potential, which is one of the commonly used functional forms for modeling the interatomic interaction of group IV semiconductor materials, such as silicon, is used in the present study. The functional form of Tersoff potential is given by Tersoff[13]:

$$V = \sum_i v_i = \frac{1}{2}\sum_{i \neq j} v_{ij},$$

$$v_{ij} = f_c(r_{ij})\left[f_R(r_{ij}) + b_{ij}f_A(r_{ij})\right].$$

(7-1)

where V is the total potential energy of the system. f_R and f_A are repulsive and attractive pair potentials, respectively, and f_C is a cutoff function given by

$$f_R(r_{ij}) = A_{ij}e^{(-\lambda_{ij} r_{ij})},$$

$$f_A(r_{ij}) = -B_{ij}e^{(-\mu_{ij} r_{ij})},$$

(7-2)

$$f_C(r_{ij}) = \begin{cases} 1, & r_{ij} < R \\ \dfrac{1}{2} + \dfrac{1}{2}\cos\left[\pi\dfrac{(r_{ij}-R)}{(S-R)}\right], & R < r_{ij} < S \\ 0 & r_{ij} > S \end{cases}$$

where r_{ij} is the bond distance between atoms i and j, S is the cutoff radius, R is the inner cutoff radius, and $\lambda_{ij} = \lambda$, and $\mu_{ij} = \mu$ are potential parameters. The strength of each bond depends upon the local environment. It is lowered when the number of neighbors is relatively high. This dependence is expressed by the parameter b_{ij}

$$b_{ij} = \chi_{ij}\left(1 + \beta_i^{n_i}\,\zeta_{ij}^{n_i}\right)^{-1/2n},$$

$$\zeta_{ij}\sum_{k\neq i,j} f_C\left(r_{ik}\right)\omega_{ik}\,g\left(\theta_{ijk}\right),$$

$$g\left(\theta_{ijk}\right) = 1 + \frac{c_i^2}{d_i^2} - \frac{c_i^2}{d_i^2 + \left(h_i - \cos\left(\theta_{ijk}\right)\right)^2}, \tag{7-3}$$

which can diminish the attractive force relative to the repulsive force. Here,

$$\lambda_{ij} = \frac{\lambda_i + \lambda_j}{2}; \quad \mu_{ij} = \frac{\mu_i + \mu_j}{2}. \tag{7-4}$$

The term ζ_{ij} defines the effective coordination number of atoms taking into account the relative distance between two neighbors $(r_{ij},\ r_{ik})$ and the bond angle θ_{ijk}. The function $g(\theta)$ has a minimum at $h = \cos(\theta)$. The parameter d determines sensitivity of the potential to the angle and c expresses the strength of the angular effects. The parameters R and S are not optimized but chosen so as to include first neighbors only for several selected high symmetry structures, such as graphite, diamond, and simple cubic and face-centered cubic lattice structures.

The parameter x_{ij} strengthens or weakens heteropolar bonds in multi-component systems. $x_{ii} = 1$, and $x_{ij} = x_{ji}$. Also, $\omega_{ii} = 1$. The parameters chosen to fit theoretical and experimental data were for cohesive energy, lattice constants, and bulk modulus obtained from realistic and hypothetical configurations. Table 7-1 lists the parameters of the Tersoff potential for silicon.[13] These values are adequate for computing bulk parameters of silicon at room temperature. However, they do not agree with the potentials computed based on first principle calculations at different temperatures and for isolated configurations. The parameters in Table 7-1 also do not predict the correct melting point for silicon.[241] As a result, the application of the Tersoff potential with the parameters given in Table 7-1 may lead to erroneous MD computations.

Using internal atomic coordinates is generally the most efficient way to specify the potential energy (see Equations (7-1)–(7-3)). Internal coordinates are specified using bond distances, bond angles, and dihedral (torsional) angles. Figure 7-1 shows variables involved in internal coordinates. The dihedral angle is the angle between two planes passing through atoms i-j-k and j k l, respectively. It is measured as the angle between the vectors normal to these planes. However, since the internal coordinates are highly coupled, it is convenient to solve the equations of motion during MD simulation in the Cartesian coordinate system, and convert Cartesian coordinates to corresponding internal coordinates after imposing certain constraints (e.g., first neighbor represented as atom 2 in GAUSSIAN 03 software).[141]

Table 7-1 Parameters of Tersoff Potential[13] for Silicon.[72] (Reprinted with permission from Physical Review of the American Physical Society.)

Tersoff Parameters	Values
$A\ (eV) \times 10^3$	1.8308
$B\ (eV) \times 10^2$	4.7118
$\lambda\ (°A^{-1})$	2.4799
$\mu\ (°A^{-1})$	1.7322
$\beta \times 10^{-6}$	1.1
n	0.787
$c \times 10^5$	1.0039
D	16.217
H	−0.59825
$R\ (°A)$	2.7
$S\ (°A)$	3.0

7.1.5. GA-NN Approach

Genetic algorithms (GAs) and other nonlinear function approximation routines based on a stochastic gradient search require the specification and evaluation of an objective or fitness function to determine the quality of fit. Problems involving the repetitive evaluation of the objective function (such as mean squared error in fitting a complicated function) become computationally time-consuming. In this chapter, we present an approach based on using the function approximation capability of NNs for computation of the objective function. The method is applied for parameter determination of the interatomic potential for silicon using the Tersoff functional form.[13]

The objective function $O[P]$ was defined as the mean squared error (MSE) between the potential energy calculated using a set of parameters for a Tersoff

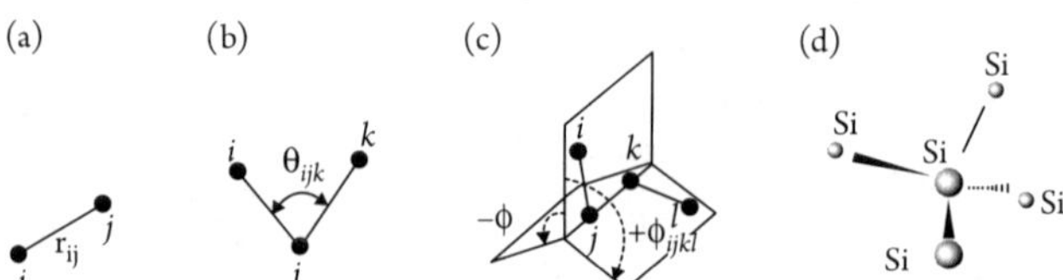

Figure 7-1 Internal coordinates: (a) bond distance, (b) bond angle, (c) dihedral angle, and (d) configuration of 5 silicon atoms[72] (reprinted with permission from Physical Review of the American Physical Society).

functional form and the actual energy values obtained from first principle calculations. The potential energy $[V]$ was computed for each of the N_c configurations (x_k), $k = 1, 2, \ldots, N_c$, using one of the N_p sets of parameters

$$O[P] = \frac{\sum_{k=1}^{} \left(V_k - \check{V}_k[P] \right)^2}{N_c} \tag{7-5}$$

where $[P]$ is the set of parameter vector values for the chosen potential function, V_k is the potential computed from the chosen function with parameter set at $[P]$, and $\check{V}_k$ is the actual potential values to be fitted, obtained here from Tersoff or *ab initio* calculations.

As shown in Figure 7-2, the gradients of $O[P]$ are steep for small values of the parameters (β, d, and η). As the values of the parameters increase, the gradient of the objective function decreases until it reaches a large plateau region marked by near-zero gradients. Conventional gradient-based optimization methods tend to be inefficient for objective functions of such form because gradients tend to be computationally cumbersome to evaluate. Ill-conditioned numerical structures, slow convergence, and suboptimal solutions are known to manifest whenever gradient search algorithms are employed for multi-variable combinatorial optimization with such objective functions. GAs are more effective compared to traditional gradient-based methods. Our approach to NN-accelerated GA is summarized in Figure 7-3.

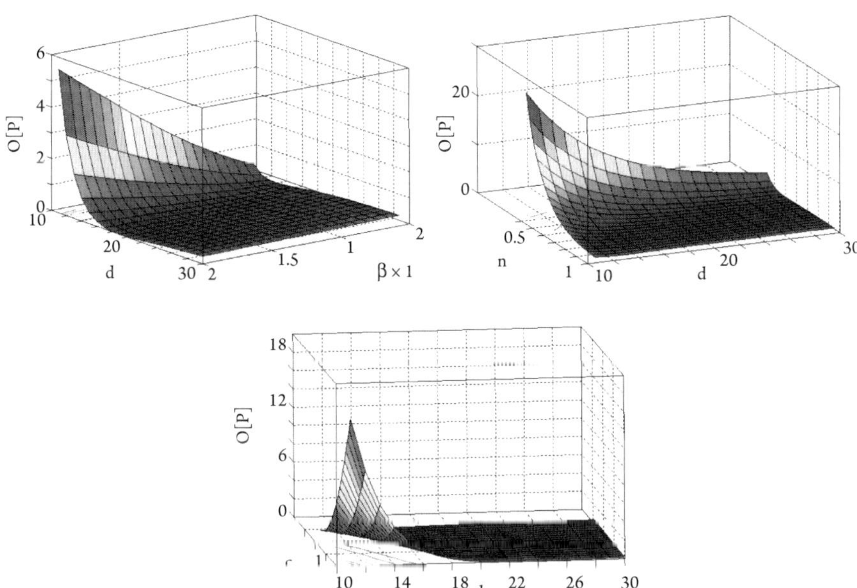

Figure 7-2 Variation of the objective function $O[P]$ with (a) β and d (b) n and d and (c) c and d, respectively, showing large plateau regions[72] (reprinted with permission from Physical Review of the American Physical Society).

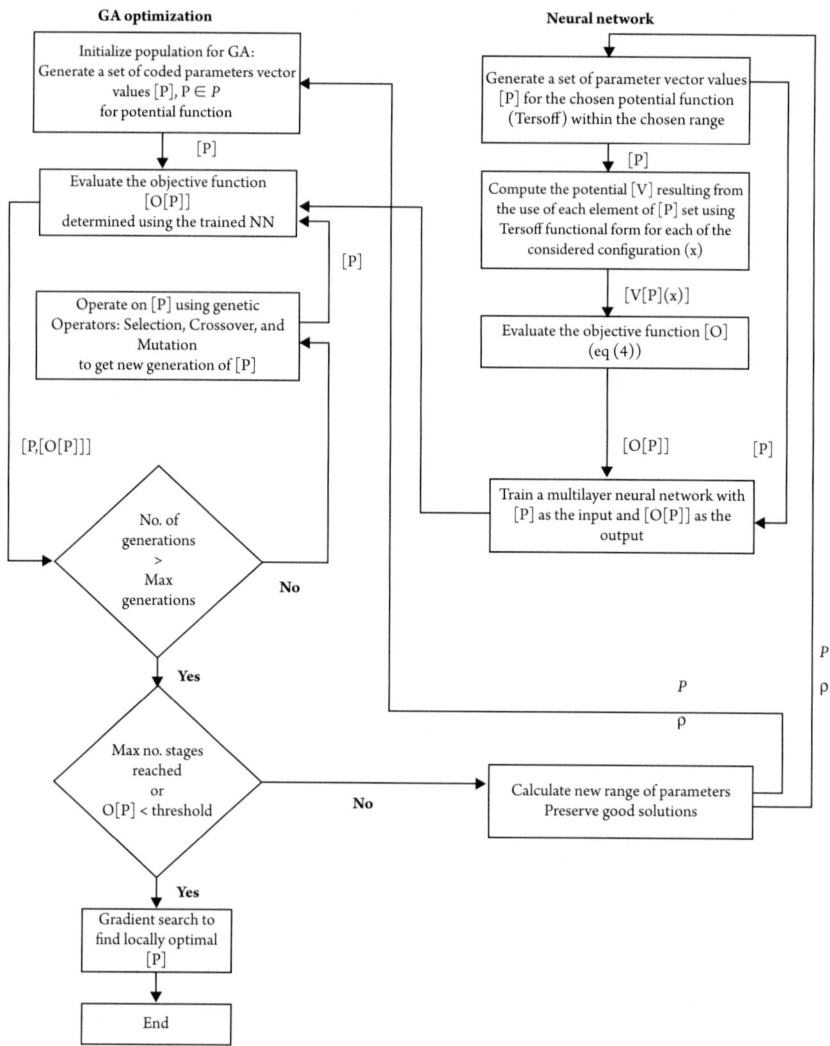

Figure 7-3 Flowchart of the approach for GA optimization[72] (reprinted with permission from Physical Review of the American Physical Society).

The approach involves off-line training of an NN to capture the relationship between theparameter $[P]$ and the objective function $[O[P]]$, and using the NN estimates of $[O[P]]$ as surrogates for GA optimization. The entire procedure of parameter optimization is carried out in multiple stages. All initial stages consist of two major components, namely, off-line NN training and GA parameterization, as shown in Figure 7-3. The procedure begins with the generation of an initial set of parameter vector values $[P]$, and computation of corresponding potential $[V[P]]$ and hence, the objective function as described in Equation (7-4). A multilayer NN is trained with potential energy parameters $[P]$ as inputs and corresponding objective function $O[P]$ as the output. Regularization and early stopping are used during training to avoid overfitting the training data. Next, the GA is executed based on NN estimates of $O[P]$. The GA consists of the following steps:

1. *Initialization of population*: A set of parameter vector values $[P]$ of the considered potential functional form[13] is generated within a specified range. The dimensionality of the parameter vector P is 9 corresponding to the number of parameters used in the Tersoff potential. The set $[P]$ contains 50 such elements. Each parameter within a set is coded in a binary string of length 25.

2. *Generating a new set of parameter values $[P]$ using genetic operators*: The objective function set $[O[P]]$ corresponding to a set of parameters generated $[P]$ is evaluated using an NN trained off-line. Genetic operators, such as crossover and mutation,[234] are used to arrive at a new set of parameters for Tersoff potential functional form. The new set of parameters for Tersoff potential is selected based on the fitness. In other words, the smaller the value of $O[P]$, the higher the probability for a parameter vector value P to be selected. The GA terminates after the maximum number of iterations is reached, or if the GA solutions appear to have converged. If the fit at the end of the phase is below a threshold level, subsequent phases consisting of NN retraining and GA parameterization are undertaken. However, the range set for the parameter vector P is reduced and the length of binary string used to encode each of the parameters is increased, for example, from 25 to 50 bits to improve the resolution. Once the fit improves above a threshold level, it is assumed that the parameter values lie in the basin of the global minimum. A gradient based method is used to determine the best fit.

7.1.6. Application to Fitting of PES of Si

First, let us apply the approach to approximate actual Tersoff potential parameters. The results can help us benchmark the approximation capability of the GA-NN approach. Next, we apply this approach to approximate *ab initio* potential values with a Tersoff functional form.

A two-layered feedforward NN is trained for mean squared error (MSE), that is, between the potential calculated using Tersoff parameters and GA parameters for $N = 1{,}000$ configurations of 5-atom Si clusters. The NN consists of 40 neurons in the hidden layer and 1 neuron in the output layer. The input exemplar patterns for training the NN consist of 4,000 arbitrarily chosen realizations of the 9-dimensional parameter vector $[P]$ of Tersoff functional form and the corresponding $[O[P]]$ obtained from Equation (7.1). Architecture of such a network is commonly denoted as (9-40-1). The trained NN is tested using 500 element sets of $[P]$ and $[O[P]]$. The standard deviations of the actual outputs $[O[P]]$ and those computed using an NN $[O[P]]$ are 0.0636 eV^2 for Stage 1, and 0.0036 eV^2 for Stage 4.

The NN-estimated $[O[P]]$ values are used to evaluate the fitness of various input sets $[P]$ during the GA. The nine parameters for the Tersoff potential, namely, A, B, λ, μ, β, η, c, d, and h (see Table 7-1) are first encoded in a binary string. During the initial stages 25 binary bits, and during final stages 50 bits are used to represent each variable. Optimization of parameters is carried out in four stages. During each stage, the range for the parameter set is narrowed to improve the accuracy in successive stages. The mutation rates are gradually decreased from 0.92 to 0.12 during each stage using a functional form $\left[1-k_1\left(1+k_2 e^{-n/N}\right)^{-1}\right]$ where n is the current iteration and N is the total number of specified iterations, and k_1 and k_2 are constants chosen for each stage to yield the mutation rate variation profiles shown in Figure 7-4(a). The crossover rate is varied from 0.9 in Stage I to 0.62 in Stage IV as shown in Figure 7-4(b). The mutation rate affects the convergence of the final solution. During the initial stages, a large value for mutation rate facilitates in searching through a large parameter space, while a smaller value for mutation rate toward the end promotes convergence to the optimal solution.

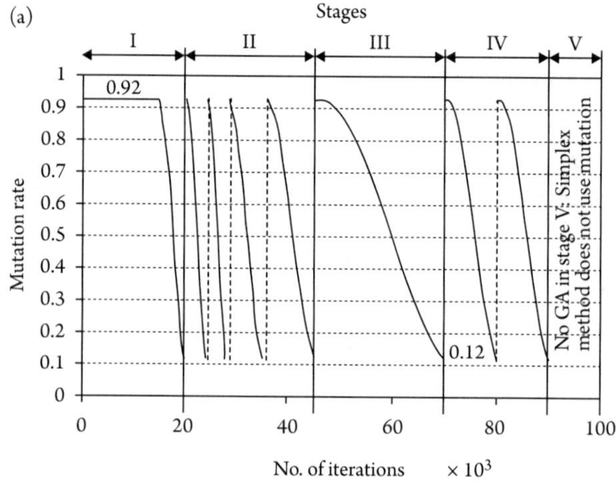

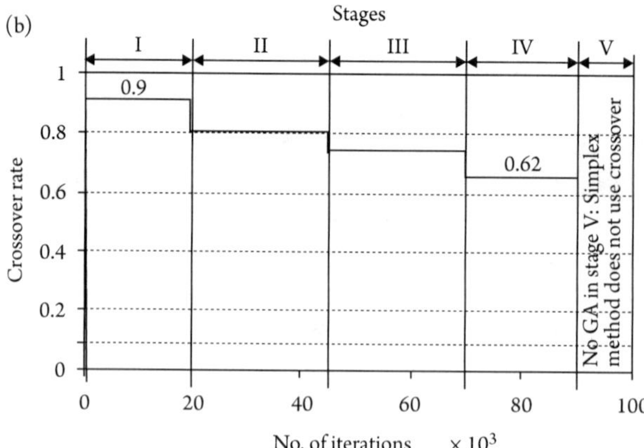

Figure 7-4 Variation of (a) mutation rate and (b) cross over rate during each stage with number of iterations[72] (reprinted with permission from Physical Review of the American Physical Society).

The convergence pattern of MSE between the GA parameterized Tersoff potential value and actual Tersoff potential value taken over all the configurations is summarized in Figure 7-5 and Table 7-2. Evidently, during the initial stages of the GA, the use of multipoint crossover provides a faster convergence as opposed to single-point crossover.[242] This could be because multipoint crossover tends to produce more variation among individuals during successive iterations, which helps in narrowing down the range of parameters during initial stages. On the other hand, single-point crossover along with high resolution (number of bits used in binary representation) helps to improve the accuracy.

It may be noted that it takes five stages to make the GA optimized Tersoff parameters values match the actual values of the Tersoff potential given in Table 7-1.[13] At this point the MSE drops below $10^{-5}\,\mathrm{eV}^2$. Among all the Tersoff parameters, the potential appears to be most sensitive to variations in A and B. Consequently, the GA is able to determine the near-optimal settings of these values during the early stages itself. Significant training efforts spanning multiple stages of GA are needed to optimize the other, less sensitive parameters (e.g., λ and h). The advantage of using an NN trained for MSE is that it obviates the need to calculate Tersoff potential energy for each set of parameters and for each configuration. The use of an NN has reduced the computational time from ~1 week without the NN to ~8 hrs with the NN.

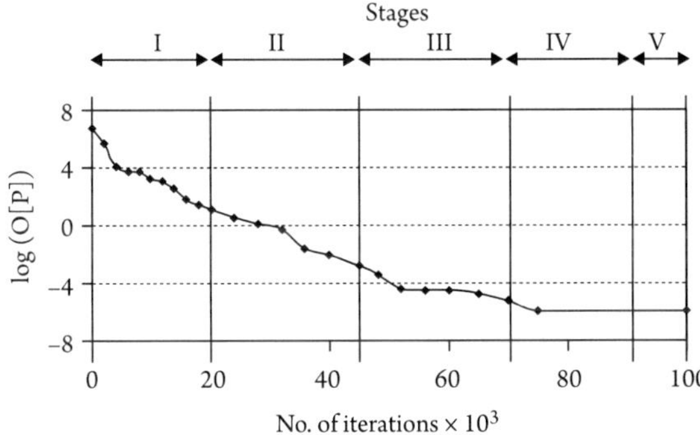

Figure 7-5 Variation of MSE with the number of iterations at various stages for parameterization of Tersoff potential[73] (reprinted with permission from Physical Review of the American Physical Society).

Table 7-2 Variation of MSE at different stages of GA fitting of Tersoff potential.[72] (Reprinted with permission from Physical Review of the American Physical Society.)

Stage No.	Stage I		Stage II		Stage III			Stage IV		Stage V
MSE (eV²)	0.01	0.0019	4.72×10^{-4}	5.45×10^{-5}	4.32×10^{-5}	3.57×10^{-5}	2.53×10^{-5}	7.65×10^{-6}	1.43×10^{-6}	1.32×10^{-6}

The next application to demonstrate the GA-NN approach involved the fitting of *ab initio* potentials of 5 atom Si-clusters using a Tersoff functional form. Bukkapatnam et al.[72] used 10,000 configurations of Si_5 clusters to compute the target objective function for training and testing. The configurations and their potentials were obtained by performing an MD simulation of orthogonal machining of a silicon workpiece using a single point cutting tool at a +15° rake angle and 10° clearance angle, 1 Å cutting depth, 54.3 Å cutting width, at a cutting speed of 491.2 m s^{-1} (see Figure 7-6).[140] A Tersoff potential with parameters corresponding to the bulk Silicon (Table 7-1) was employed in the initial step of the iterative MD simulation process. At every time step of the MD simulations, the configurations of silicon clusters that are present in front of the tool, in the chip, and within a few unit cell distances in the workpiece beneath the tool are stored. It was observed from an examination of a histogram of atomic configurations of Si in the primary and in the secondary deformation zones that out of about 39,000 configurations identified, over 60% of the Si atoms evolve into clusters of five atoms. The histogram also shows that Si_7 (20%) followed by Si_6 and Si_4 clusters (<10%) occur a significant number of times in the deformation zones during the machining of an Si workpiece.

The database obtained from the cutting simulations was augmented with 10,000 additional Si_5 configurations near equilibrium. These configurations were obtained by placing thermal energy corresponding to a temperature of 300 K in the silicon workpiece and then following the vibrational motions of the lattice using MD with the Tersoff empirical potential for bulk silicon. Subsequent to the MD calculations, the energies and force fields for each of the 28,000 stored configurations were computed using density functional theory (DFT) with a 6-31G** basis set and the B3LYP procedure for incorporating correlation energy. Next, the 10,000 configurations needed for GA training and testing were randomly selected from among the available configurations in the database.

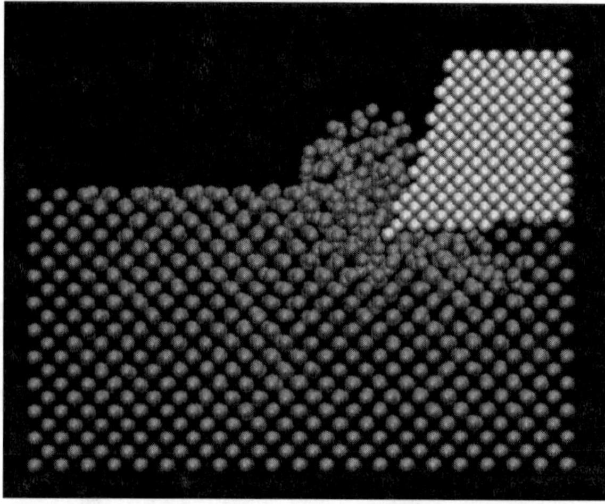

Figure 7-6 Atomic configuration of Si workpiece and a diamond tool used in MD simulation of Si machining[72] (reprinted with permission from Physical Review of the American Physical Society).

The objective function was computed for every one of the randomly generated 9,000 parameter vector realizations. These results were used to train a two-layer NN to approximate the objective function $[O[P]]$ as described in Section 7.1.3. Figure 7-7 (a)–(i) shows the histogram plots of the number of configurations as a function of nine internal coordinates used in the calculation of the *ab initio* energies, using GAUSSIAN 03 software (Frisch et al.[141b]). The standard deviation of the errors between the objective functions was 0.00847 eV^2 during Stage I and 0.00358 eV^2 in the final stage (Stage IV). The low values of the standard deviation indicate the closeness of the NN-estimate of $O[P]$ to the actual $O[P]$, whose computation is extremely cumbersome. Next, the GA was trained for five stages, and the NN was re-trained at the end of Stages I through III. In the final step, a Nelder-Mead simplex algorithm[127,243] was used to arrive at the optimal set of parameters. The specific GA parameters are used as follows:

Stage I: 100 individuals, 15 bits to represent each variable, 4 point crossover
Stage II: 50 individuals, 20 bits to represent each variable, 2 point crossover

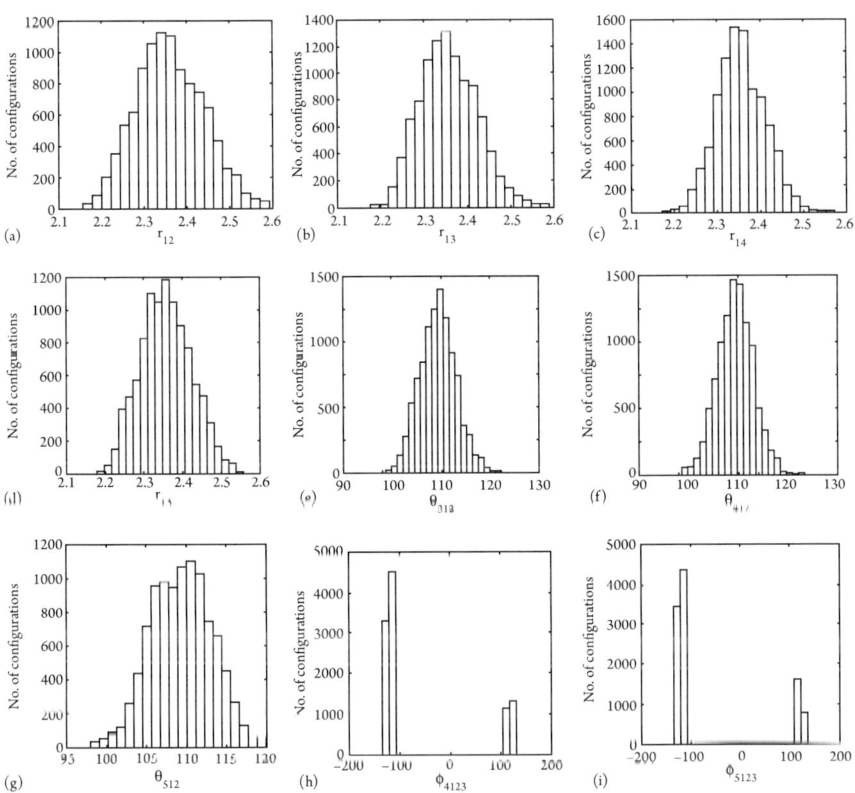

Figure 7-7 (a)-(i). Histograms of the number of configurations as a function of internal coordinates, (a)-(d) bond distances (r), (e)-(g) bond angles (θ) and (h)-(i) dihedral angle (ϕ)[72] (reprinted with permission from Physical Review of the American Physical Society).

Stage III: 50 individuals, 50 bits to represent each variable, 2 point crossover
Stage IV: 25 individuals, 75 bits to represent each variable, single point crossover
Stage V: Nelder-Mead simplex method (does not use genetic operators)

The variation of the best fitness function ($O[P]$) values at the end of each stage is summarized in Figure 7-8. Table 7-3 summarizes the parameter values for Tersoff functional form that seem to adequately fit the *ab initio* potentials of 5-atom Si clusters. The MSE of the GA-fit was reduced from 0.74 eV^2 to about 0.28 eV^2 during four stages of GA execution. The final stage of Nelder-Mead simplex reduced the errors by two orders of magnitude to 0.0026 eV^2. In contrast, the MSE resulting from the use of a Nelder-Mead simplex method,[243] instead of a GA, for fitting has yielded an MSE of 0.28 eV^2 and those from the application of a simplex method (with Stage I GA results as starting points) has yielded an average MSE of about 0.12 eV^2. These results suggest that it would be necessary to have the GA converge to the best possible solution (leading to the identification of the correct valley containing the global minima) prior to the application of a Nelder-Mead simplex.[243] The GA-parameterized Tersoff functional fit to the potentials of 28,200 configurations was tested. These configurations included those not used in the NN for fitting the GA. The results show a small dispersion around the 45° line, corresponding to a standard deviation of 0.0507 eV. Most of the points lie along the 45° line indicating that the GA-parameterized model estimates the *ab initio* potentials accurately for a vast majority of the configurations.

Finally, the extrapolation capability of the GA-estimated potential function was evaluated. Figure 7-9 shows that both NN and GA optimized Tersoff potential fit *ab initio* energy very well within a range of data that was used for the fitting. The NN fit is accurate for bond distances up to 3 Å, whereas GA-Tersoff is accurate up to 2.7 Å. This can be attributed to the fact that for GA-Tersoff, one starts with an assumed functional form based on some physical intuition involved in the bonding

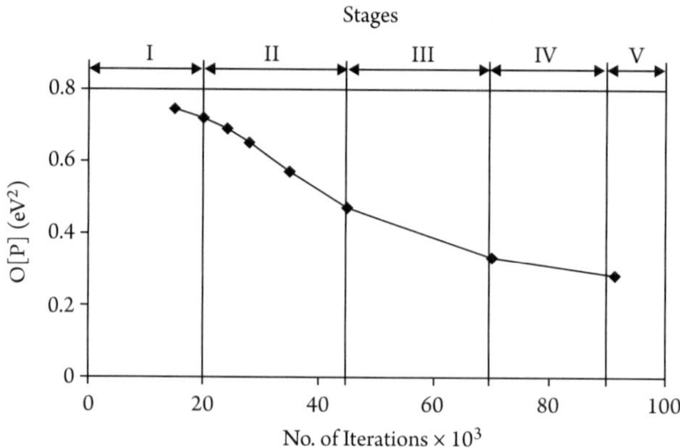

Figure 7-8 Variation of objective function $O[P]$ (here, MSE) at various stages of GA fitting of *ab initio* potentials to Tersoff functional form[72] (reprinted with permission from Physical Review of the American Physical Society).

Table 7-3 Variation of mean square error (MSE) and Tersoff potential function parameter estimates at different stages of GA fitting of *ab initio* energies.[72] (Reprinted with permission from Physical Review of the American Physical Society.)

	Stage I		Stage II			Stage III	Stage IV		Stage V
MSE (eV^2)	0.74552	0.71826	0.68913	0.6532	0.57253	0.4693	0.3331	0.28162	0.002567
$A \times 10^3$	1.8345	5.2328	9.7774	2.3203	2.3042	4.8184	5.2688	4.6293	3.5831
$B \times 10^2$	4.5224	5.8889	1.2736	5.8423	1.3211	4.8441	4.2669	3.8348	4.1021
λ	12.1868	17.5864	8.2376	10.3244	19.2320	3.2461	2.3431	2.2423	2.9402
μ	2.7139	5.5381	8.38749	12.2139	2.3149	2.3429	1.2320	2.0831	1.5392
β	0.5875	0.2436	0.1335	0.2201	0.0087	0.0018	0.0003	0.0006	9.38E-05
η	0.18947	0.01201	0.31941	0.2349	0.7923	0.5121	0.6903	0.77428	0.66311
$c \times 10^3$	3.2817	3.1344	2.7214	3.2899	1.1947	1.0265	1.03791	1.0335	1.0686
d	11.9426	31.6264	66.3297	54.2320	19.2322	14.3446	20.2361	22.1385	18.5568
h	0.1294	0.2957	0.21987	0.2342	−0.02923	−0.7038	−0.6613	−0.8343	−0.8205

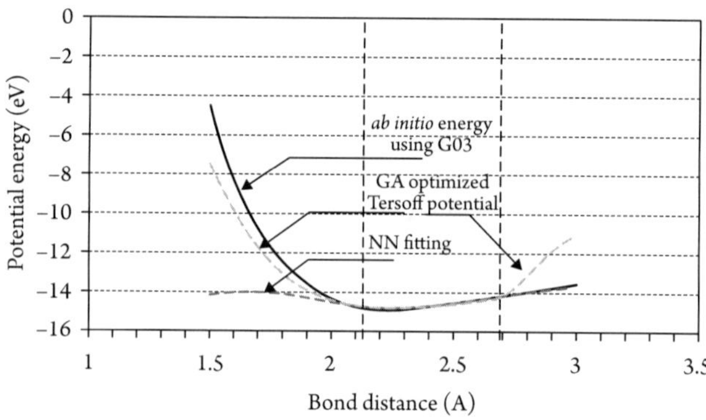

Figure 7-9 Variation of NN- and GA-Tersoff-estimates of *ab initio* potential with bond distances compared against the variation of actual *ab initio* potential[72] (reprinted with permission from Physical Review of the American Physical Society).

environment and then adjusts parameters to fit *ab initio* energy using the functional form. As a result, the accuracy of the fit critically depends on the underlying functional form used, whereas in the case of NN fitting, no functional form is assumed and the accuracy of the fit depends on the number of neurons used. It should be noted that NN fit is good only over a range of data that was used for fitting, that is, an NN fit is good only for interpolation and not for extrapolation. An additional advantage of a GA-Tersoff method is that it can be used for extrapolation. As seen in Figure 7-9, for bond distances < 2.1 Å, both GA-Tersoff and NN fit start deviating from *ab initio* energy. While the energy predicted by the NN fit remains almost constant, the energy predicted by GA-Tersoff correctly follows the trend of *ab initio* energy with the magnitudes deviating somewhat from the *ab initio* energy. This is important for MD/MC simulations involving non-equilibrium configurations, in which case it is important that the fit be reasonably accurate even for extrapolation. The loss of accuracy in GA-Tersoff extrapolations beyond 2.7 Å may be attributed to the fact that the Tersoff functional form in Equation (7-1) uses $R = 2.7$ Å as the inner cutoff on the bond distance, and it may be possible that the functional form used to attenuate the interaction effects beyond $r > R$ may not adequately capture the actual atomic interactions in 5-atom Si clusters.

Although the Tersoff functional form with the parameters given in Table 7-3 provides an excellent fit to the Si_5 *ab initio* energies obtained from density functional theory (DFT), we would not expect this potential to accurately represent the energies for smaller or larger silicon clusters. The nature of the bonding changes significantly when a silicon atom is removed from an Si_5 cluster to form Si_4. Therefore, we would expect a corresponding significant change in the potential parameters required to describe smaller clusters. The bonding changes are probably less significant as the size of the cluster is increased. Consequently, we would expect the parameters for the Si_5 system to more accurately describe Si_6 or Si_7 than Si_4.

Also, it may be possible to develop a robust potential for Si_n ($n > 3$) clusters by employing the present GA-NN method with *ab initio* DFT energies to obtain the

Tersoff parameters for a series of clusters. Subsequently, these data could be used to train a neural network to permit interpolation of the parameter values for any given cluster size. As n increases, we would expect the interpolated values to approach those for bulk silicon. We are currently investigating this possibility. Also as stated in the foregoing, the present study focuses on five-atom clusters because the majority of Si-clusters ahead of the tool in a machining simulation experiment consisted of five atoms. Future work will investigate the parameterization of potential functions for other clusters.

7.1.7. Summary

A genetic algorithm (GA) approach is presented to fit atomic potentials to a physically meaningful form. A GA-based parameterization process requires a repetitive computation of the objective function, which in this case is expressed as the mean squared error (MSE) taken over all configurations of interest (here, about 28,000 configurations of five-atom Si clusters were used) between the potentials determined from the *ab initio* calculations and those estimated using the GA-selected parameter vector values. An NN is used to learn the relationship connecting the parameter vector and the objective function. The NN thus trained provides surrogate estimates of the objective function, thereby obviating the need to compute the objective function for every parameter value selected during the GA fitting process. In fact, the use of an NN is the key to reducing the overall computational time for parameterization from a few weeks to a few hours. The GA has been found to quickly establish the correct "range" of the parameter values to optimize a nonlinear and perhaps combinatorial objective function. Since the objective function has significant slope changes and long plateaus, a simplex (Nelder-Mead) search[243] is used during the final stage of the parameterization process to ensure faster convergence to the optimal $[P]$ values. It has been shown that the approach can correctly identify the parameters used to fit Tersoff potential to within +/−0.0023 eV (corresponding to a ~95% confidence limit) accuracy. The results also show that it is possible to fit *ab initio* potentials of five-atom Si clusters to a Tersoff functional form within +/−0.0507 eV accuracy. More important, it appears that the GA-parameterized fit to Tersoff functional form seems to have good extrapolation capabilities compared to the use of NNs.

7.2. INTERNAL ENERGY TRANSFER CALCULATIONS USING NN METHODS

7.2.1. Introduction to Internal Energy Transfer Calculations

The study in the dynamics and rates of intramolecular vibrational energy transfer (IVR) is fundamental to the development of the understanding of unimolecular reaction rate processes. Because of the fundamental importance of this process, there have been hundreds of experimental and theoretical investigations of IVR.

Questions of central importance related to IVR are the pathways of intramolecular energy flow and whether these pathways are dependent upon the prepared initial state of the molecule. The rates at which such processes occur are of central importance in determining whether it is possible to produce mode-specific rate enhancement of chemical reactions and in assessing the accuracy of theories that assume that a rapid internal distribution of energy will be present.

Theoretical MD studies of IVR are impeded by a number of technical problems. One of these is the development of a sufficiently accurate PES to ensure that the mode couplings are properly represented. Another is the computational time required to execute a sufficient number of trajectory calculations to obtain meaningful statistical averages over all the internal variables present. A third problem involves the development of a reasonable means of monitoring the energy flow from one mode to another. This is complicated by the fact that the vibrational modes are coupled so that there is no exact method available by which a unique "mode energy" can be computed.

In most computations, the vibrational motions of the molecule are described in terms of normal coordinates.[244,148] Internal, bond, and angle coordinates are also often useful. Both of these coordinate systems have frequently been employed in MD studies of IVR.

Since trajectories are most conveniently computed using Cartesian coordinates and momenta, to monitor IVR dynamics using either normal mode or internal bond-angle coordinates, it is necessary to transform the Cartesian velocities to the corresponding quantities in terms of bonds, angular bending, and dihedral velocities. These are generally denoted as curvilinear coordinates and velocities.

At any point during a trajectory, the Cartesian velocities can be transformed to curvilinear velocities, $\mathbf{v}_r$, based on a Wilson et al.'s[244] treatment of molecular vibrations. In this method, the kinetic mode energies are expressed in the form

$$2T = \mathbf{v}_r \, \mathbf{K} \mathbf{v}_r, \tag{7-6}$$

where $\mathbf{K}$ is the kinetic-energy matrix. $\mathbf{v}_r$ can be computed by transforming the Cartesian velocities, $\mathbf{v}_x$, using

$$\mathbf{v}_r = \mathbf{B} \mathbf{v}_x, \tag{7-7}$$

where $\mathbf{B}$ is the transformation matrix from Cartesian to curvilinear coordinates. McIntosh and Michaellion[245] have shown that $\mathbf{K}$ is the inverse of Wilson's[244] $\mathbf{G}$ matrix,

$$\mathbf{G} = \mathbf{B} \mathbf{M}^{-1} \mathbf{B}^{T}, \tag{7-8}$$

where $\mathbf{M}^{-1}$ is a diagonal matrix of the reciprocal atomic masses and $\mathbf{B}^{T}$ is the transposed $\mathbf{B}$ matrix. This permits the kinetic energy to be written in the form

$$2T = \mathbf{v}_r \, \mathbf{G}^{-1} \, \mathbf{v}_r = \Sigma \, (v_r)_i \, G_{ij}^{-1} \, (v_r)_i. \tag{7-9}$$

It is common practice to approximate the mode kinetic energies as the diagonal terms of the above expression. That is,

$$T_i = 0.5\, (v_r)_i\, G_{ij}^{-1}\, (v_r)_i. \tag{7-10}$$

The off-diagonal terms, which include the mode couplings between the curvilinear velocities, are usually neglected. The error thus introduced into the calculations can be reduced by scaling the sum of the computed kinetic mode energies to the total Cartesian kinetic energy, which is exact. The size of the scaling factor provides an estimate of the error.

While these computations are straightforward, they can be computationally time-consuming. For example, to obtain the temporal dependence of the mode kinetic energies in a single trajectory lasting a few picoseconds, the $\mathbf{B}$, $\mathbf{G}$, and $\mathbf{G}^{-1}$ matrices must be computed several thousand times for each trajectory. As the number of atoms present in the molecular system increases, this can become a computational bottleneck in IVR studies.

Sumpter et al.[246] have shown that this bottleneck can be essentially removed by employing an NN to execute the transformation from Cartesian velocities to velocities and mode kinetic energies in the curvilinear system. The spirit of their approach is the same as that described in Section 7.1 of this chapter for GA calculations. In GA calculations, the repeated computation of the objective function produces a computational bottleneck. By training an NN to perform this function for any set of empirical PES parameter values, the bottleneck is effectively removed. In the present case, Sumpter et al.[246] employ an NN to convert the Cartesian coordinates and momenta to curvilinear velocities and mode kinetic energies, thereby removing the computational bottleneck of having to execute the repeated calculations of the $\mathbf{B}$, $\mathbf{G}$, and $\mathbf{G}^{-1}$ matrices.

7.2.2. Application of NNs for Acceleration of IVR Calculations: H_2O_2 System

7.2.2.1. NN for Conversion of Cartesian Velocities and Kinetic Mode Energies

Sumpter et al.[246] have trained a three-layer NN to convert Cartesian coordinates and momenta to the corresponding kinetic mode energies in terms of curvilinear coordinates thereby avoiding the repeated execution of the computations represented by Equations (7-6) through (7-10). As a set of first examples, they chose the four-atom H_2X_2 systems, where, in principle, X is any atom that forms the molecule H-X-X-H.

In terms of local, curvilinear coordinates, the H_2X_2 molecules exhibit six internal vibrational modes that comprise two H-X stretching modes, one X-X stretch, two H-X-X bending modes, and one four-body torsional mode that alters the dihedral angle. The objective is, therefore, to devise and train an NN that accepts the

12 Cartesian coordinates, the corresponding 12 conjugate Cartesian momenta, and the four masses as input variables and produces the corresponding kinetic energies of the six vibrational modes without actually solving Equations (7-6) through (7-10).

Based on empirical testing, Sumpter et al.[246] found that the optimum architecture for such an NN is a three-layer structure with 28 elements in the input vector, 38 and 12 neurons in the two hidden layers, and six neurons in the output layer. This (28-38-12-6) NN contains a total of 1,648 weight and bias parameters that are adjusted iteratively by backpropagation methods as described in Section 3.5.

Since the transformation from Cartesian coordinates and momenta to the kinetic mode energies for H_2X_2 systems depends only upon the values of the Cartesian coordinates and momenta along with the masses of the four atoms present, the NN, once trained with a sufficient range of masses, coordinates, and momenta will be able to execute the transformation for any tetra-atomic molecule with the general formula H_2X_2.

Sumpter et al.[246] trained the NN using data from trajectories computed for the H_2O_2 system on a PES obtained from *ab initio* data.[247,248] The sampling trajectory initially excited one of the OH stretching modes to the $v = 6$ vibrational state with all other modes being in their ground vibrational state with only zero-point energy present. The equations of motion were integrated for 3 ps during which the Cartesian coordinates and momenta at 2,000 integration points were employed along with Equations (7-6) through (7-10) to compute the kinetic mode energies.

These 2,000 sets of coordinates and momenta along with the H_2O_2 masses were used as the training input for the NN with the six mode kinetic energies, all scaled between 0.1 and 0.9, being the targets for the iterative training of the NN. In order to force the NN to learn the relationships between the input and target values rather than sense the correlations existing between adjacent points in the data set, the 2,000 examples in the database were randomized before being presented to the NN. The NN was trained for 9,400 epochs after which it had a maximum error relative to the training data of 0.07 and a root mean square error of 0.0075.

The ability of the NN to produce kinetic mode energies from Cartesian coordinates and momenta was examined by Sumpter et al.[246] in several ways. These included (a) choosing random points from the H_2O_2 trajectory used to train the network and comparing the NN predictions with results obtained from Equations (7-6) to (7-10), (b) integrating the training trajectory for times greater than 3 ps and comparing NN results with computed mode energies at randomly selected points, and (c) executing the same comparison using a reactive H_2O_2 trajectory in which hydrogen peroxide dissociated two OH radicals.

Comparison of NN predictions with computed mode kinetic energies using MD results indicated that the predictions of the NN closely resembled those obtained by algebraically transforming the MD results. The relative percentage error (RPE) of the NN predictions defined by Sumpter et al.[246] is

$$RPE = \frac{\left| E(t)_{MD} - E(t)_{NN} \right|}{E(t)_{MD}} \times 100, \tag{7-11}$$

where $E(t)_{MD}$ and $E(t)_{NN}$ are the kinetic mode energies at time t for a particular vibrational mode during a trajectory computed using algebraic transformation of the MD results and the NN predictions, respectively. The average values of the RPE for the six H_2O_2 modes were

$$<RPE(OH^*)> = 8.2\%, <RPE(OOH^*)> = 8.2\%, <RPE(OO)> = 2.5\%$$

$$<RPE(OH)> = 5.4\%, <RPE(OH)> = 12.0\%, \text{ and } <RPE(torsion)> = 5.6\%.$$

These average were computed for integration times between 3 and 10 ps, which were not used in the training of the NN.

Sumpter et al.[246] suggested that the average RPE is higher for the OOH bends and the OH stretches because of the rapid and effective energy exchange that occurs between an initially excited OH mode and the adjacent OOH bend which makes their mode kinetic energies behave like a highly fluctuating or chaotic function. Such chaotic motion is expected to present difficulties to an NN that relies on sensing the underlying correlations between the variables for accuracy.

Recent experience suggests that when fitting analytic functions such as those represented by Equations (7-6) to (7-10), NNs should exhibit high fitting accuracy provided the database is sufficiently large to ensure that the NN is not being required to extrapolate into regions where it has not been trained. The average RPEs for the prediction of the kinetic mode energies of H_2O_2 are generally below this expected accuracy. Consequently, it appears likely that the training to data points taken from the first 3 ps of a single H_2O_2 trajectory was probably not sufficient to allow the NN to generalize accurately to other unknown trajectories. If this is true, it is a problem that is easily corrected. Nevertheless, the results obtained by Sumpter et al.[246] clearly establish the ability of a trained NN to execute this type of analysis.

Figures 7-10 and 7-11 show a comparison of the scaled, algebraically computed, O-O stretching mode energies with those obtained from the NN for a set of 2,000 examples given in arbitrary order. Thus, the abscissa of the plots is just a numerical listing of the points from 1 to 2,000. The points in Figure 7-10 were taken from a nonreactive H_2O_2 trajectory whereas those shown in Figure 7-11 are from a reactive trajectory that dissociates to 2 OH radicals.

Sumpter et al.[246] demonstrated that the same NN that was trained using data taken from nonreactive trajectories can make similarly accurate predictions about mode kinetic energies in reactive trajectories. While it might seem that such predictions would require extrapolation by the NN, this is not the case. The equations connecting the mode kinetic energies and the Cartesian coordinates and momenta are the same whether the trajectory is reactive or nonreactive. Figure 7-11 shows a comparison of the kinetic mode energies in the O-O mode, which is very close to the reaction coordinate for the dissociation, computed algebraically and using the NN. The similarity between the two curves is obvious. The average RPE for all the vibrational modes in H_2O_2 was predicted to within 12% for the reactive trajectory.

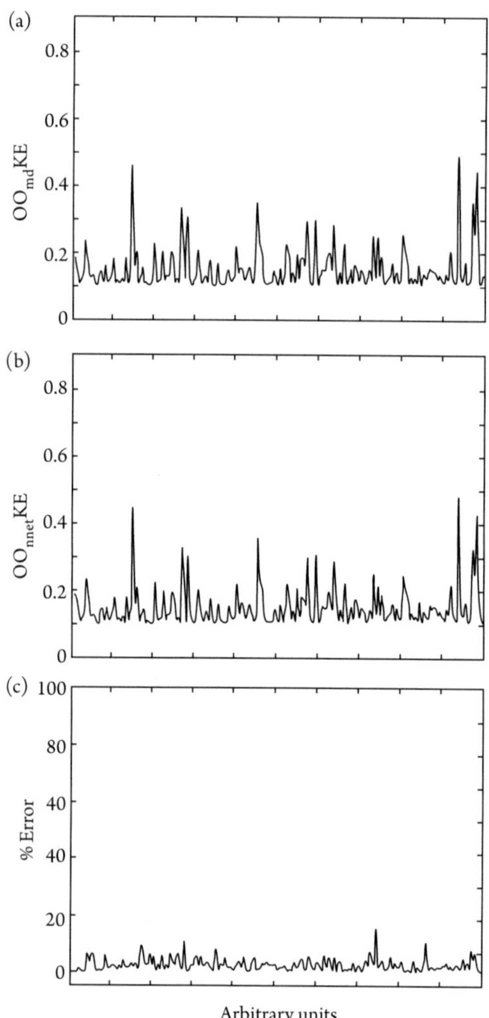

Figure 7-10 The mode kinetic energy for the O-O stretching mode of H_2O_2 at 2000 randomly selected points taken from a nonreactive H_2O_2 trajectory in which one of the OH stretching modes is excited to the $v = 6$ state with all other modes being in their ground vibrational states. (a) Algebraically computed using Eqs. (7-6) to (7-10). (b) Predicted by the NN. (c) Average RPE (see Eq. (7-11) in text)[246] (reprinted with permission from the American Institute of Physics and the authors).

As noted previously, the transformation equations between Cartesian coordinates and momenta and kinetic mode energies depend only upon the masses of the atoms and not upon the details of the PES. Therefore, if the NN is properly trained to learn the mass dependence, the same NN should be able to predict mode energies for a wide variety of tetra-atomic molecules. Sumpter et al.[196] examined this possibility by training the NN using example points from two trajectories, each integrated

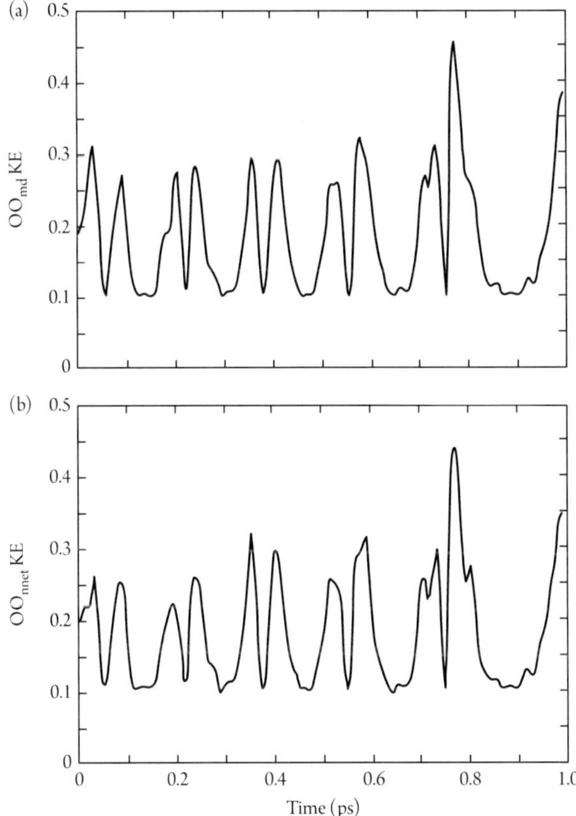

Figure 7-11 Same as that for Fig. 7-10 except the sampling points were taken randomly from a reactive H_2O_2 trajectory that dissociates to 2 OH radicals[246] (reprinted with permission from the American Institute of Physics and the authors).

for 3 ps for H_2O_2 and H_2Se_2. The training was continued until the NN could predict the energies of 4,000 examples from each system to a maximum network error of 0.1 in scaled units. Subsequently, the NN was employed to predict kinetic mode energies along unknown trajectories for various H_2X_2 tetra-atomic molecules. Table 7-4 shows typical results.

The results shown in Table 7-4 make it clear that NNs are able to obtain model kinetic energies for a variety of tetra-atomics with accuracy levels similar to those obtained for hydrogen peroxide even when only two masses were used in the NN training. More extensive training would undoubtedly improve the accuracy still further.

Finally, Sumpter et al.[246] demonstrated that NNs have the capability to predict the energy flow out of an excited CH mode in a polyethylene molecule modeled using a single chain of 300 atoms. The NN input comprised the level of CH excitation, temperature of the system, and the integration time and desired mode. The output was the energy of the desired mode. Figure 7-12 shows the resulting predictions

187

Table 7-4 Average relative percent errors between NN predictions and algebraic evaluation of mode kinetic energies from MD results for the various tetraatomic molecules studies.

| | % Error[a] | | | | | |
Molecule	HX*	HXX*	XX	HXX	HX	HXXH
H_2C_2	10	9	5	10	10	6
H_2N_2	6	7	14	12	13	5
H_2O_2[b]	4	4	6	7	10	2
H_2Si_2	3	12	5	10	15	3
H_2S_2	15	20	10	17	12	10
H_2Se_2[b]	10	12	9	10	10	5

(a) See Eq. (7-11). Averages were taken over 2,000 examples from trajectories not employed in the training of the NN.

(b) Molecules used to train the NN. The examples, however, were taken from trajectories not used in the training.

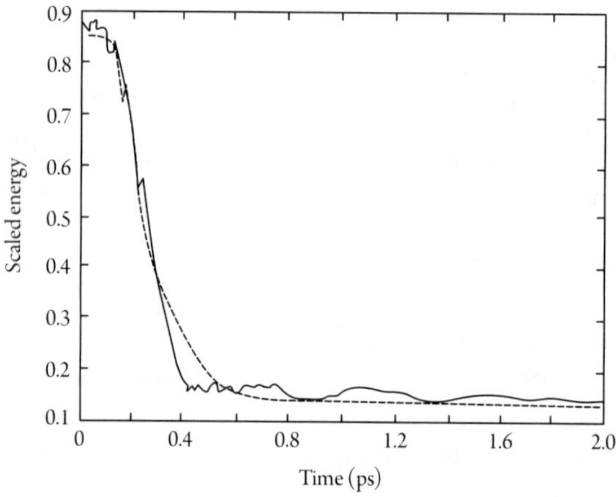

Figure 7-12 Energy in a CH stretching mode of a polyethlyene molecules as a function of time for an initial excitation of $v_{CH} = 7$ and a backbone temperature of 20 K. The solid curve is the ensemble average of five classical trajectories while the dashed curve is the NN prediction. The input to the NN was the time sequence (0 to 2 ps), the initial excitation level and the temperature. The NN output is the scaled energy[246] (reprinted with permission from the American Institute of Physics and the authors).

of the trained NN for the case in which the CH mode was initially excited to the $v = 7$ vibrational state and $T = 20$ K. The solid curve is the MD trajectory result while the dashed curve gives the predicted energy obtained from the NN. Clearly, the agreement is very good. Similar accuracy was obtained for CH excitations between 0 and 14 and system temperatures between 0 and 350 K.

The general conclusion is that NNs can be a powerful tool for calculating and monitoring energy flow in molecular systems.

8

EMPIRICAL POTENTIAL-ENERGY SURFACES FITTING USING FEEDFORWARD NEURAL NETWORKS

8.1. FITTING TO *AB INITIO* ELECTRONIC STRUCTURAL DATA

8.1.1. Introduction

When the system of interest becomes too complex to permit the use of *ab initio* methods to obtain the system potential-energy surfaces (PES), empirical potential surfaces are frequently employed to represent the force fields present in the system under investigation. In most cases, the functional forms present in these potentials are selected on the basis of chemical and physical intuitions. The parameters of the surface are frequently adjusted to fit a very small set of experimental data that comprise bond energies, equilibrium bond distances and angles, fundamental vibrational frequencies, and perhaps measured barrier heights to reactions of interest. Such potentials generally yield only qualitative or semiquantitative descriptions of the system dynamics.

Several research groups have significantly improved the accuracy of the values of the experimental properties computed using empirical potential surfaces by fitting the chosen functional form for the potential to the force fields obtained from trajectories using *ab initio* Car-Parrinello[17] molecular dynamics simulations. The fitting to the force fields is usually done using a least-squares fitting approach. This method has been employed by Izvekov et al.[18] to obtain effective non-polarizable three-site force fields for liquid water. Carré et al.[19] have employed such a procedure to obtain a new pair potential for silica. In their investigation, the vector of potential parameters was fitted using an iterative Levenberg-Marquardt[117,135,249] algorithm. Tangney and Scandolo[20] have also developed an interatomic force field for liquid SiO_2 in which the parameters were fitted to the forces, stresses, and energies obtained from *ab initio* calculations. Ercolessi and Adams[21] have used a quasi-Newtonian procedure to fit an empirical potential for aluminum to data obtained from first-principals computations.

Empirical potentials can be improved by making the parameters parameterized functions of the coordinates defining the instantaneous positions of the atoms of the system. This approach has been successfully employed by numerous investigators.[16,250] The difficulty with this procedure is that the number of parameters that must be adjusted increases rapidly. Appropriate fitting of these parameters requires a much more extensive database. Finally, the actual fitting process can often be tedious, difficult, and time-consuming.[16,250h]

Extremely large databases to which potentials may be fitted can be obtained using *ab initio* electronic structure calculations. When this approach is adopted, robust methods must be developed to ensure that the important regions of the multi-dimensional configuration space are adequately sampled in the *ab initio* calculations and efficient techniques must be developed to efficiently fit the parameters of the potential to the database. Several robust and powerful sampling methods have developed to achieve this objective. Many of these methods have been discussed and described in Chapter 4 and elsewhere in this monograph.

This section presents a novel method developed by Malshe et al.[22] that reduces or eliminates many of the problems associated with fitting empirical potentials, particularly when some of the parameters are functions of the system configuration. First, the method completely obviates the problem of selecting the form of the functional dependence of the parameters upon the system's coordinates. This form is, in effect, determined automatically by a neural network. Second, the entire fitting procedure is automated so that months of effort are replaced with a few days of computer computation. Third, the method provides a procedure to avoid local minima in the multi-dimensional parameter hyperspace.

The basic approach used to achieve these objectives is to generalize NN fitting procedures to fit not only the weight and bias matrices of the network but also the parameters of the empirical potential to a database of *ab initio* energies. It is shown that the potential parameters may easily be made functions of the atomic structure of the system with the parameters defining those functions being fitted by the NN so that the overall procedure is efficient and robust. As an example, the method is applied to the fitting of the DFT energies of Si_5 clusters using a Tersoff[13] functional form. The Si_5 database is the same as that described in Chapter 6, Section 6.3, and Chapter 7.

8.1.2. General Method

Let us assume that there exists an appropriate parameterized empirical potential function $V(q_i; A, B, C, \ldots)$, where the q_i are the coordinates specifying the configuration of the system, and $A, B, C, \ldots$ are the adjustable parameters contained within the empirical function. The problem is to obtain the best fit possible to the database by adjustment of the parameters, where some of the parameters may be functions of the q_i.

The first step is to partition the parameter set into two parts, those that are to be regarded as functions of the q_i and those whose values are independent of the q_i. Let $\mathbf{A}(q_i)$ be a column vector containing the M_1 parameters functionally dependent upon the q_i, where $A_n(q_i)$, $(n = 1, 2, \ldots, M_1)$, are the elements of $\mathbf{A}(q_i)$. The remaining M_2 parameters, B_j $(j = 1, 2, \ldots, M_2)$ are independent of the q_i. They form a column vector $\mathbf{B}$. The objective is to adjust the functional dependence of the parameters in $\mathbf{A}$ upon the q_i and the parameters in $\mathbf{B}$ so as to minimize the difference between the potential given by $V(q_i; \mathbf{A}, \mathbf{B})$ and the target values for the potential.

Figure 8-1 illustrates the overall NN method for the adjustment of the parameters of $V(q_i; \mathbf{A}, \mathbf{B})$. The input vector, $\mathbf{P}_k$, of the NN comprises the set of q_i that specifies the kth configuration of the system. Typically, the elements of $\mathbf{P}_k$ are interatomic distances, bond angles, and dihedral angles. The elements of $\mathbf{P}_k$ are combined with the weight and bias matrices of the network, $\mathbf{W}$ and $\mathbf{b}$, respectively, to produce the corresponding elements of $\mathbf{A}_k$. These results are then used along with the present values of elements of $\mathbf{B}$ to compute $V(\mathbf{P}_k; \mathbf{A}_k, \mathbf{B})$. This potential is combined with the target potential, $V_{T,k}$, for the kth configuration, which is generally obtained from *ab initio* calculations at some level of accuracy, to compute the interpolation error e_k for the kth input vector $\mathbf{P}_k$:

$$e_k = V_{T,k} - V(\mathbf{P}_k, \mathbf{A}_k, \mathbf{B}). \tag{8-1}$$

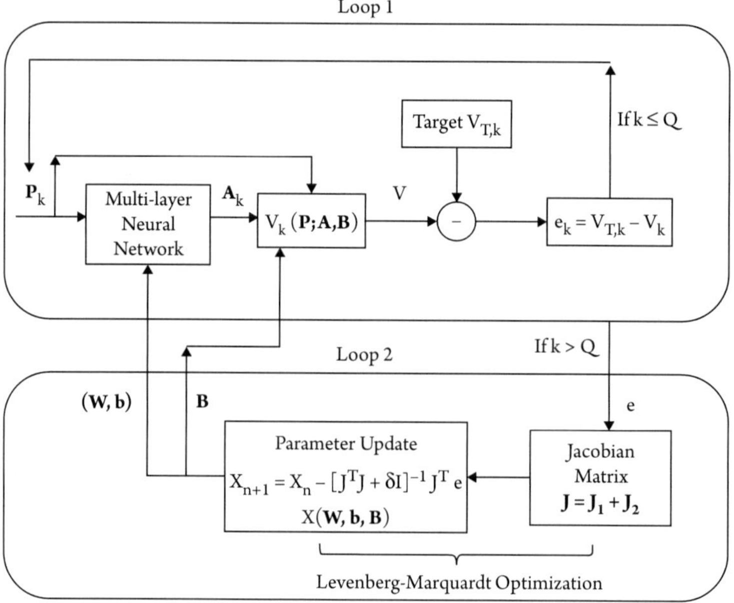

Figure 8-1 Flow diagram for the operation of the NN procedure for empirical parameter adjustment to a database. Q is the number of points in the database. See the text for definitions of the remaining variables[22] (reprinted with permission from American Institute of Physics).

As indicated in Loop 1 of Figure 8-1, this procedure is repeated for each of the Q configurations present in the database.

In the general case, the NN for a system containing n atoms will usually be a two-layer network comprising one hidden layer and an output layer along with the input vector. The input vector will have at least $(3N-6)$ elements to specify the system configuration. However, in some cases it may be convenient to over-specify the configuration. For example, a common choice is the ensemble of interatomic distances or inverse distances. If this choice is made, the input vector will contain $n(n-1)/2$ elements. The number of neurons in the hidden layer, K, is arbitrary. The optimum value is generally determined by empirical investigation to find the minimum value of K required to achieve sufficient interpolation accuracy. The output layer contains M_1 neurons that produce the M_1 required values of the elements in $\mathbf{A}_k$.

In the next step, the Levenberg-Marquardt algorithm[117,135,249] is employed to update the elements of B and the weight and bias matrices of the NN so as to minimize the error. This procedure requires the computation of the Jacobian matrix which consists of partial derivatives of the error ($\mathbf{e}$) with respect to the parameters of the potential. Since the parameters in $\mathbf{A}_k$ are treated differently from the parameters in $\mathbf{B}$, the computation of the Jacobian matrix is split into two parts. First, the Jacobian $\mathbf{J}_1$ is computed. $\mathbf{J}_1$ comprises the partial derivatives of the e_k $(k = 1, 2, 3, \ldots, Q)$ with respect to the weights and biases of the NN that determine the elements of $\mathbf{A}_k$. In the second step, the derivatives of the e_k $(k = 1, 2, 3, \ldots, Q)$ with respect to the elements of $\mathbf{B}$ are computed to obtain $\mathbf{J}_2$. The total Jacobian, $\mathbf{J}$, is a $Q \times (N+M_2)$ matrix, where N is the total number of elements in $\mathbf{W}$ and $\mathbf{b}$. $\mathbf{J}$ is obtained from the concatenation of $\mathbf{J}_1$ and $\mathbf{J}_2$. This overall procedure is illustrated diagramatically in Loop 2 of Figure 8-1.

After the execution of one epoch of optimization using the Levenberg-Marquardt procedure, the updated weights and biases along with the updated $\mathbf{B}$ matrix are returned to Loop 1 for execution of the next cycle of optimization. The iterative procedure is continued until convergence is attained using either an early-stopping procedure with a validation set[105,118] or some other suitable method.

In order to avoid being trapped in local minima of the multi-dimensional parameter hyperspace and to average out statistical errors present in the fitting, the NN procedure is repeated several times using different initial guesses for $\mathbf{W}$, $\mathbf{b}$, and $\mathbf{B}$ and a different partitioning of the database into training, testing, and validation sets to generate a committee of networks. The final predicted value for the potential can be taken as the average of the committee of networks, or the result from the single NN producing the best fit can be used.

8.1.3. Application to Fitting an *Ab Initio* Database for Si$_5$ Clusters

In this illustrative example, Malshe et al.[22] fit a Tersoff empirical potential[13] to a database comprising the DFT energies of 10,202 Si$_5$ structures observed in MD simulations of the nanometric cutting of a silicon workpiece with a single-point cutting tool having a +15° rake angle, a 10° clearance angle, a 1 Å cutting depth, a 54.3 Å cutting width, and a 491.2 m s⁻¹ cutting speed.[140]

Electronic structure calculations[141b] using DFT methods with a 6-31G** basis set were executed for each of the 10,202 Si_5 configurations present in the database. For convenience of interpretation, the resulting energies were converted to eV relative to the energy of five separated, ground-state silicon atoms.

To apply the general NN fitting method to Si_5 with a Tersoff potential, we take the NN input vector $\mathbf{P}_k$ to be a (10×1) column vector whose elements are the 10 Si_5 interparticle distances.

For an arbitrary system, the Tersoff function,[13] E, is given by

$$E = \sum_i E_i = \sum_{i \neq j} V_{ij} \tag{8-2}$$

where

$$V_{ij} = f_c(r_{ij}) \left[F_R(r_{ij}) + b_{ij} F_A(r_{ij}) \right], \tag{8-3}$$

$$F_R(r_{ij}) = C_{ij} \, exp(-\lambda_{ij} \, r_{ij}), \tag{8-4}$$

$$F_A(r_{ij}) = - D_{ij} \, exp(-\mu_{ij} \, r_{ij}), \tag{8-5}$$

The cutoff function, $f_c(r_{ij})$, is given by Equation (8-6). Its use produces a potential E that is continuous with continuous first derivatives. However, the second derivatives are discontinuous at the cutoff points, R_{ij} and S_{ij}.

$$f_C(r_{ij}) = \begin{cases} 1, & \text{if } r_{ij} < R_{ij} \\[2mm] 0.5 + 0.5\cos\left[\dfrac{\pi\left(r_{ij} - R_{ij}\right)}{\left(S_{ij} - R_{ij}\right)}\right] & \text{if } R_{ij} < r_{ij} \leq S_{ij} \\[2mm] 0, & \text{if } r_{ij} > S_{ij} \end{cases} \tag{8-6}$$

$$b_{ij} = \left(1 + \beta_i^{n_i} \, \zeta_{ij}^{n_i}\right)^{-1/2n_i}, \tag{8-7}$$

$$\zeta_{ij} = \sum_{k \neq i,j} fc\left(\left(r_{ij}\right)\omega_{ik}\, g\left(\theta_{ijk}\right)\right), \tag{8-8}$$

and

$$g\left(\theta_{ijk}\right) = 1 + c_i^2 d_i^2 - c_i^2 \Big/ \left[d_i^2 + \left(h_i - \cos\left(\theta_{ijk}\right)\right)^2\right]. \tag{8-9}$$

For a general system, the Tersoff parameters differ for each unique bonding pair. In the present case, all pairs correspond to Si-Si bonds. Consequently, the i-j subscripts on the potential parameters can be dropped. This simplification leaves a total

194

of 12 parameters, $C, D, \lambda, \mu, R, S, \beta, n, \omega, c, d$, and h. In this illustration, the value of ω is set to unity and those for R and S to 2.850 Å and 3.000 Å, respectively. This decision leaves 9 parameters to be adjusted to the *ab initio* Si_5 database.

Physical considerations and preliminary calculations suggest that the C and D parameters in the two-body terms should be strongly dependent upon the Si_5 configuration. These multiplicative parameters play a key role in determining the Si-Si bond strength, which is very different in Si_3 than is the case for Si_2. Therefore, C and D are placed in matrix $\mathbf{A}_k$ that contains the parameters whose values are to be functions of the configuration of the system. The $\mathbf{B}$ matrix comprises the remaining seven parameters. With these assignments, $\mathbf{A}_k$ becomes a (2×1) column vector while $\mathbf{B}$ is a (7×1) vector.

For illustrative purposes, the simple (1-2) NN shown in Figure 8-2 is used to compute C_{ij} and D_{ij}. When the $\mathbf{P}_k$ are presented to the network in Loop 1 of Figure 8-2, each of the three elements of $\mathbf{P}_k$ are input into the NN one at a time to produce the corresponding values of C and D for that particular element. This simple choice for the NN makes the values of the two-body C and D parameters linear functions of the corresponding interparticle distance:

$$C_{ij} - W_1 \, r_{ij} + b_1 \text{ and} \tag{8-10}$$

$$D_{ij} = W_2 \, r_{ij} + b_2 , \tag{8-11}$$

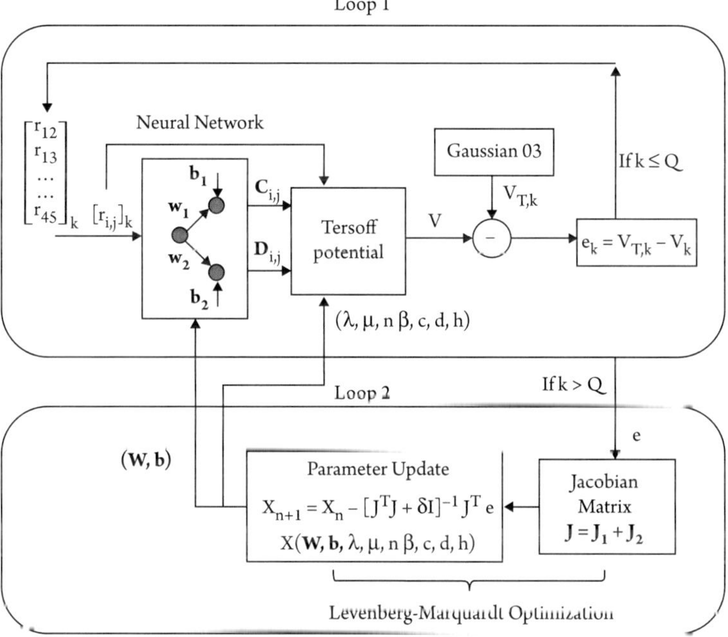

Figure 8-2 Flow diagram for the operation of the NN procedure for empirical parameter adjustment to a database for a Tersoff potential with C and D treated as functions of the two-body interparticle distances. Q is the number of points in the database, which in this case is 10,202. See the text for definitions of the remaining variables[22] (reprinted with permission from American Institute of Physics).

where C_{ij} and D_{ij} are the C and D parameters that are associated with the two-body, Tersoff term whose input element is r_{ij}.

After computation of the error vector $\mathbf{e}$ whose elements are the errors for all Q configurations in the database, control is transferred to Loop 2 in Figure 8-2 for the execution of one training epoch using the Levenberg-Marquardt[117,135] algorithm. The required $\mathbf{J}_1$ matrix has dimensions $Q \times (N+M_2)$, where N is the number of weights and biases in the NN. For the present illustrative example, $N = 4$ and $M_2 = 7$. The elements of $\mathbf{J}_1$ are the derivatives of the errors with respect to the weights and biases of the NN. Therefore, the $\mathbf{J}_1$ matrix is,

$$\mathbf{J}_1 = \begin{bmatrix} \dfrac{\partial e_1}{\partial W_1} & \dfrac{\partial e_1}{\partial W_2} & \dfrac{\partial e_1}{\partial b_1} & \dfrac{\partial e_1}{\partial b_2} & 0 & 0 & 0 & \cdots & 0 \\[2.5ex] \dfrac{\partial e_2}{\partial W_1} & \dfrac{\partial e_2}{\partial W_2} & \dfrac{\partial e_2}{\partial b_1} & \dfrac{\partial e_2}{\partial b_2} & 0 & 0 & 0 & \cdots & 0 \\[2.5ex] . & . & . & . & 0 & 0 & 0 & \cdots & 0 \\[1ex] . & . & . & . & 0 & 0 & 0 & \cdots & 0 \\[2.5ex] \dfrac{\partial e_Q}{\partial W_1} & \dfrac{\partial e_Q}{\partial W_2} & \dfrac{\partial e_Q}{\partial b_1} & \dfrac{\partial e_Q}{\partial b_2} & 0 & 0 & 0 & \cdots & 0(10202 \times 11) \end{bmatrix} \tag{8-12}$$

The elements of $\mathbf{J}_1$ are most conveniently computed using a standard chain rule. That is,

$$\frac{\partial e_J}{\partial W_1} = \frac{\partial(V_{T,J} - V_J)}{\partial W_1} = -\left[\frac{\partial V_J}{\partial C_{1\,2}} \frac{\partial C_{1\,2}}{\partial W_1} + \frac{\partial V_J}{\partial C_{1\,3}} \frac{\partial C_{1\,3}}{\partial W_1} + \frac{\partial V_J}{\partial C_{2\,3}} \frac{\partial C_{2\,3}}{\partial W_1} \right] \tag{8-13}$$

$$\frac{\partial e_J}{\partial W_2} = \frac{\partial(V_{T,J} - V_J)}{\partial W_2} = -\left[\frac{\partial V_J}{\partial D_{1\,2}} \frac{\partial D_{1\,2}}{\partial W_2} + \frac{\partial V_J}{\partial D_{1\,3}} \frac{\partial D_{1\,3}}{\partial W_2} + \frac{\partial V_J}{\partial D_{2\,3}} \frac{\partial D_{2\,3}}{\partial W_2} \right] \tag{8-14}$$

with analogous expressions for the derivatives of e_J with respect to b_1 and b_2.

The $\mathbf{J}_2$ Jacobian is the matrix of derivatives of the errors with respect to the potential parameters contained in vector $\mathbf{B}$. $\mathbf{J}_2$ is given by Equation (8-15) below:

$$\mathbf{J}_2 = \begin{bmatrix} 0 & 0 & 0 & 0 & \dfrac{\partial e_1}{\partial \lambda} & \dfrac{\partial e_1}{\partial \mu} & \dfrac{\partial e_1}{\partial \beta} & \dfrac{\partial e_1}{\partial n} & \dfrac{\partial e_1}{\partial c} & \dfrac{\partial e_1}{\partial d} & \dfrac{\partial e_1}{\partial h} \\[2.5ex] 0 & 0 & 0 & 0 & \dfrac{\partial e_2}{\partial \lambda} & \dfrac{\partial e_2}{\partial \mu} & \dfrac{\partial e_2}{\partial \beta} & \dfrac{\partial e_2}{\partial n} & \dfrac{\partial e_2}{\partial c} & \dfrac{\partial e_2}{\partial d} & \dfrac{\partial e_2}{\partial h} \\[2.5ex] . & . & . & . & & & & & & & \\[1ex] . & . & . & . & & & & & & & \\[2.5ex] 0 & 0 & 0 & 0(10202 \times 11) & \dfrac{\partial e_Q}{\partial \lambda} & \dfrac{\partial e_Q}{\partial \mu} & \dfrac{\partial e_Q}{\partial \beta} & \dfrac{\partial e_Q}{\partial n} & \dfrac{\partial e_Q}{\partial c} & \dfrac{\partial e_Q}{\partial d} & \dfrac{\partial e_Q}{\partial h} \end{bmatrix} \tag{8-15}$$

The gradients of the Tersoff potential with respect to the parameters contained in the **B** vector are

$$\frac{\partial V}{\partial \lambda_{ij}} = -r_{ij} f_c\left(r_{ij}\right) f_R$$

$$\frac{\partial V}{\partial \mu_{ij}} = -r_{ij} b_{ij} f_c\left(r_{ij}\right) f_A$$

$$\frac{\partial V}{\partial \beta_i} = f_c\left(r_{ij}\right) f_A \left[-\frac{1}{2} \frac{b_{ij} \zeta_{ij}^{n_i} \beta_i^{n_i-1}}{\left(1+\zeta_{ij}^{n_i} \beta_i^{n_i}\right)} \right]$$

$$\frac{\partial V}{\partial n_i} = \frac{1}{2} f_c\left(r_{ij}\right) f_A b_{ij} \left[\frac{ln\left(1+\zeta_{ij}^{n_i} \beta_i^{n_i}\right)}{n_i^2} - \frac{\zeta_{ij}^{n_i}\left(ln\zeta_{ij} + ln\beta_i\right)}{n_i\left(1+\zeta_{ij}^{n_i} \beta_i^{n_i}\right)} \right]$$

$$\frac{\partial V}{\partial c_i} = -f_c\left(r_{ij}\right) f_A b_{ij} \frac{\zeta_{ij}^{n_i-1} \beta_i^{n_i}}{\left(1+\zeta_{ij}^{n_i-1} \beta_i^{n_i}\right)} \left[\sum_{k(\neq i,j)} f_c\left(r_{ik}\right)\left(\frac{c_i}{d_i^2} - \frac{c_i}{d_i^2 + \left(h_i - cos\theta_{ijk}\right)^2} \right) \right]$$

$$\frac{\partial V}{\partial d_i} = -f_c\left(r_{ij}\right) f_A b_{ij} \frac{\zeta_{ij}^{n_i-1} \beta_i^{n_i}}{\left(1+\zeta_{ij}^{n_i-1} \beta_i^{n_i}\right)} \left[\sum_{k(\neq i,j)} f_c\left(r_{ik}\right)\left(-\frac{c_i^2}{d_i^3} - \frac{c_i^2 d_i}{\left(d_i^2 + \left(h_i - cos\theta_{ijk}\right)^2\right)^2} \right) \right]$$

$$\frac{\partial V}{\partial h_i} = -f_c\left(r_{ij}\right) f_A b_{ij} \frac{\zeta_{ij}^{n_i-1} \beta_i^{n_i}}{\left(1+\zeta_{ij}^{n_i-1} \beta_i^{n_i}\right)} \left[\sum_{k(\neq i,j)} f_c\left(r_{ik}\right)\left(-\frac{c_i^2\left(h_i - cos\theta_{ijk}\right)}{\left(d_i^2 + \left(h_i - cos\theta_{ijk}\right)^2\right)^2} \right) \right]$$

A good quantitative measure of the fitting accuracy may be obtained from the computed RMS error at each of the 10,202 points in the database. The best result obtained by Malshe et al.[22] using three different initial estimates of the fitting parameters is an RMS error of 0.0148 eV (1.43 kJ mol^{-1}) for Solution-1 in Table 8-1. Before optimization using the flow diagram shown in Figure 8-2, the RMS error using the initial guess for Solution-1 was 14.083 eV. This large fitting error is reduced to by a factor of 951.6 in about 800 iterations (epochs) of the loop shown in Figure 8-2. These iterations required about 80 hours of CPU time on a single processor with a 2.6 GHz clock speed.

The distribution of errors at the 10,202 points in the database is shown as a histogram in Figure 8-3. As can be seen, the large majority of errors lie between −0.02 eV and +0.02 eV.

Table 8-1 Parameters for the modified Tersoff potential with parameters C and D treated as linear functions of the two-body interatomic distance.[22] (Reprinted with permission from American Institute of Physics.)

Parameter	Values		
	Solution-1	Solution-2	Solution-3
W_1 (eV Å^{-1})	0.0000	−243.721813	−243.77000676
W_2 (eV Å^{-1})	−0.0014	0.135643281	−0.472394167
b_1 (eV)	713.4128	398.4742375	398.5871931
b_2 (eV)	410.1866	−0.892099763	0.727862883
λ (Å^{-1})	2.0815	1.637737984	1.637775465
μ (Å^{-1})	1.3273	1.637755192	2.023039578
β	0.3077	0.001906998	0.001906998
n	0.0098	0.009909063	0.009909061
c	1068.6001	10686.00999	10686.00999
d	18.5451	18.553194	18.553194
h	−1.2943	0.294251	0.294251
rms error(eV)	0.0148	0.0153	0.0153

In all parameters sets, R = 2.8500 Å, S = 3.0000 Å, and ω = 1.000. In all cases, the database is the same 10,202 Si_5 DFT energies.

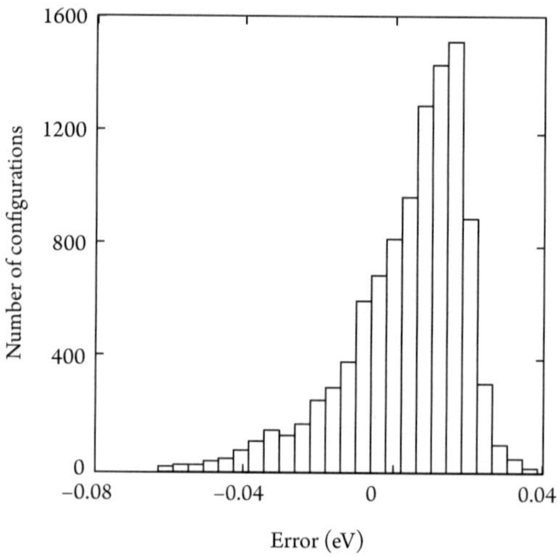

Figure 8-3 Distribution of errors for Solution-1 in Table 8-1. Total number of points in the database is 10,202. The rms error for the distribution is 0.0148 eV (1.43 kJ mol^{-1})[22] (reprinted with permission from American Institute of Physics).

Solution-1 in Table 8-1 is close to the best fit possible using a Tersoff potential form[13] for the Si_5 system where C and D are assumed to be linear functions of the Si-Si distance involved in the two-body term. If more of the parameters were made simultaneous functions of all the elements of $\mathbf{P}_k$ with a more elaborate NN employed to adjust the weights and biases, the results would be substantially improved. However, even the present simple treatment yields fitting accuracy comparable to or better than most previously reported generalized fitting methods.

It should be noted that the operation of the general method is independent of the choice of empirical function to be fitted. It provides the means to rapidly obtain the best possible fit of the chosen form to the database. If the empirical function employed is well judged, the fitting results will be excellent. If the opposite is true, the fitting accuracy will be substantially reduced.

The power of the present method derives from three considerations. First, it obviates the problem of selecting the form of the functional dependence of the parameters upon the system's coordinates by employing a neural network. If this network contains a sufficient number of neurons, it will automatically find something close to the best functional form. This is the case since Hornik et al.[108] have shown that two-layer NNs with sigmoid transfer functions in the first hidden layer and linear functions in the output layer are universal approximators for analytic functions. Second, the entire fitting procedure is automated so that excellent fits are obtained rapidly with a minimum of human effort. Third, the neural network method provides a procedure to avoid local minima in the multi-dimensional parameter hyperspace. The Si_3 system is just a simple illustration of the technique. The real advantages will be realized when the method is applied to much more demanding systems.

8.2. FITTING EMPIRICAL POTENTIALS TO VIBRATIONAL SPECTRAL DATA

8.2.1. Introduction

When empirical potential parameters are to be fitted to a database comprising *ab initio* electronic structural data, the computational difficulties reside primarily in executing the electronic structure calculations at a sufficiently high level of accuracy and at a sufficient number of molecular configuration points to ensure that the important regions of configuration space are adequately represented in the database. Once such a database is obtained, the fitting process is relatively straightforward using the NN methods described in Section 8.1 of this chapter.

Since the PES generally underlies the determination of most experimental data on any molecular system of interest, it should, in principle, be possible to obtain the PES or the best parameters for a chosen empirical PES from measured experimental data. If this intriguing possibility can be realized, it would obviate the need to compute extensive databases of *ab initio* electronic structural data. In addition, since the experimental measurements come equipped with the correct PES, it might be reasonably argued that a PES determined from experimental data will be more accurate than one obtained from first principles electronic structure calculations since the latter are necessarily approximate.

This concept is not new. There have been many attempts made to obtain potential surfaces using various types of experimental data. Simple examples include using x-ray and microwave data to obtain equilibrium bond distances, Raman and IR spectra to find the fundamental vibrational frequencies, and the temperature dependence of a reaction rate to extract activation energies and then incorporating that information into the empirical PES by adjustment of the potential's parameters. While such procedures are straightforward, they generally do not suffice to determine all the parameters of a PES, particularly if the functional form of the PES is complex. In addition, in many cases, the parameters are coupled in a complex manner that makes adjustment to experimental data difficult and laborious.

Since it is generally true that the topology of the PES is correlated with the experimental observations, an NN is an ideally suited tool to unravel and map these often hidden correlations and thereby obtain the parameters of the PES. In 1992, Sumpter and Noid[74] demonstrated how this might be done using the measured IR vibrational spectra for macromolecules. The following section describes this work and the results obtained.

8.2.2. Application to Macromolecules

Sumpter and Noid[74] demonstrated the utility of NNs in inverting experimental spectral data to obtain the parameters of an empirical PES by application to a model of polyethylene that comprised a single chain of 100 backbone carbon atoms and 200 substituent hydrogen atoms. The PES they chose to represent the internal forces and couplings in this system was the sum of bonded and nonbonded interactions.

$$V = V(bonded) + V(nonbonded). \tag{8-16}$$

The bonded and nonbonded interactions were written as many-body expansions about the equilibrium position.

$$V(bonded) = \Sigma\, V_{CC} + \Sigma\, V_{CH} + \Sigma\, V_{CCC} + \Sigma\, V_{HCH} + \Sigma\, V_{HCC} + \Sigma\, V_{CCCC} \tag{8-17}$$

and

$$V(nonbonded) = \Sigma\, V_{CC} + \Sigma\, V_{CH} + \Sigma\, V_{HH}. \tag{8-18}$$

The individual functions were chosen to yield a relatively accurate description of the vibrational motion of the molecule. They were

(A) Morse functions for the bonded two-body or stretching interactions

$$V(r_{ij}) = \Sigma_{ij}\, D_{ij}\,[1 - exp\{-\,\alpha_{ij}\,(r_{ij} - r^o_{ij})\}]^2$$

with $\qquad\qquad r_{ij} = r_{CH}, r_{CC}. \tag{8-19}$

In Equation (8-19) r_{ij} denotes the bond distance between the ith and j^{th} atom of the polymer chain.

(B) Harmonic functions for the bending interactions

$$V(\theta_{ijk}) = \Sigma_{ijk} K(\theta_{ijk}) \, [\theta_{ijk} - \theta^{o}_{ijk}]^2 \, ,$$

with $\theta_{ijk} = \theta_{CCC}, \theta_{HCC}$, or θ_{HCH}. (8-20)

(C) Threefold well potentials for the backbone torsional or dihedral angles, τ

$$V_{CCCC}(\tau) = \Sigma_{ijkl} [- \alpha' \, cos(\tau_{ijkl}) + \beta' cos^3(\tau_{ijkl})].$$ (8-21)

(D) Exp-6 potentials for the nonbonded interactions

$$V(R_{ij}) = \Sigma_{ij} \alpha_{ij}/R^6_{ij} + \beta \, exp(- \gamma R_{ij}),$$ (8-22)

where R_{ij} is the distance between atoms i and j that are not directly bonded.

Equations (8-16) through (8-22) contain a total of 23 adjustable parameters. Sumpter and Noid have given these parameter values for a polyethylene molecule.[74]

In order to train an NN to predict the values of the 23 parameters contained in Equations (8-16) to (8-22), Sumpter and Noid[74] constructed a set of hypothetical IR spectra for different sets of parameter values. This was done by first executing a normal-mode analysis[244] for a given set of parameters to determine the 894 fundamental vibrational frequencies of their 300-atom macromolecular model. From these results, they constructed a vibrational spectrum $g(\omega)$, where $g(\omega)$ is defined to be the number of normal mode frequencies between ω and $\omega + d\omega$, by simply making histograms of the various fundamental frequencies as a function of ω.

Figures 8-4a and 8-4b show typical results for $g(\omega)$. The spectrum shown in Figure 8-4a was obtained using the parameter set appropriate for polyethylene (see Reference 74). The one shown in Figure 8-4b was obtained for a different set of potential parameters. Comparison of the two figures shows that changes in the potential parameters produce significant alterations in the computed $g(\omega)$ spectrum. This sensitivity to the parameter values suggests that an NN will be able to sense the underlying correlations between the parameter values and the details of $g(\omega)$ and be able to predict the best set of parameters from the measured spectrum.

The NN employed by Sumpter and Noid to predict the potential parameters given the vibrational spectrum was a two-layer network whose input vector contained 426 elements that provided the value of $g(\omega)$ at equally spaced frequencies along the calculated spectrum. The hidden layer contained 7 neurons, and the output layer 18. Since Equations (8-16) to (8-22) contain 23 parameter values, apparently

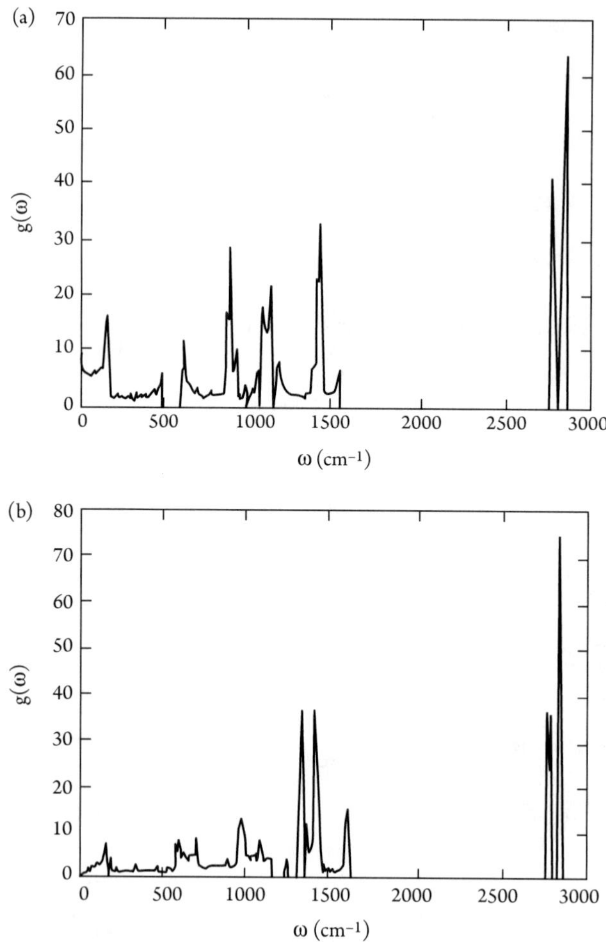

Figure 8-4 $g(\omega)$ spectra for a 300-atom model of the polyethylene molecule. (a) Calculated spectra (see text) for the potential parameters fitted to polyethylene. (b) Calculated spectra for a different set of parameters in Eqs. (8-16) through (8-22) (reprinted with permission from Elsevier and the authors; see ref. 74).

Sumpter and Noid fixed five of these and varied the other 18. The predicted values of those parameters are produced by the 18 neurons of the output layer.

The (426-7-18) NN was trained using spectra computed from 51 different sets of parameter values. Training was continued for 20,000 epochs, which required about one CPU hour of computational time on a CRAY-XMP/14. After training, the prediction of parameter values given a new spectrum as input required less than 2 CPU seconds.

Sumpter and Noid[74] assessed the predictive accuracy of the trained NN in terms of the absolute relative percentage error (RPE) between the actual potential parameters producing a test spectrum and those predicted by the NN. Specifically, the RPE is defined by

$$RPE = \frac{\left| P_{known} - P_{NN} \right|}{P_{known}} \times 100, \tag{8-23}$$

where P_{known} and P_{NN} are the values of one of the known parameters and the corresponding parameter predicted by the NN, respectively.

For the sets of parameters used to generate the training data, the average RPE for a given parameter was found to range from zero to about 1.4% with the majority of the parameters having RPEs less than 0.5%.

The ability of the NN to make predictions using parameter sets that were not employed in the training was investigated by computing $g(\omega)$ spectra for several such sets and calculating the average RPE for each of the parameters using Equation (8-23). In this case, the maximum RPE was 1.0% with the majority of values being near zero provided the parameter sets used for testing were within the training range for each parameter. When the NN was forced to extrapolate to parameters whose values were outside the training range, the maximum error increased to 3.9%, but the majority of RPE values were still near zero. Since NNs usually do not extrapolate accurately, this is a very happy, but unexpected, result.

9

NEURAL NETWORK METHODS FOR DATA ANALYSIS AND STATISTICAL ERROR REDUCTION

9.1. INTRODUCTION

The use of neural networks (NNs) to predict an outcome or the output results as a function of a set of input parameters has been gaining wider acceptance with the advance in computer technology as well as with an increased awareness of the potential of NNs.[105,249,251-253] A neural network is first trained to learn the underlying functional relationship between the output and the input parameters by providing it with a large number of data points, where each data point corresponds to a set of output and input parameters.

Sumpter and Noid[74] demonstrated the use of NNs to map the vibrational motion derived from the vibrational spectra onto a PES with relatively high accuracy. In another application, Sumpter et al.[246] trained an NN to learn the relation between the phase-space points along a trajectory and the mode energies for stretching, torsion, and bending vibrations of H_2O_2. Likewise, Nami et al.[254] demonstrated the use of NNs to determine the TiO_2 deposition rates in a chemical vapor deposition (CVD) process from the knowledge of a range of deposition conditions.

In view of the success achieved in obtaining interpolated values of the PESs for multi-atomic systems using an NN trained by the *ab initio* energy values for a large number of configurations, it is reasonable to ask whether we can successfully compute the results of an MD trajectory for a chemical reaction using an NN trained by the data obtained by previous MD simulations. If this can be done successfully, it becomes possible to execute a small number of trajectories, M, and then utilize the results of these trajectories as a database to train an NN to predict the final results of a very large number of trajectories N, where $N >> M$, that can be used to increase the statistical accuracy of the MD calculations and to further explore the dependence of the trajectory results upon a wide variety of variables without actually having to perform any further numerical integrations. In effect, the NN replaces the computationally laborious numerical integrations.

9.2. INTERACTION OF CARBON (C_2) DIMER WITH DIAMOND—MD SIMULATIONS

The above intriguing possibility was investigated by Agrawal et al.[75] using the interaction of a carbon C_2 dimer with a diamond (100) surface as a test system. MD simulations of C_2 collisions with a model (100) diamond surface at a given surface temperature were executed to compute the probabilities of C_2 chemisorption, scattering, and desorption probabilities as a function of the direction (θ,ϕ) of the initial C_2 velocity vector, impact parameter (b), translational energy (E_{trans}), and rotational energy (E_{rot}). These data were then used as a database to train an NN to predict these probabilities for various input conditions.

The methods for conducting MD simulations of gas-surface processes have been described in numerous journal articles.[255-261] Therefore, only the essential details of the MD investigation are given here.

A total of 324 atoms are used to model the system. Out of these, 282 atoms of diamond substrate are used to model the (100) crystalline face with 40 atoms of hydrogen on the top layer of the diamond surface and 2 atoms in the C_2 dimer. The BRENNER MD code used employs the potential given by Brenner et al.[262] with van der Waals interactions incorporated using a Lennard-Jones (LJ) potential. The (x, y) coordinates of the atoms on the top three layers are as shown in Figure 9-1(a) for atoms of top layer, $z = 0$. Each of the carbon atoms on the top layer, except the central atom and boundary atoms, is capped by a hydrogen atom. The equations of motion are integrated using Gear's predictor-corrector method[263] with a time step size Δt of 0.5 fs.

The results of a given trajectory depend upon a multitude of input variables. These include b, θ, ϕ, E_{rot}, E_{trans}, the initial orientation of the C_2 dimer, the angle defining the C_2 rotational plane, the initial C_2 vibrational energy and its phase, the temperature of the system, and all of the variables that define the vibrational phases of the diamond surface. Here, θ denotes the angle of incidence, i.e., the angle between the direction of the initial translational velocity vector of the C_2 dimer and the perpendicular on the surface (Z direction). The impact parameter b is defined as the distance between the location of the central atom C (see Figures 9-1 (a) and (b)) and the point of intersection P of the initial velocity vector and the diamond surface, as shown in Figure 9-1(b). ϕ represents the angle between the line CP and the X axis, as shown in Figure 9-1(b), and E_{rot} and E_{trans} refer to the rotational and translational kinetic energies of the dimer, respectively.

9.3. STATISTICAL DATA ANALYSIS AND RESULTS

Agrawal et al.[75] focused their efforts on determining the dependence of probabilities for chemisorption, scattering, and desorption upon b, θ, ϕ, E_{rot}, E_{trans}. The initial

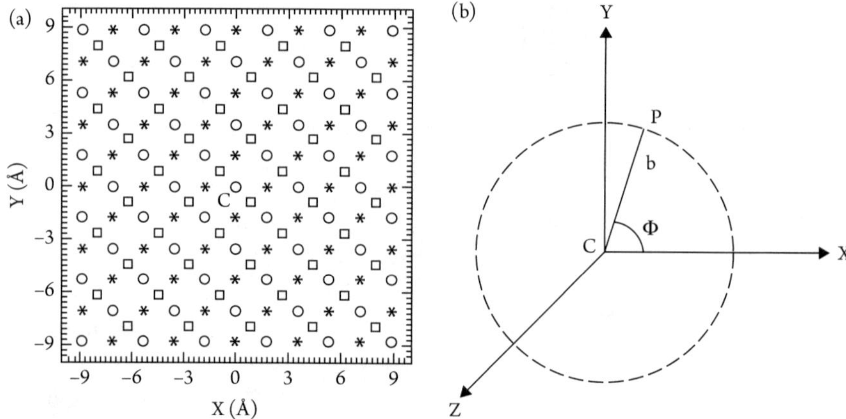

Figure 9-1 (a) The projection of the top three layers of carbon atoms of the diamond (100) lattice on the XY plane: first layer (O), second layer (□), third layer (*). The central atom **C** denotes the radical site. (b) Schematic diagram to depict parameters b and ϕ. **P** represents the point of intersection of the initial C_2 velocity vector and the diamond surface[75] (reprinted with permission from American Institute of Physics).

C_2 vibrational energy is set equal to the zero-point energy and the temperature of the lattice is maintained constant at 600 K using the Berendsen thermostat procedure.[264] For a given input set of b, θ, ϕ, E_{rot}, E_{trans}, the probabilities of different reactions are determined by running 50 trajectories to effect the averaging over the remaining variables.

Using MD simulations, Agrawal et al.[75] first computed the reaction probabilities, $P_x(a;MD)$ (x = C, S, and D, which correspond to chemisorptions, scattering, and desorption, respectively), for a given set of input parameters, $a = a(b,\theta,\phi,E_{rot},E_{trans})$. By running such calculations for different values of a, N sets of values of $P_x(a;MD)$ were computed. After scaling all input and output data using Equation (2-34), these N sets of data points are used to train a two-layer, feedforward NN illustrated in Figure 9-2.

The NN employed by Agrawal et al.[75] used a tan-sigmoid transfer function in the hidden layer and a linear transfer function for the output layer. The architecture of the NN was chosen to be (5-50-3). The five elements of the input vector are b, θ, ϕ, E_{rot}, and E_{trans}.

The three reactions of the C_2 dimer with the radical site on the diamond (100) surface of interest are

$$\blacklozenge + C_2 \rightarrow \blacklozenge - C_2 \text{ (chemisorption)} \qquad \text{(Reaction } R_1\text{),}$$

$$\blacklozenge + C_2 \rightarrow \blacklozenge + C_2 \text{ (scattering)} \qquad \text{(Reaction } R_2\text{), and}$$

$$\blacklozenge + C_2 \rightarrow \blacklozenge - C_2 \rightarrow \blacklozenge + C_2 \text{ (adsorption and desorption)} \qquad \text{(Reaction } R_3\text{).}$$

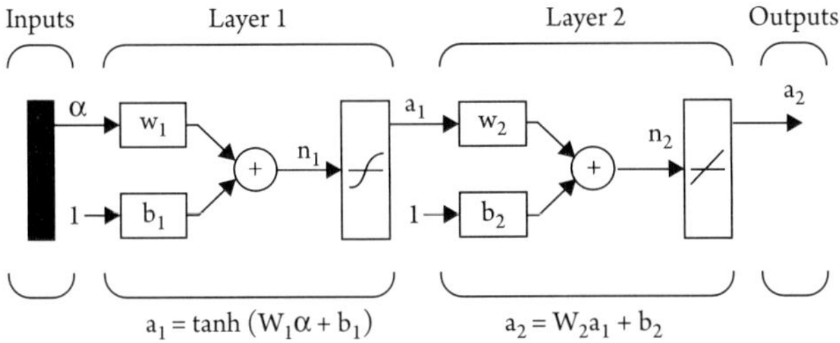

Figure 9-2 Schematic illustration of the three-layer NN used to present the reaction probabilities[75] (reprinted with permission from American Institute of Physics).

The notation ♦ refers to the central carbon atom on the top diamond (100) surface that is not bonded to a hydrogen atom. It is, therefore, a radical site. Agrawal et al.[75] also observed a reaction in which C_2 is inserted into two surface carbon atoms:

$$\text{H--}\blacklozenge' + C_2 + \blacklozenge''\text{--H} \rightarrow \text{H--}\blacklozenge' - C - C - \blacklozenge''\text{--H} \qquad \text{(Reaction R}_4\text{)}$$

Here, $\blacklozenge'$ and $\blacklozenge''$ denote two neighboring carbon atoms on the diamond surface, and H denotes a hydrogen atom bonded to that carbon atom. Since reaction R_4 was found to occur with very low probability, it was ignored in the NN analysis.

The probabilities, $P_x(\alpha;\text{MD})$ [$(x=C, S, D)$ corresponding to reactions (R1)–(R3)], of a reaction for a given set of input parameters $\alpha = \alpha(b,\theta,\phi,E_{rot},E_{trans})$ are computed by running 50 trajectories to average over the other variables that include the initial phases of vibration of C_2 and the lattice, and the initial orientation of the C_2 bond vector. The vibrational energy of C_2 and temperature of the lattice are kept fixed in all the calculations. $P_x(\alpha;\text{MD})$ was computed for 1,900 different values of α. Out of these 1,900 data points, 1,500 data points were used for training, 200 for validation, and 200 for testing. The NN shown in Figure 9-2 was trained to predict the probabilities of reactions for all values of α, within the range of α present in the training set. Early stopping procedures with a validation set were employed to prevent overfitting.[117,118]

The initial success of the NN training is shown in Figure 9-3 (top plots), where the 1,500 values of $P_x(\alpha;\text{NN})$ given by the NN are plotted versus the corresponding probabilities, $P_x(\alpha;\text{MD})$, given by the trajectory calculations. Next, for 200 values of α in the testing set, $P_x(\alpha;\text{NN})$ was computed and compared with $P_x(\alpha;\text{MD})$ obtained from the MD simulations (see bottom plots of Figure 9-3). The results of the testing set are very similar to those for the training set.

For the purpose of the following discussion, it is appropriate to describe the relationships among three related reaction probabilities: $P_x(\alpha;\text{MD})$, $P_x(\alpha;\text{NN})$, and $P_x(\alpha)$. For brevity, when the specification of some labels is not essential, one or more of the labels among x, α, MD, and NN may be ignored. $P_x(\alpha;\text{MD})$ is the reaction probability for process x computed from the trajectory results using

$$P_x(\alpha;\text{MD}) = \frac{Nx}{NT}, \qquad (9\text{-}1)$$

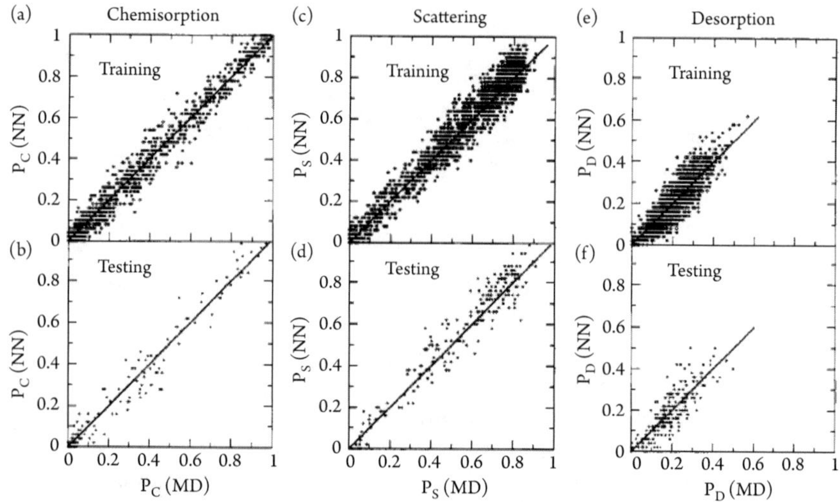

Figure 9-3 Reaction probabilities given by the NN vs. the corresponding probabilities given my MD calculations for the training and testing sets. Here, [(a) and (b)] chemisorption, [(c) and (d)] scattering, and [(e) and (f)] desorption[75] (reprinted with permission from American Institute of Physics).

where Nx is the number of trajectories that resulted in reaction x out of a total of N_T trajectories computed. $P_x(\alpha)$ is the underlying true, but unknown, classical reaction probability for process x on the potential-energy surface used in the calculations. These two probabilities are related by

$$P_x(\alpha) = \lim_{N_T \to \infty} P_x(\alpha; \mathrm{MD}) \,. \tag{9-2}$$

When N_T is finite, there will be a random statistical fluctuation present in $Px(\alpha;\mathrm{MD})$ that will cause its value to deviate from $P_x(\alpha)$. This fluctuation will be such that ~ 95% of the time, we expect to have

$$P_x(\alpha;) - 2\Delta \le P_x(\alpha; \mathrm{MD}) \le P_x(\alpha) + 2\Delta, \tag{9-3}$$

where

$$\Delta = [P_x(\alpha)\{1 - P_x(\alpha)\}/N_T]^{1/2}. \tag{9-4}$$

Finally, $P_x(\alpha; \mathrm{NN})$ is the reaction probability predicted by the NN for process x.

As previously noted, Figures 9-3 (a)–(f) show plots of $P_x(\alpha, \mathrm{NN})$ versus $P_x(\alpha; \mathrm{MD})$ for chemisorptions, scattering, and desorption, respectively, with $N_T = 50$ in the training and testing sets. If the agreement between $P_x(\alpha, \mathrm{NN})$ and $P_x(\alpha, \mathrm{MD})$ were perfect, all points would lie on the 45° lines shown in the plots. In the analysis

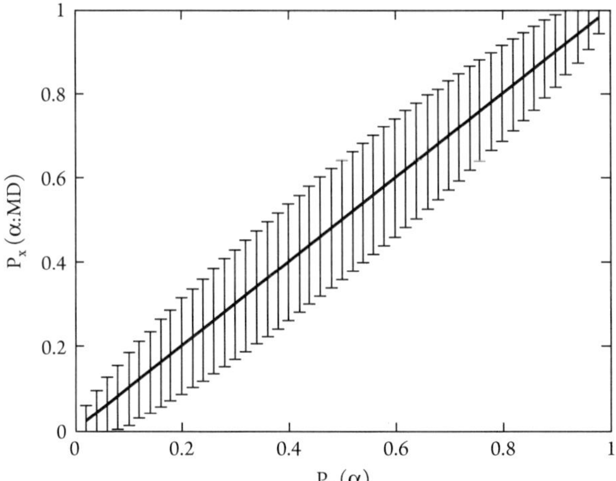

Figure 9-4 The expected statistical spread of the probabilities $P_x(\alpha{:}MD)$ as given by Eq. (9-3) vs. the true probabilities $P_x(\alpha)$[75] (reprinted with permission from American Institute of Physics).

that follows, it is demonstrated that most of the deviation of the plotted points from the 45° line is due to the statistical variations present in $P_x(\alpha, MD)$ rather than to inaccurate interpolation of the neural net. This conclusion is based on the variation of $P_x(\alpha, MD)$ with $P_x(\alpha)$ computed using Equations (9-3) and (9-4). The results are shown in Figure 9-4.

A comparison of the spread of data from the ideal 45° line in Figure 9-3(a) or 9-3(b) with that shown in Figure 9-4 shows that they match; 95% of the points in these curves are within distances of 0.08, 0.10, and 0.11, respectively, from the 45° line. This shows that the expected statistical spread of $P_x(\alpha, MD)$ values is very close to that observed in the plots of $P_x(\alpha, MD)$ versus $P_x(\alpha, NN)$ shown in Figure 9-3. In effect, $P_x(\alpha, NN)$ is fitting $P_x(\alpha)$ and not the statistical fluctuations in $P_x(\alpha, MD)$. This observation indicates that neural net fitting of the MD results provides a rapid and effective means of averaging out the statistical fluctuations present in $P_x(\alpha, MD)$ when N_T is small.

The NN with an early stopping procedure[117,118,265] tends to fit the trend and not the statistical fluctuations. This observation is not a new one. It has been shown that the number of training iterations corresponds to the complexity of the neural network function.[266] Therefore, as training continues, the complexity of the network function increases and it becomes more likely that the network will begin to fit statistical fluctuations rather than the underlying true probability function. By determining when the error on the validation set begins to increase, we can stop the training at an optimal point, before overfitting occurs. It is trivial to visualize that with overfitting, in Figure 9-3, the spread in the training plots would shrink and that in the testing plots would increase.

To check further that the spreads in the plots of Figure 9-3 are mainly due to the statistical variability associated with the results of the MD calculations, $P_x(\alpha, MD)$ has been computed by running 500 trajectories instead of 50, for each

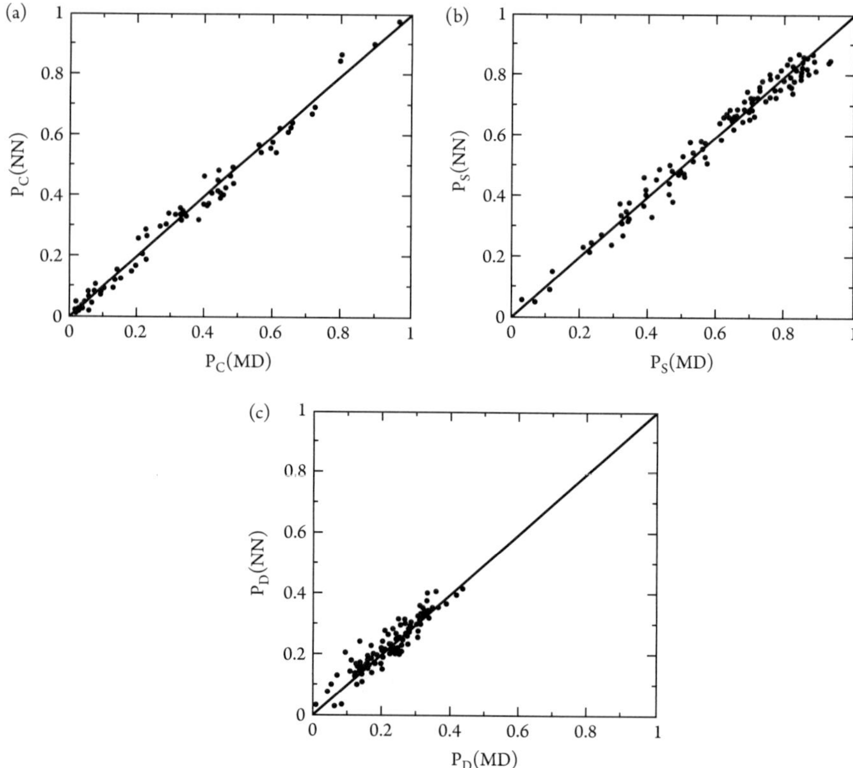

Figure 9-5 The probabilities of the various reactions vs the corresponding probabilities obtained from MD calculations using sets of 500 trajectories. (a) chemisorption, (b) scattering, and (c) desorption[75] (reprinted with permission from American Institute of Physics).

value of α. This is expected to decrease the statistical fluctuations in $P_x(\alpha, MD)$ by a factor of 3.2. The plots of such probabilities, $P_x(\alpha, MD)$ vs $P_x(\alpha, NN)$, are shown in Figure 9-5. In this case, the $P_x(\alpha, MD)$ values have been determined by using the same sets of NN weights and biases used to obtain Figure 9-3. The decreased spread of the points around the 45° line suggests that the same NN, which was obtained by fitting the less accurate MD data, is in better agreement with the more accurate MD data. This point again indicates that the NN fitting procedure involving early stopping fits the trend of the output data and not the statistical fluctuations associated with the output data. This implies that the statistical error present in the MD results may be most conveniently reduced using NNs rather than by computing additional trajectories.

A comparison of the spread in Figures 9-3 and 9-5 shows that there has been a decrease in the spread by a factor of 2.0. The reduction in the spread by a factor of 2.0, instead of 3.2, is indicative of the effect of error, ε_{NN}, associated with the neural network fitting.

To estimate the deviation associated with the NN fitting, the same 1,900 data points were employed 50 times to obtain 50 different NNs such that each network has been trained with different random initial weights and a different partitioning of the database between training, validation, and testing sets. In the two-sigma limit, the variation of such 50 values of $P_x(\alpha;\text{NN}\_)$ from the mean value for each of the 1,900 values of α was determined. The average of such variations is found to be equal to 0.04.

It is useful to consider two limiting cases. When the statistical error, ε_{MD}, in the input MD data is very small as compared to ε_{NN}, we would expect some spread in figures similar to Figures 9-3 (a)-(f) due to the error in NN fitting, ε_{NN}. When ε_{MD} is very large as compared to ε_{NN}, then the probabilities predicted by the NN are expected to be more accurate than those given by MD, and the spread of points from the 45° line in a figure similar to Figure 9-3 would be close to ε_{MD}. The results from Figures 9-3 to 9-5 show that, for the present application, we have $\varepsilon_{MD} \gg \varepsilon_{NN}$.

To improve the network accuracy, the 1,900 data points previously described were used to train an ensemble of f networks. In this investigation, Agrawal et al.[75] used f = 50. Each network in the ensemble was trained using a different random partitioning of the data into 1,600 training data and 300 validation data. In addition, each network was trained with different random initial weights. After all networks in the ensemble have been trained, the same input is applied to each network and the average of all responses is computed. The ensemble average produces a more accurate response than any one member of the ensemble because the committee averages out the variations in individual network responses due to randomness in sampling of the data and randomness in the choice of initial weights and biases provided by training.[267]

The results obtained for the C_2-diamond (100) surface system are examples of the advantage provided by analysis of the data using an NN committee. This procedure has been discussed in Chapter 3, Section 3.8.1. Equations (9-3) and (9-4) show that the two-sigma limit statistical uncertainty in values of $P_x(\alpha;\text{MD})$ is ±28.3% when $P_x(\alpha)$ is 0.50 and N_T is 50. In contrast, the corresponding statistical uncertainty present in the results obtained using the NN committee is about ± 8%. This statistical error reduction of a factor of 3.5 is achieved with very little additional computational effort.

A major advantage of NNs can be realized by determining the probability of $P_x(\alpha;\text{NN})$ corresponding to any value of α within the range of b, θ, ϕ E_{rot}, and E_{trans} for which the NN has been trained. For example, in almost negligible time (as compared to the time required to run tens of thousands of trajectories) the dependence of $P_x(\alpha;\text{NN})$ on any variable, say b, for any fixed values of θ, ϕ E_{rot}, and E_{trans} can be found. As an example, Figure 9-6 shows such variations of P_C, P_S, and P_D as a function of b for $\theta = 17°$, $\phi -310°$, $E_{rot} - 0.052$ eV, and $E_{trans} - 0.06$ eV given by the NN ensemble. For comparison, the results obtained by trajectory calculations are also shown in the figures. As can be seen, the agreement between the NN and MD results is very good. Similarly, the computed MD results and NN predictions of variation of P_C, P_S, and P_D as a function of E_{trans} and E_{rot} for fixed values of other parameters are compared in Figure 9-7, and the results shown in Figure 9-8 give the corresponding

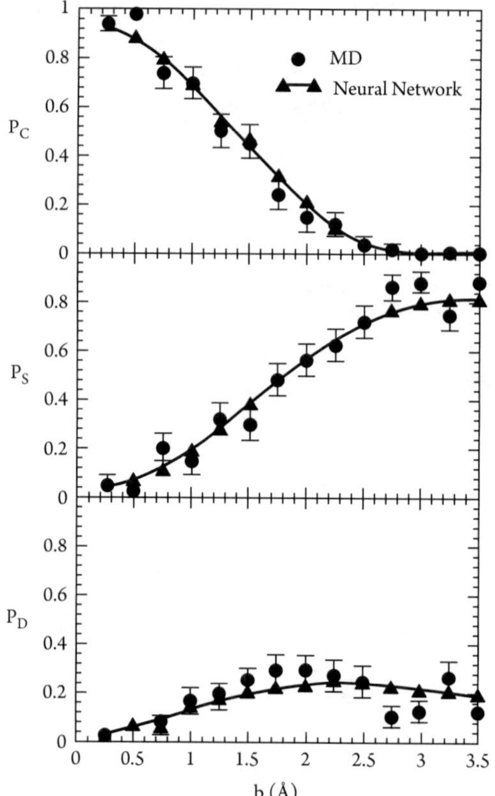

Figure 9-6 The variation of the chemisorptions probability (P_C), scattering probability (P_S), and desorption probability (P_D) as a function of impact parameter (b), for E_{trans} = 0.06 eV, E_{rot} = 0.052 eV, θ = 17°, and ϕ = 310°. The triangles (Δ) joined by the line denote the NN ensemble predictions and the circles ($\bullet$) represent the MD data. The error bars associated with the MD points correspond to one-sigma limit[75] (reprinted with permission from American Institute of Physics).

curves when θ and ϕ are varied. All these curves show that the agreement between the MD results and NN predictions is very good.

It is trivial to exhibit the utility of such a trained NN. As an example, Figure 9-9 shows the variation of P_C, P_S, and P_D with angle θ predicted by the NN ensemble for different values of impact parameters for some fixed values of ϕ, E_{rot}, and E_{trans}. It is easy to visualize that any number of such curves for the dependence of P_C, P_S, and P_D on any of the parameters can similarly be obtained in a CPU time that is negligible compared to the time required to obtain the same results by explicit MD calculations.

9.4. CONCLUSIONS

On the basis of the results discussed, Agrawal et al.[75] reached the following conclusions:

1. By using a large number of data, the reaction probabilities predicted by a trained NN committee will be in very good accord with those provided by MD calculations.

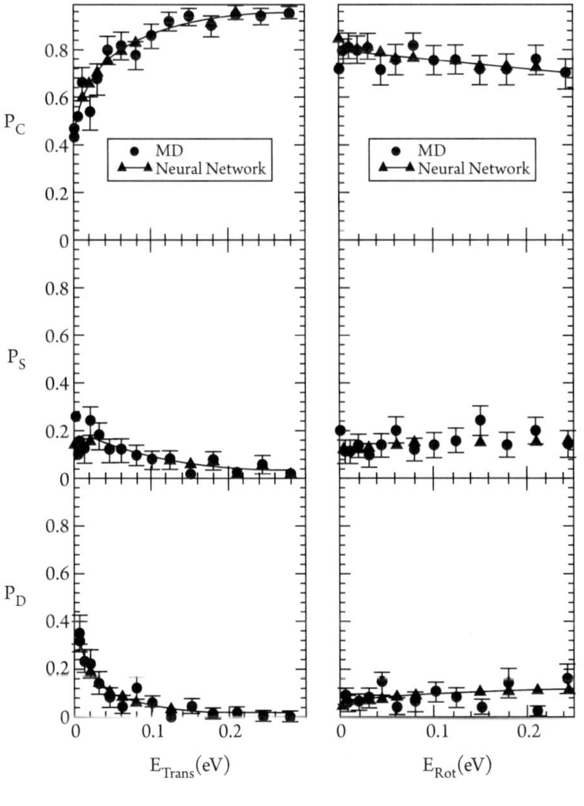

Figure 9-7 Same as Figure 9-6 except that these curves show variation with E_{trans} for E_{rot} = 0.052 eV, b = 1.0 Å, θ = 11°, and ϕ = 110° (left curves), and that with E_{rot} for E_{trans} = 0.06 eV, b = 1.0 Å, θ = 17°, and ϕ = 110° (curves on the right)[75] (reprinted with permission from American Institute of Physics).

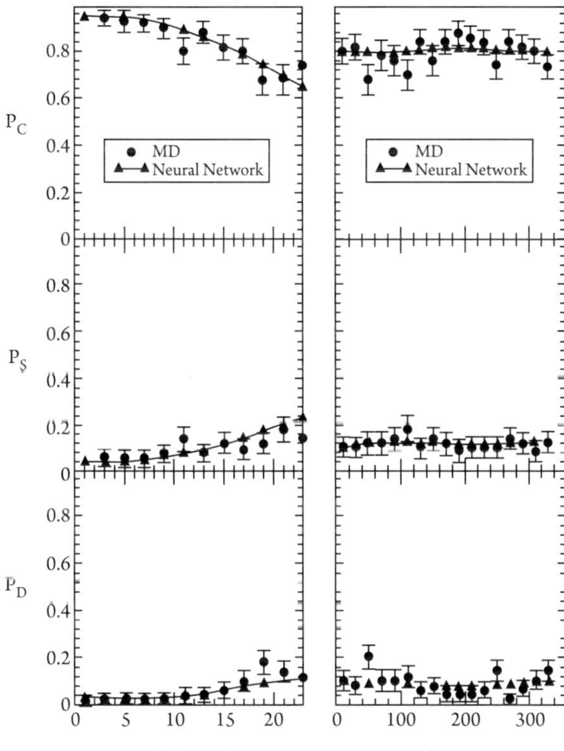

Figure 9-8 Same as Figure 9-6 except that these curves show variation with θ for E_{trans} = 0.124, for E_{rot} = 0.052 eV, b = 1.0 Å, and ϕ = 110° (curves on the left), and that with ϕ for E_{trans} = 0.06 eV, F_{rot} = 0.052 eV, b = 1.0 Å, and θ = 11°[75] (reprinted with permission from American Institute of Physics).

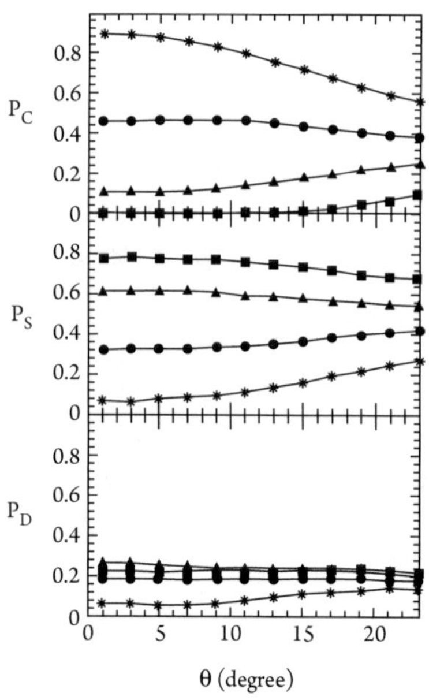

Figure 9-9 The variation of the chemisorption probability (P_C), scattering probability (P_S), and the desorption probability (P_D) predicted by the NN committee as a function of angle θ for different values of the impact parameter b = 1.0 Å (*), 1.5 Å (•), 2.0 Å (▲), 2.5 Å (■) for E_{trans} = 0.06 eV, E_{rot} = 0.052 eV, and ϕ = 110°[75] (reprinted with permission from American Institute of Physics).

2. After training an NN committee, determining the interpolated reaction probabilities as a function of the input parameters involves only the evaluation of simple analytical expressions. Thus, interpolation by using an NN committee can significantly reduce the burden of performing a large number of expensive MD calculations.

3. NN fitting employing early stopping will fit the trend of the input data and not the statistical fluctuations associated with the input data. Thus, NNs can provide a computationally convenient method of averaging out statistical variations inherent in the MD calculations.

These observations demonstrate that NN committees can be an extremely powerful tool for interpolating reaction probabilities as a function of different variables in chemical reactions thereby avoiding computational expense of integrating coupled sets of differential equations to compute MD trajectories.

10

OTHER APPLICATIONS OF NEURAL NETWORKS TO QUANTUM MECHANICAL PROBLEMS

10.1. SOLVING THE MOLECULAR VIBRATIONAL SCHRÖDINGER EQUATION

10.1.1. Introduction and General Theory

The usual method for solving the vibrational Schrödinger equation to obtain molecular vibrational spectra and the associated wave functions generally involves the expansion of the vibrational wave function, $\psi_k(\mathbf{y})$, in terms of a linear combination of a set of basis functions. That is,

$$\psi_k(\mathbf{y}) = \sum_{i=1}^{N} c_{ik}\chi_i(\mathbf{y}). \tag{10-1}$$

In Equation (10-1), the $\chi_i(\mathbf{y})$ are predetermined basis functions that depend upon the coordinate vector $\mathbf{y}$ that specifies the nuclear configuration of the system. The linear coefficients, c_{ik}, are usually determined by a variational-type calculation that minimizes the variational energy, λ_k, with respect to the expansion coefficients where

$$\lambda_k = \frac{\langle \psi_k(\mathbf{y})|H|\psi_k(\mathbf{y})\rangle}{\langle \psi_k(\mathbf{y})|\psi_k(\mathbf{y})\rangle}. \tag{10-2}$$

and H is the vibrational Hamiltonian operator for the system.

The major computational problem with this approach is the size of N in Equation (10-1). In most cases, N must be large if accurate results are to be obtained. When the molecular system of interest is large, the Hamiltonian operator will contain many terms, and this fact, coupled with a large value of N, means the minimization of λ_k in Equation (10-2) will require the evaluation of many integrals and involve handling extremely large matrices.

The primary reason that N must be large in Equation (10-1) is that, in most cases, the $\chi_i(\mathbf{y})$ are predetermined and, therefore, have a fixed functional form. To achieve accurate results, the approximate wave function in Equation (10-1) must be close to the true, but unknown, wave function in all the important regions of configuration space. Since the basis set functions have fixed functional forms, the only flexibility $\psi_k(\mathbf{y})$ has derives from variations of the linear expansion coefficients. If the basis set $\{\chi(\mathbf{y})\}$ is complete, we are guaranteed that if N is made sufficiently large, $\psi_k(\mathbf{y})$ will converge to the true wave function for sufficiently large N, but in most cases, convergence will be achieved only for very large N.

If, however, we were to replace the right-hand side of Equation (10-1) with an NN that has a totally flexible functional form such that it is a universal approximator for any analytic function that spans the same space, we might expect to see $\psi_k(\mathbf{y})$ converge to the true wave function using a number of neurons, N, that is much smaller than the number of basis functions needed in the linear expansion.

This concept of using an NN to represent the vibrational wave function was first explored by Lagaris et al.[174] in 1997 and later by Nakanishi and Sugawara[176] and Sugaware[175] in 2000 and 2001. In this initial work, the method was applied only to model systems, and vibrational wave functions along with the associated energies were computed one at a time. Recently, Manzhos and Carrington[76,77] have improved the method. Most important, they have shown how it can be applied to real molecular systems[77] in a manner that permits many vibrational states and their energies to be obtained in a single calculation. In this section, we will describe their approach.

In the Manzhos-Carrington method,[76,77] the NN representing the vibrational wave functions of the system is taken to be a single radial basis function neural network (RBFNN), which has been shown to be a universal approximator for analytic functions.[268,269] In such a network, the usual sigmoid transfer functions are replaced with Gaussian-type functions, which have the right asymptotic behavior for wave functions without modification.

Consider a two-layer RBFNN with input vector $\mathbf{y}$ with d elements, a hidden layer with N neurons, and an output layer with M neurons. The output from neuron i, a_i, in such a network is

$$a_i = \left(\frac{b_i}{\sqrt{\pi}}\right)^d \exp\left(-b_i^2 \sum_{j=1}^{d}\left(w_{ij} - y_j\right)^2\right) . \tag{10-3}$$

As usual, each input element to neuron i is associated with a weight, w_{ij}, and each neuron has a bias, b_i. The argument for the Gaussian transfer function is the negative product of the summation over all input elements of the squares of the differences between w_{ij} and y_j and the square of the bias for the neuron. The functional form of

a_i shows that the w_{ij} locate the center or the position of the Gaussian while the bias parameters determine the width of the Gaussians.

The input to the kth output node of the RBFNN is the summation of the a_i over all neurons multiplied by a weight factor c_k. That is, the output of node k, O_k, is

$$O_k = \sum_{i=1}^{N} c_{ik} a_i = \sum_{i=1}^{N} c_{ik} \left[\left(\frac{b_i}{\sqrt{\pi}} \right)^d \exp\left(-b_i^2 \sum_{j=1}^{d} \left(w_{ij} - y_j \right)^2 \right) \right]. \tag{10-4}$$

Comparison of the right-hand sides of Equation (10-4) and Equation (10-1) shows that the output from the neurons of the RBFNN play the role of the basis functions of Equation (10-1) while the weights assigned by the RBFNN to each of the neuron outputs into the output neurons play the role of the linear expansion coefficients of Equation (10-1). Therefore, the output from the kth node in the output layer plays the role of the k-state vibrational wave function. That is, $\psi_k(\mathbf{y}) \equiv O_k$. The big difference is that the "basis functions" produced by the RBFNN have very flexible form rather than the fixed form of the $\chi(\mathbf{y})$ in Equation (10-1). Therefore, we may expect convergence of $\psi_k(\mathbf{y})$ to the true wave function for much smaller values of N in Equation (10-4) than in Equation (10-1).

It should also be recognized that if the input elements are the coordinates of the molecular system under consideration, the number of input elements into the RBFNN will be the dimensionality of the molecular system. The problem is now to develop a method that permits the best set of parameters appearing in Equation (10-4) to be obtained.

In order to solve for energy levels and wave functions of a set of nuclei vibrating in the force field produced by the motion of the electrons of the system, we must first solve the electronic structure problem to obtain the electronic potential energies at a discrete set of configuration points whose derivatives provide the nuclear force field.

If the usual variational methods are then employed to solve the molecular vibrational problem, it will be necessary to fit these data to an analytic function so that the required integrals can be computed. The various methods for executing such fitting have been discussed in detail in Chapters 2 through 7 of this monograph. One of the great advantages using NNs to solve the vibrational problem is that the method obviates this fitting problem since no integrals are needed.

Let us assume we have executed the required electronic structure calculations to obtain the electronic potential energies at a set of L distinct configuration points. If an accurate empirical potential is available, these computationally intense calculations can be replaced with simple calculations of the potential at the L selected points. We will denote the computed potential at point n as $V_n = V(^n\mathbf{y})$.

The initial efforts to use NNs to solve the vibrational Schrödinger equation did so by using the set $\{V_n\}$ to compute the vibrational energies and wave functions one state at a time.[174-176] While this is not as efficient as the method developed by Manzhos and Carrington,[76,77] it is easier to visualize what is being done. Therefore, we will examine this problem first.

If $\psi_k(\mathbf{y})$ were the exact k^{th}-state vibrational wave function and E_k the corresponding exact vibrational energy, then we would have

$$\left(\mathcal{H} - E_k\right)\psi_k(\mathbf{y}) = \left(T + V(\mathbf{y}) - E_k\right)\psi_k(\mathbf{y}) = 0 \tag{10-5}$$

at every point in configuration space. In Equation (10-5), T represents the kinetic energy operator for the system of interest. More specifically, Equation (10-5) would be satisfied at all of the L configuration points in our selected database. That is, we would have

$$\left(T + V(^n\mathbf{y}) - E_k\right)\psi_k(\mathbf{y})\Big|_n = 0 \quad \text{for } \left(n = 1, 2, \ldots, L\right), \tag{10-6}$$

where the left-hand side of Equation (10-6) is evaluated at $\mathbf{y} = {}^n\mathbf{y}$ after operation by T on $\psi_k(\mathbf{y})$.

If, however, $\psi_k(\mathbf{y})$ and E_k are not the exact k^{th}-state wave function and energy, then we will have

$$\left(T + V(^n\mathbf{y}) - E_k\right)\psi_k(\mathbf{y})\Big|_n \neq 0 \quad \text{for } \left(n = 1, 2, \ldots, L\right). \tag{10-7}$$

In some cases, the left-hand side will be greater than zero, and in others, it will be less.

Now consider the sum of the k-state residuals, R_k, which are represented by the left-hand side of Equation (10-7) at each of the L points in the database. That is,

$$R_k = \frac{\displaystyle\sum_{n=1}^{L}\left|\left(T + V(^n\mathbf{y}) - E_k\right)\psi_k(\mathbf{y})\Big|_n\right|^2}{\displaystyle\sum_{n=1}^{L}\left|\psi_k(^n\mathbf{y})\right|^2} > 0 \tag{10-8}$$

where the absolute squares on the right-hand side of Equation (10-8) ensure that the sum will always be greater than zero. However, we know from Equation (10-6) that the closer R_k is to zero, the closer the wave functions and energies will be to their exact values.

Equation (10-8) provides the means by which the parameters of the wave function, c_{ik}, b_i, and w_{ij}, along with the energy E_k may be determined. All that need be done is to execute a non-linear minimization of R_k with respect to these variables. The results will then provide the best approximate values of the energy and associated wave function.

It should be noted that this procedure requires no integrations. All that is needed are the potentials $V(^n\mathbf{y})$ at a finite set of discrete points. Consequently, it is not necessary to fit the computed energies in the database to an analytic PES. Moreover, these points need not be equally spaced. They can be randomly chosen so long as $\psi_k(^n\mathbf{y})$ is not negligibly small at any of the selected points.

In addition to using an RBFNN instead of the usual transfer functions, Manzhos and Carrington[76,77] have put the above procedure into matrix form so that the wave functions and associated energies for multiple states are obtained in a single calculation. They have also introduced some powerful methods for the execution of the minimization and a new way to increase accuracy without significantly increasing the computational difficulty in effecting the non-linear minimization.

Equation (10-5) is written in matrix form as

$$\left(\boldsymbol{M}-E_k\boldsymbol{S}\right)c_k =0, \tag{10-9}$$

where c_k is the kth column of a matrix of eigenvectors, $\boldsymbol{c}$ and $\boldsymbol{M}$ and $\boldsymbol{S}$ are rectangular matrices of dimension $L \times N$ whose elements depend upon the values of the potential and the neurons at the L different points in configuration space. These elements are

$$S_{ni} =\left(\frac{b_i}{\sqrt{\pi}}\right)^d \exp\left(-b_i^2 \sum_{j=1}^{d}\left(w_{ij} -^n y_j\right)^2\right) \tag{10-10}$$

and

$$M_{ni} = T_{ni}S_{ni} +V_n S_{ni}. \tag{10-11}$$

The form of the elements T_{ni}, depend upon the form of the kinetic energy operator which is system dependent.

Finally, instead of executing a non-linear minimization of Equation (10-8), Manhzos and Carrington minimize

$$R = \frac{1}{M}\sum_{k=1}^{M}\frac{\left\|\left(\boldsymbol{M}-E_k\boldsymbol{S}\right)\mathbf{c}_k\right\|^2}{\left\|\boldsymbol{S}\mathbf{c}_k\right\|^2}, \tag{10-12}$$

where M is the number of neurons in the output layer of the RBFNN, or the number of wavefunctions one wishes to compute.

The numerical procedures employed by Manzhos and Carrington[76,77] simultaneously optimize the non-linear parameters **w** and **b** along with the linear parameters **c**. This is done by solving the rectangular "eigenvalue" problem represented by Equation (9-9) for the linear NN coefficients **c**. Because the $\boldsymbol{M}$ and $\boldsymbol{S}$ matrices are rectangular, an approximate solution is sought by using an algorithm developed by Boutry et al.[270] that minimizes

$$\min\frac{\left\|\left(\boldsymbol{M}-E_k\boldsymbol{S}\right)\mathbf{c}_k\right\|^2}{\left(1+\left|E_k\right|^2\right)}. \tag{10-13}$$

The Boutry et al.[270] algorithm suggests that an approximate right eigenvector is the right eigenvector of $(M - E_k S)^H (M - E_k S)$ corresponding to the smallest eigenvalue where the superscript "H" denotes the Hemitian conjugate. Once the approximate linear coefficients are obtained, the energies, E_k, and the non-linear parameters $\mathbf{w}$ and $\mathbf{b}$ are optimized by prefitting with the Generalized Pattern Search Positive Basis 2N algorithm and then optimized with the Levenberg-Marquardt algorithm, both of which are available in MATLAB.

In order to increase the accuracy of the procedure without significantly increasing the difficulty of executing the non-linear optimization, Manzhos and Carrington utilize "composite" neurons of variable width. As noted in the discussion following Equation (10-3), the location (center) of the neuron is determined by the w_{ij} whereas the width of the Gaussian is determined by the bias b_i. Thus, "composite neurons" are a group of neurons all with the same center but with different widths. Since for each neuron there is only one bias parameter whereas there are $d\,w_{ij}$ parameters, the composite procedure allows additional neurons to be inserted with an increase of only one non-linear parameter instead of $d + 1$ if both location and width of the neurons are allowed to vary.

10.1.2. Application to H$_2$O

Manzhos and Carrington[77] have applied the general method outlined in Section 10.1.1 to compute the first five vibrational levels of H$_2$O. They employed the PES developed by Jensen[271] to provide the discrete set of potential points $V(^n\mathbf{y})$. The sampling of this surface was done randomly using a sampling weight that favored the selection of points with lower potential energies. Specifically, a randomly selected point $\mathbf{y}$ was accepted if

$$\frac{V_{max} - V(^n\mathbf{y})}{V_{max}} > \xi \tag{10-14}$$

where V_{max} was taken to be 15,000 cm^{-1} and ξ is a random number whose distribution is uniform on the interval $[0,1]$.

The Schrödinger equation was solved in symmetrized Radau coordinates, R, r, $c \equiv \cos\theta$. In atomic units, the kinetic energy operator, T, in these coordinates is

$$2m_H T(r,R,c) = \frac{\partial^2}{\partial R^2} + \frac{\partial^2}{\partial r^2} + \left(R_1^{-2} + R_2^{-2}\right)\frac{\partial}{\partial c}\left(1 - c^2\right)\frac{\partial}{\partial c}, \tag{10-15}$$

where $R = \left(R_1 + R_2\right)/\sqrt{2}$ and $r = \left(R_1 - R_2\right)/\sqrt{2}$ with R_1 and R_2 being the unsymmetrized Radau coordinates. Application of T to Equation (10-4) results in the following expression for the matrix element T_{ni} in Equation (10-11)

$$T_{ni} = \left(2b_i^2/m_H\right)\left(1 - b_i^2\left(w_{i1} - {}^nR\right)^2 - b_i^2\left(w_{i2} - {}^nr\right)^2 + 0.5\left({}^nR_1^{-2} + {}^nR_2^{-2}\right)\right)$$

$$\cdot \quad (10\text{-}16)$$

$$x\left[\,2\,{}^nc\left(w_{i3} - {}^nc\right) + \left(1 - {}^nc^2\right) - 2\left(1 - {}^nc^2\right)b_i^2\left(w_{i3} - {}^nc\right)^2\,\right]$$

The procedures employed by Manzhos and Carrington[77] to perform the RBFNN calculations were as follows: The programming was done in MATLAB. The weight matrix **w** was initialized by setting each row to the coordinate vector of a randomly chosen data point. The ith component of the initial vector **b** was set to the inverse of the distance between the initial position of neuron i (computed from **w**) and the position of the closest initial neuron. The initial b_i of the additional γ_{max} composite neurons with the same center were derived from the original b_i $(i = 1, 2, \ldots N)$ using

$$b(\gamma-2)N+i = \frac{b_i(\gamma_{max}-1)}{1+0.5(\gamma-1)} \quad \text{for } \gamma = 2, 3, \ldots, \gamma_{max}. \qquad (10\text{-}17)$$

The E_k were randomly initialized in the window $(E_k)_{exact} \pm 10$ cm^{-1}, where the $(E_k)_{exact}$ were taken from the computed energy spectrum in Reference 272 using the same PES.[271]

The results obtained by Manzhos and Carrington[77] for H$_2$O are shown in Table 10-1. As can be seen, the accuracy of the RBFNN solution of the Schrödinger equation improves as the number of sampling points is increased. When L reaches 6,000, the average fitting accuracy for the energies reaches about 0.025%. The errors in the wavefunctions when ≈ 25 neurons are present is about 1%.

10.1.3. Summary

When NNs are used as wavefunctions to obtain an approximate solution to the Schrödinger equation, the neurons of the NN play the role of the basis set expansion functions in terms of the usual methods. The most important conclusion resulting from these initial attempts[174-176,76,77] to execute such calculations is that the number of neurons required to obtained an accurate solution is much less than the number of basis set functions that would be required in the usual variational calculation. This reduction is realized because of the flexibility of the NN that is not present in the usual basis set expansion functions.

The work of Manhzos and Carrington[76,77] has shown that use of an RBFNN is preferable to an NN using sigmoid transfer functions because an RBFNN ensures that the boundary conditions are satisfied and thus obviates the need for additional multiplier functions to achieve this objective.

In addition to the reduction in the number of basis functions required, other advantages of the method include the following:

1. No integrations are required.
2. Because no integrations are executed, there is no need to fit a PES to a discrete of energies computed by electronic structure calculations. Only the energies for the molecular configurations in the database are needed. No derivatives of these energies need be computed.

Table 10-1 Quality of the RBFNN fits to the Schrödinger equation for the first five vibrational levels of water. Energies differences between those obtained from the RBFNN fit and the values given in Ref. 272 are expressed in cm^{-1}. The square root of R in Eq. (10-12) is a measure of the accuracy of the fitted wave functions. The fourth column of the table gives this value relative to the system energy in terms of percentage. Data are taken from Ref. 77. The notation NC means the iterative method to obtain the non-linear parameters did not converge.

L No. Points	N, No. Neurons	γ_{max}	$100(\sqrt{R}/E)$	$<|\Delta E|>$
750	10	1	NC	NC
		2	3.67	6.45
		3	3.90	4.22
	20	1	1.74	4.43
		2	0.99	6.80
		3	NC	NC
	40	1	0.76	3.44
	60	1	NC	NC
1500	10	1	NC	NC
		2	3.69	6.11
		3	3.19	4.89
	15	1	2.68	7.38
		2	2.26	5.76
		3	1.98	2.74
	20	1	2.02	7.28
		2	1.19	4.68
	30	1	1.13	3.71
3000	10	1	NC	NC
		2	3.85	4.67
	15	1	NC	NC
		2	2.46	5.28
		3	1.80	3.23
	20	1	2.74	4.38
		2	1.37	3.84
		3	0.96	3.51
	30	1	1.43	3.66
		2	0.66	2.87
	40	1	1.15	2.68
6000	10	1	NC	NC
		2	3.79	4.69
	20	1	2.06	4.58
		2	1.44	2.77
	30	1	1.02	2.46
		2	1.44	2.77
	50	1	0.84	1.82

3. The introduction of composite neurons significantly reduces the computational effort required to fit the non-linear parameters.
4. The computational effort required for parameter determination is further reduced by solving a rectangular matrix problem[270] to obtain the linear parameters.

In their paper,[77] Manhzos and Carrington conclude that "for H_2O, the ideas enable one to obtain a small group of levels at reasonable accuracy from a very small matrix. For small molecules, it might be possible to get better accuracy. For large molecules, it should be possible to achieve the sort of accuracy obtained in this paper by using small basis sets. That will enable one to compute low resolution vibrational spectra or approximate wavefunctions of large molecules."

10.2. PREDICTION OF HIGH-LEVEL ELECTRONIC STRUCTURE ENERGIES FROM HARTREE-FOCK ENERGIES

10.2.1. Introduction

The first nine chapters of this monograph have repeatedly stressed the central importance of obtaining accurate potential surfaces in any MD, MC, or quantum scattering investigation. Over the last decade, it has become possible to obtain the PES and the corresponding force field directly from the results of *ab initio* electronic structure calculations provided the system under investigation is not too large or complex. In order to achieve this goal, it has been necessary to address four, distinct, difficult problems. The first three of these problems are the development of methods for (a) determining the subset of the total configuration space that is important in the reaction dynamics, (b) finding the number and identity of configuration points required to obtain a sufficiently converged PES, and (c) accurately fitting an analytical PES to the database of *ab initio* energies or forces. The NN and other approaches and the sampling methods used to obtain solutions to these three problems have been discussed in detail with numerous examples in Chapters 1 through 9. The fourth problem is the real bottleneck to the execution of *ab initio* investigations of chemical reaction dynamics. There must be sufficient computational power available to execute the required electronic structure calculations at the required level of accuracy. This fourth problem is the focus of attention in this section of Chapter 9.[78,79]

To date, at least five methods have been introduced to obtain accurate analytic fits to a converged database of *ab initio* electron structure energies and gradients. They are (a) moving Shepard interpolation (MSI) techniques,[23-25,86,89-93,144-146] (b) reproducing kernel Hilbert space (RKHS),[26,27] (c) interpolating moving least squares (IMLS)[28-32] and (d) invariant polynomials (IP),[35-37] and (e) neural network (NN) methods. These latter methods have been extensively reviewed and discussed in previous chapters of the monograph.

The level of fitting accuracy achieved using these methods varies with the complexity of the system and the number of fitting points in the database. As the volume of the important regions of configuration space increases, the fitting accuracy generally decreases. Thus, the accuracy for three- and four-body systems restricted to regions near equilibrium and not undergoing chemical reactions is much greater than for such systems undergoing two-center bond scission reactions. If three- or four-center reactions occur, the fitting accuracy will decrease still further. When five- or six-atom systems undergoing multiple simultaneous reactions are involved, fitting generally becomes much more difficult regardless of the fitting method employed. As the number of fitting points in the database decreases, the global accuracy of the fitting will generally decrease.

In Section 10.2, we discuss NN methods that address the last of the difficulties previously cited, namely, the computational time required to execute the required *ab initio* electronic structure calculations at each point in the database. When the system under investigation contains only three or four atoms all lying in the first two periods of the periodic table, high-level *ab initio* calculations can be conducted without excessive difficulty. If a cluster of high-speed workstations is brought to bear, all calculations can usually be completed in a few days or less. However, as the number and the size of the atoms present increases and as the number of reactions and their complexity also increases, the computational requirements for the electronic structure calculations become very large. At present, this is the factor that limits the complexity of reactions whose dynamics can be investigated using *ab initio* methods.

Malshe et al.[78] have shown that it is possible to obtain configuration energies equivalent to those computed at highlevel with extended basis sets by employing NNs and lower-level Hartree-Fock calculations along with higher-level energies computed for a small subset of configurations. In test cases, it was found that the accuracy of the predicted higher-level *ab initio* energies is greater than the NN fitting accuracy for the database of higher-level energies. Although these investigations are in their infancy, the initial results suggest that this may provide a powerful method to substantially circumvent the computational bottleneck to obtaining databases of high-level electronic structure results for many-atom systems undergoing simultaneous, complex reactions. In the same spirit, Balabin and Lomakina[79] have shown that reasonably accurate equilibrium DFT energies can be obtained from lower-level HF calculations for a wide variety of molecules by using appropriately trained NNs.

Section 10.2.2 describes the procedure and its rationale. Section 10.2.3 gives the results of an application of the method to the unimolecular dissociation of the six-atom, vinyl bromide molecule. As has been previously discussed in other chapters, this dissociation is very complex in that it involves four, distinct two-center bond scission reactions and two distinct three-center reactions. Consequently, the configuration space that must be adequately represented by the *ab initio* database is very large and complex. The results for several cases are presented to investigate how the predictive error varies with the parameters of the procedure. Section 10.2.4 describes an NN method that permits equilibrium DFT energies for a wide variety of molecules to be obtained from HF energies using NNs that have been appropriately trained. Section 10.2.5 presents a discussion and evaluation of the method proposed. The results are summarized in Section 10.2.6.

10.2.2. Concepts and General Procedures

The basic problem that impedes the computation of *ab initio* databases that can be employed with MSI, IMLS, NN, or other methods to obtain potential surfaces for conducting MD or MC studies of reaction dynamics is that the number of configurations whose energy must be computed increases rapidly as the number of atoms, their complexity, the total energy, and the complexity of the reactions the system is undergoing increase. This is simply a consequence of the fact that the number of electrons in the system and the volume of configuration space that must be sampled to obtain a converged PES increase rapidly as these factors increase

Table 10-2 gives the total number of configurations required for convergence in some of the systems that have been investigated to date. As can be seen, when only

Table 10-2 Database sizes required for convergence of some systems whose dynamics have been investigated using *ab initio* methods.[78] (Reprinted with permission from American Institute of Physics.)

System	N^a	*Reference*
H_3	550	144
H_2Br	3,000	122
$BeH_2 + H$	438	91
$OH + H_2$	400	86
$NH + H_2$	600	23
$H_3^+ + O$	1,000	90
BeH_3	500–1,000	93
CH_3^+	500–1,000	93
$HOOH$	6,489	94
H_2CN	830	31
$HONO$	21,584	125
$HONO$	20,192	147
$HOOH$	25,608	124
$HOOH$	$> 10^4$	125
$H_2O_3^+$	66,965	34
$HOONO$	55,471	35
$H + CH_4$	20,728	39
H_5^+	105,888	37
C_2H_3Br	71,969	126
$H_5O_2^+$	48,189	36

(a) Number of configuration points employed in the *ab initio* database.

three- or four-atom systems undergoing only a single two-center bond dissociation reaction are studied, the database usually contains less than the energies for 1,000 configurations. The largest database for such systems seems to be 6,489 configurations employed by Kawano et al.[94] to investigate the O-O bond dissociation dynamics for HOOH. When a four-atom system is undergoing simultaneous two-center and four-center reactions, as is the case for HONO, the size of the required database rises to about 20,000.[125,147] Huang et al.[34] had to employ a database containing the energies of 66,965 $H_3O_2^-$ molecular configurations to study the vibrational structure of that five-atom molecule. To investigate the association reaction of OH with NO_2 to form HOONO, Chen et al.[35] used a database comprising 55,471 configurations. Xie et al.[37] used a database containing 105,888 points to obtain an extremely accurate PES for H_5^+. When Malshe et al.[126] investigated the dissociation dynamics of vinyl bromide with 4 two-center and 2 four-center reactions occurring, the database required for convergence contained the energies of 71,969 configurations. It is reasonably clear that *ab initio* MD or MC studies of complex reaction dynamics in systems containing seven or more atoms will require databases containing the energies of more than 10^5 configurations.

If the electronic structure calculations need to be conducted at high level with large basis sets, the computational requirements will be very large. The CPU time required for computation of the database for vinyl bromide serves as an illustration. The electronic structure calculations[141b] for this system were carried out at the UHF-Stable = Opt/MP4(SDQ) level of accuracy using a 6-31G(d,p) basis set for the carbon and hydrogen atoms and Huzinaga's (4333/433/4) basis set augmented with split outer s and p (43321/4321/4) orbitals along with a polarization f orbital for the bromine atom.[126] Using a workstation with a 1.66 GHz clock speed, the calculation of the energy for each configuration took an average of 82 s. For the entire 71,969 database, the total CPU time was about 1,639 hours. The most accurate fits to *ab initio* data are achieved by fitting the gradients of the potential as well as the potential itself.[122,124] If the gradients for vinyl bromide are computed at each point, the required CPU time per configuration rises to 220 s. The entire database, therefore, requires 4,398 hours of CPU time when the CPU clock speed is 1.66 GHz.

Since some systems need to be investigated using even higher levels of computational accuracy and more elaborate basis sets than the MP4(SDQ) calculations discussed here, the magnitude of the computational bottleneck is obvious. Even if a cluster of 10 workstations each with clock speeds of 3.3 GHz is brought to bear on the vinyl bromide database, all the computational resources of the entire cluster would be working for 220 hours to complete the computations of the gradients. Even more time would be needed for more complex systems requiring even higher levels of *ab initio* accuracy.

If the same basis set is employed, HF calculations of the energy of a particular vinyl bromide configuration requires only 14.6% as much CPU time as the corresponding UHF-Stable = Opt/MP4(SDQ) computation. For gradient calculations, this percentage is 13.6% using the GAUSSIAN-03 suite of programs.[141b] If HF calculations could be substituted for the UMP4(SDQ) calculations, the computational requirements would drop by a factor between 6.9 and 7.3. For higher-level calculations, this factor would be correspondingly greater. Moreover, if HF calculations using much smaller basis sets could be substituted for UMP4(SDQ) or higher-level computations with much larger basis sets, the decrease in computational requirements would easily exceed an order of magnitude.

Unfortunately, such a substitution cannot be executed without sacrificing much, if not all, of the accuracy of the final PES and that of the computed dynamics. Since we cannot simply replace higher-level electronic structure calculations with HF results, we need to be able to accurately predict what the higher-level results would be based on a knowledge of the HF energies and forces. If the correlation corrections to HF results produced by UMP4(SDQ) calculations or ones at even higher levels of accuracy were independent of the configuration of the molecular system being investigated, such prediction would be simple. Of course, this is not the case, as shown by the results in Figure 10-1.

Figure 10-1 shows a plot of the computed UHF-Stable = Opt/MP4(SDQ) energies versus the corresponding HF energies for 68,308 different configurations of vinyl bromide using the basis set described in detail in Chapter 5, Section 5.6. If the correlation correction given by the higher-level calculation were dependent only upon the HF energies, all results would fall on a single smooth curve. This is obviously not the case. The corrections are not only a function of the HF energy but also of the configuration of vinyl bromide. For example, when the HF energy is -16.0 eV, Figure 10-1 shows that the UMP4(SDQ) energies vary from about -21 eV to -19 eV depending upon the molecular configuration. At lower energies near equilibrium, this variation decreases because the possible variations in the configuration near equilibrium are much smaller than is the case at higher energies. This behavior is clearly seen in Figure 10-1.

Although no one-to-one correlation between HF and UMP4(SDQ) results exists, the HF results are, nevertheless, clearly highly correlated to the higher-level energies as nearly all 68,308 points fall on a near triangular wedge in Figure 10-1. If there were no correlation, the distribution of points would be random. Malshe

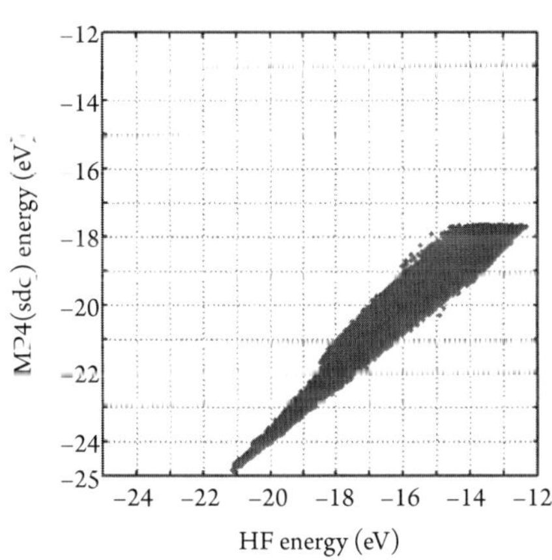

Figure 10-1 Variation of the MP4(sdq) energies as a function of the HF energy in eV for the 68,308 configurations in the vinyl bromide database[78] (reprinted with permission from American Institute of Physics).

et al.[78] have suggested that this behavior will be observed regardless of what higher-level method is employed to obtain the configuration energies and forces.

If this supposition is correct, then all that is required to accurately predict higher level electronic structure results from HF calculations is the implementation of a method that can sense and map the complex, non-linear, underlying correlations that must exist between the higher-level results and the variables describing the molecular configuration along with the HF results. Hornik et al.[108] have shown that two-layer, feedforward NNs with sigmoid transfer functions in the hidden layer and a linear one in the output layer provide such a generalized method since they are universal approximators for analytic functions.

Based on this concept, Malshe et al.[78] have proposed a general procedure for the development of large *ab initio* databases for use in the development of accurate *ab initio* potential-energy surfaces and force fields for an N-atom system.

With N atoms present, the internal configuration of the system can be specified using a minimum of $3N-6$ generalized coordinates. However, when using an NN, there is no requirement that the minimum set be used. Sometimes, it may be convenient to employ the complete set of bond distances or inverse bond distances. Such a redundant set would comprise $N(N-1)/2$ coordinates. For a set of K coordinates that specifies the molecular configuration, we denote those coordinates as p_i ($i = 1, 2, 3, \ldots, K$). For a given basis set, let the computed HF energy of the configuration specified by the p_i ($1 \leq i \leq K$) be p_{K+1}. For a molecular system whose configuration is to be specified by K coordinates, the gradient at the HF level of computational accuracy will contain K components, $p_{K+1+n} = (\partial V / \partial p_n)$ for ($n = 1, 2, 3, \ldots, K$), where V is the computed HF energy of the molecular configuration. Therefore, we define the input vector to our two-layer, feedforward NN as $\mathbf{P}$ where the elements of $\mathbf{P}$ are $(p_1, p_2, p_3, \ldots, p_{K+1})$ if we intend to predict only the higher-level energy. If it is desired to predict the higher-level gradient as well as the energy, $\mathbf{P}$ becomes a $2K+1$ column vector whose elements are $(p_1, p_2, p_3, \ldots p_{2K+1})$.

The number of neurons in the hidden layer of the NN is adjusted to fit the complexity of the problem. We denote its value by S. The second layer is the output layer. It will contain a single neuron that will provide the predicted higher-level *ab initio* energy of the configuration. If both the energy and the gradients are desired, the output layer will contain $K+1$ neurons.

Let us assume that using sampling trajectories combined with novelty sampling or some other suitable sampling method,[25,136, 124,147] we have determined that the *ab initio* database needs to contain the high-level energies and perhaps the corresponding gradients for M previously selected configuration points. The NN is trained by first selecting the *ab initio* computational level and basis set desired. A fraction f of the M configuration points at which electronic structure calculations are executed at the chosen level with the selected basis set is then selected. These calculations will produce both the HF and higher-level energies and gradients. This set of $f \times M$ results is divided into training, validation, and testing sets in the usual manner[105,117,118] and the NN is trained using any appropriate method to fit the target higher-level *ab initio* energies and perhaps gradients.

Subsequent to training the NN, the higher-level energies and perhaps the gradients of the remaining $(1\text{-}f)M$ configurations are predicted by computing the HF results for these points to obtain the required input vector for the NN. The NN is then employed to predict the higher-level results for these $(1\text{-}f)M$ configurations.

As described, the effect of the method is to replace the high-level electronic structure calculations at $(1\text{-}f)M$ of the configurations with HF calculations using the same basis set. It may be equally effective to train the NN using HF calculations executed with a much smaller basis set to reduce the computational requirements still further. This approach will require HF calculations using the smaller basis set to be executed at all M points in the database. When the NN is trained, these HF energies and gradients are employed rather than the ones obtained with the larger basis set.

The above method will be effective in substantially reducing the computational requirements associated with obtaining large, high-level *ab initio* databases provided two conditions are met. The first of these is obvious. The high-level energies and gradients predicted using the NN and the computed HF energies must be accurate. Specifically, the error in the predictions must be less than the expected fitting error when MSI, IMLS, IP, RKHS, or NN methods are employed to fit the final database. The second condition is that the fraction f of the M configuration points must be small. If f is on the order of 0.9, the reduction of computational time will be small. If, however, f values of 0.05 to 0.20 are sufficient, very large reductions can be realized. These points are investigated in the next section by application of the general method to the database for vinyl bromide.

10.2.3. Illustrative Application to Vinyl Bromide Database

As discussed in Section 10.2.1, Malshe et al.[78] performed the initial tests of the method described in Section 10.2.2 by applying it to the vinyl bromide database described in Chapter 5, Section 5.6. This database comprises the UMP4(SDQ) and HF energies in 71,696 configurations of vinyl bromide. For purposes of testing the predictive capabilities of the method, a subset of 68,308 of these points was employed.

In this initial test of the method, Malshe et al.[78] have restricted the investigation to the configuration energies and have employed the same basis set for all HF calculations as that used for the UMP4(SDQ) computations. The input vector to the NN contains 16 elements, namely, the 15 interparticle distances and the HF energy of the configuration. In all studies, the predictive, two-layer, feedforward NN was taken to have the architecture (16-80-1).

The method was tested for eight different values of f that span the range from $f - 0.08$ to $f - 0.80$. The actual values of f examined are given in Table 10-3. For each value of f, there are $(f \times M)$ total configurational energies available for training and validation, where $M = 68,308$. Therefore, when $f = 0.20$, there are 13,600 points available for training and validation. The remaining $[(1\text{-}f) \times M]$ points are used for testing the predictive capabilities of the NN. The distribution of the $f \times M$

Table 10-3 Median error values, RMSE and MAE, using nine different random samplings for eight different values of f for the 68,308 point database for vinyl bromide. All errors are given in units of eV. In each case, f = (% train + % val)/100. The input vectors for the NN for each case comprises the 15 bond distances and the HF energy for vinyl bromide.[78] (Reprinted with permission from American Institute of Physics.)

% Train	% Val	% Test	RMSE Train	RMSE Val	RMSE Test	MAE Train	MAE Val	MAE Test
5	3	92	0.0518	0.1160	0.1247	0.0376	0.0644	0.0634
10	5	85	0.0577	0.0992	0.0994	0.0405	0.055	0.0549
15	5	80	0.0601	0.0833	0.0907	0.0398	0.0492	0.0498
20	5	75	0.048	0.0745	0.0762	0.0331	0.0425	0.0429
30	10	60	0.042	0.0609	0.0622	0.0286	0.035	0.0352
40	15	45	0.0419	0.0518	0.0545	0.0296	0.0334	0.0335
50	20	30	0.0414	0.0486	0.0518	0.0288	0.0346	0.0351
60	20	20	0.0387	0.0468	0.0480	0.0265	0.0293	0.0292

points between training and validation sets is also given in Table 10-3. In general, the percentage of the $f \times M$ points employed as the validation set varied from 20% to 37.5%.

After scaling of all input and output data using Equation (3-34), for each value of f investigated, the M-point database was randomly sampled nine times to evaluate the expected sampling error of the overall method. For each of the nine random samplings, five NNs were trained with different initial guesses for the weight and bias matrices of the network to avoid convergence to a local minima. The NN producing the most accurate testing set results of the five was selected as the NN for that particular sampling of the database.

The predictive accuracy of the (16-80-1) NNs resulting from each of the nine random data samplings are evaluated by computation of both the root mean square error (RMSE) and the mean absolute error (MAE), where

$$RMSE = N^{-1} \left\{ \sum \left[V_{NN}^{MP4} - V_{Ab\ Initio}^{MP4} \right]^2 \right\}^{1/2} \tag{10-18}$$

and

$$MAE = N^{-1} \sum \left| V_{NN}^{MP4} - V_{Ab\ Initio}^{MP4} \right| \tag{10-19}$$

where the summations run over the N data points in the testing set. V_{NN}^{MP4} and $V_{Ab\ Initio}^{MP4}$ are the MP4(SDQ) energies predicted by the (16-80-1) NN and the *ab initio* electronic structure calculations, respectively. In most cases, it will be seen that RMSE > MAE because of the added weight given to outlier points in the database that are often near the edges of the configuration space sampled where the NN errors are generally larger. Since the MAE weights all errors equally, we generally employ this as a more meaningful measure of the overall NN fitting accuracy. Typical results for each of the NNs resulting from the nine random samplings are shown in Table 10-4 for three cases when $f = 0.15$, 0.40, and 0.70.

The overall accuracy of the method is taken to be that for the NN producing the median MAE out of the nine NNs resulting from the random sampling. These results for the eight values of f investigated are given in Table 10-3. For completeness, Malshe et al.[78] reported RMSE and MAE values for training, validation, and testing sets in each case. The evaluation of the predictive accuracy of the method is measured by the results for the testing sets.

The expected sampling error can be evaluated by computing the standard deviation of the MAE of the nine NNs from the median results given in Table 10-4. Table 10-4 reports three typical examples for $f = 0.15, 0.40$, and 0.70. As can be seen, the errors due to sampling vary from 0.000792 eV to 0.001983 eV for $f = 0.70$ and $f = 0.15$, respectively. These values correspond to 0.076 kJ mol^{-1} and 0.191 kJ mol^{-1}, respectively.

Table 10-4 Typical MAE results obtained for each of nine different random samplings of the vinyl bromide database for $f = 015, 0.40$, and 0.70. All errors are given in units of eV. The standard deviations of the nine results from the median results reported in Table 10-4 are given at the bottom of each column. These deviations are a measure of the expected sampling error of the method.[78] (Reprinted with permission from American Institute of Physics.)

	F		
NN	*0.15*	*0.40*	*0.70*
1	0.0565	0.0342	0.0351
2	0.0527	0.036	0.0353
3	0.0549	0.0352	0.0349
4	0.0556	0.0350	0.0345
5	0.0509	0.0348	0.0360
6	0.0529	0.0372	0.0358
7	0.0568	0.0358	0.0346
8	0.0552	0.0347	0.0335
9	0.0531	0.0361	0.0358
Standard deviation	0.001983	0.000917	0.000792

Figure 10-2 shows the spread of the differences between the MP4(SDQ) energies predicted by the median NN with $f = 0.15$ and 10% of the database used for training and the corresponding energies obtained from MP4(SDQ) calculations using GAUSSIAN-03.[141b] The spread increases significantly at larger energies due to much wider spread of vinyl bromide configurations that have a given HF energy, each with differing amounts of correlation energy. The distribution of the NN predictive errors cannot be evaluated from Figure 10-2 because many points are superimposed. This distribution is shown in Figure 10-3. The MAE for the median NN is 0.0549 eV.

The corresponding results for NNs trained with 60% of the database and $f = 0.80$ are shown in Figures 10-4 and 10-5, respectively. Here, the spread of differences and the MAE for the median NN are much smaller due to the larger training set. The increased accuracy is, however, obtained at the expense of greatly increased computational effort.

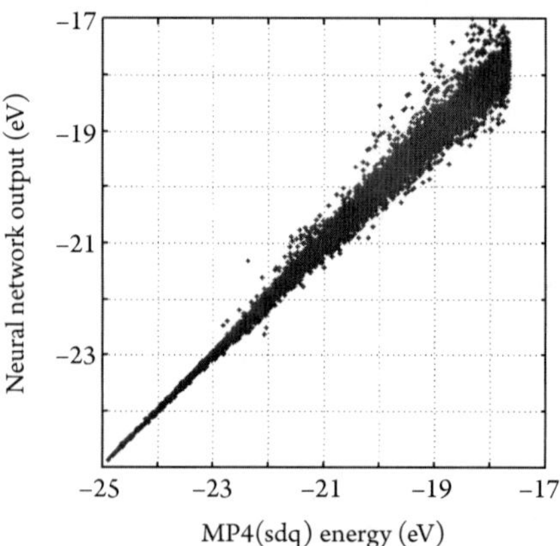

Figure 10-2 Comparison of the spread of the differences between the predicted MP4(SDQ) energies obtained from the median network with $f = 0.15$ and the computed *ab initio* MP4(SDQ) energies in eV[78] (reprinted with permission from American Institute of Physics).

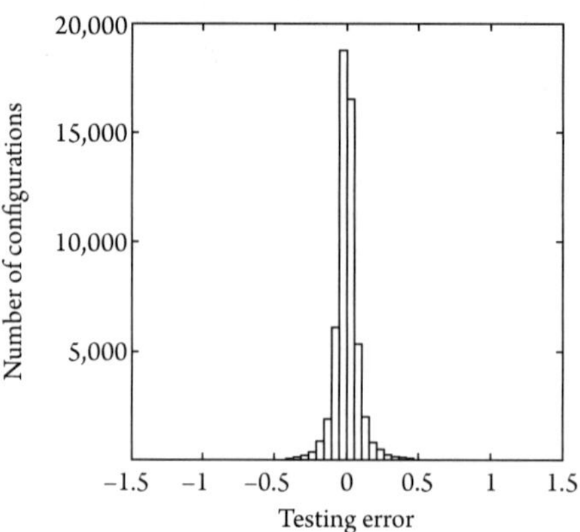

Figure 10-3 Distribution of the predicted MP4(SDQ) energies obtained from the median NN with $f = 0.15$ and the computed *ab initio* MP4(SDQ) energies in eV using GAUSSIAN-03. The MAE of the distribution is 0.0549 eV[78] (reprinted with permission from American Institute of Physics).

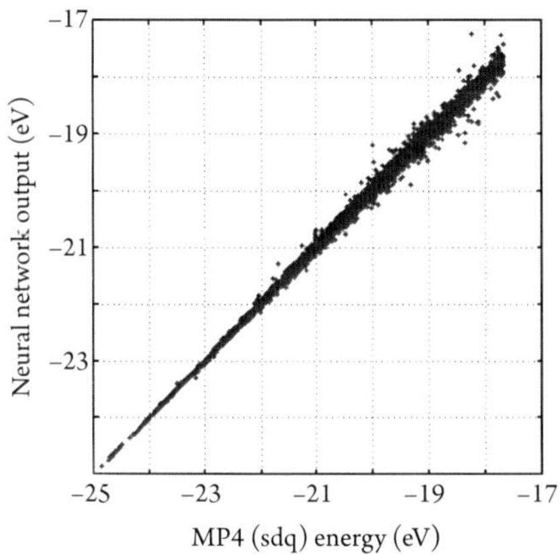

Figure 10-4 Comparison of the spread of the differences between the predicted MP4(SDQ) energies obtained from the median network with $f = 0.80$ and the computed *ab initio* MP4(SDQ) energies in eV[78] (reprinted with permission from American Institute of Physics).

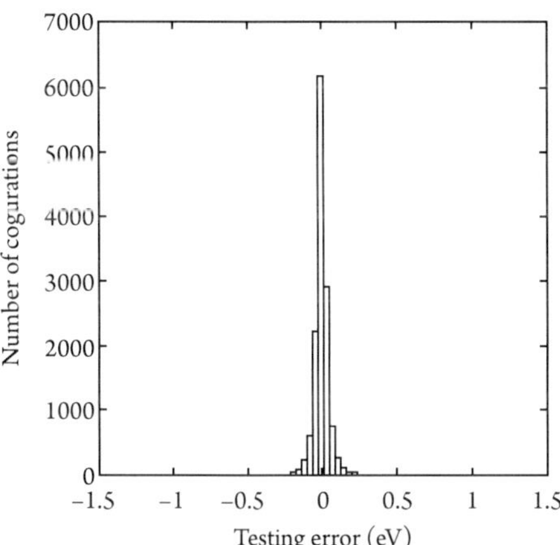

Figure 10-5 Distribution of the predicted MP4(SDQ) energies obtained from the median NN with $f = 0.80$ and a training set that comprises 60% of the database and the computed *ab initio* MP4(SDQ) energies in eV using GAUSSIAN-03. The MAE of the distribution is 0.0292 eV[78] (reprinted with permission from American Institute of Physics).

10.2.4. Application to Equilibrium Energies of Molecular Systems

The concept of obtaining higher-level electronic structure energies from lower-level calculations using NNs has been utilized by Balabin and Lomakina[79] to obtain improved results for the equilibrium structures of 208 organic molecules. Specifically, a variety of DFT results using 6-311G(3df,3pd) and 6-311G(2df,2pd) basis sets are computed using NNs with input vectors containing various molecular descriptors along with the DFT energies computed using basis sets no larger than 6-311G**. The authors call the former basis sets "large basis sets" (LBS) and the latter "small basis sets" (SBS).

These calculations are similar to those described in Sections 10.2.1 through 10.2.3 in that they employ the power of NNs to sense and map the correlations that exist between lower-level electronic structure energies and those computed with higher-level methods. In effect, the NN provides a means to deduce the missing electronic correlation energy and incorporate it into the final predicted result.

There are some differences between the studies reported by Balabin and Lomakina[79] and those reported by Malshe et al.[78] Whereas Malshe et al. utilize NNs and HF energies for vinyl bromide to predict the corresponding MP4 energies for the same system with the same basis set, Balabin and Lomakina[79] use DFT energies obtained for equilibrium structures with an SBS to predict the equilibrium DFT energies that would be obtained if an LBS were employed. The concept of using NNs to execute the predictions is the same in both cases. Neural networks can sense the underlying correlations between the two results and accurately map them. Although the concept and approach are the same in both cases, the problems that arise in the two calculations are different. These points are discussed later.

As a representative example of the method, Balabin and Lomakina attempt to predict DFT energies for 208 organic molecules using an LBS by employing an NN with one or two DFT results computed with an SBS and some molecular descriptors as the input vector. This set of molecules was restricted to contain only carbon, hydrogen, nitrogen, oxygen, and fluorine atoms or a subset of these atoms.

The first problem faced by the authors is that the appropriate nature of the molecular descriptors is not clear. This is in sharp contrast to the situation in the Malshe et al[78] studies where the input elements are a set of coordinates that completely specifies the configuration of the molecule along with the lower-level HF energy of the system. Balabin and Lomkina[79] have investigated two sets of molecular descriptors. The first set, termed CD by the authors, comprises the total number of atoms and the mole fractions of each allowed atom type for a total of six descriptors. The second group, termed QD, contained four descriptors, the HOMO-LUMO energy gap, the molecular dipole moment, the average polarizability $<\alpha>$, which was defined to be

$$\langle \alpha \rangle = \frac{\left(\alpha_{ss} + \alpha_{yy} + \alpha_{zz} \right)}{3}, \tag{10-20}$$

where the α_{ii} are the diagonal elements of the polarizability tensor, and the average quadrupole moment, $<Q>$, which is defined in a manner analogous to Equation (10-20) but using the quadrupole moment tensor. The elements of the QD input vector were computed using a 6-311G basis set.

A two-layer NN was employed in the investigation. The input vector contains the elements of either the CD or the QD descriptors along with the calculated DFT energies of the molecule computed using either a 6-311G, 6-311G*, or a 6-311G** basis set. In some calculations, two of these energies were employed as elements of the input vector. The output layer contained one neuron that gives the LBS DFT result. Hyperbolic tangent and linear transfer functions were used for the hidden and output layers, respectively. The NN training employed the Levenberg-Marquardt algorithm that minimizes the RMSE of the NN prediction relative to the training set. The 208-point database was divided into testing set (40), cross validation set (40), and training set (128). This division was randomly sampled 20 times. Average results were obtained by removing the five worst fits and averaging the remainder.

The number of neurons was optimized empirically. It was found that two to five neurons produced the best results.

The reason the NN must be chosen to contain only a few hidden neurons lies in the very restricted size of the database. With only 208 points available, if more neurons or a more extensive set of molecular descriptors is employed, overfitting will become a problem.

The relative importance of the various input parameters was investigated using cross-validation methods. As expected, the most important input elements are the SBS DFT energies. Similar results were obtained by Malshe et al.[78] These results will be described in Section 10.2.5.

Table 10-5 presents 31 benchmark results of the calculations as reported by Balabin and Lomakina.[79] In this case, all energy values are given in kcal mol^{-1}. The authors have also reported NN predictions for HF, BLYP, and BMK calculations. The reader may consult Table VII of Reference 79 for these data.

On the basis of these results, Balabin and Lomakina[79] conclude the following: (Comments by the authors of this monograph are in parentheses.)

1. The NN approach is able to provide accurate approximations to the DFT results using large basis sets, $\lfloor$6-311G(3df,3pd)$\rfloor$ or $\lfloor$6-311G(2df, 2pd)$\rfloor$ using small basis set DFT data up to 6-311G** and molecular descriptors. The average predictive accuracy is 0.6 ± 0.2 kcal mol^{-1}.
2. The use of more than one SBS DFT result significantly decreases the predictive error.
3. Comparison calculations indicate that the CD descriptors are more effective than the QD set. (This could be due to the fact that the CD set has six elements and more parameters than the QD set, which possesses only four elements.)
4. The optimum NN architecture shows almost no dependence upon the DFT functional or BS. (However, it probably would show a strong dependence upon the size of the database employed in such calculations.)

In addition to these studies,[79] several analogous investigations employing NNs to eliminate errors between quantum results and experimental data have been reported.

Hu et al.[273] and Duan et al.[274] have demonstrated that NNs can dramatically reduce the differences between DFT/B3LYP results and experimentally measured heats of formation. Using a database containing results for 350 organic molecules, they found an error reduction from 21.4 to 3.1 kcal mol^{-1}.

Wu and Xu[275] employed NNs with DFT/B3LYP results and reduced the difference between theoretically predicted heats of formation and experiment for the molecules in the G3/99 set to an RMSE of 1.34 kcal mol^{-1}. When the same procedure was applied to the computation of bond dissociation energies[276] the accuracy obtained was ± 2.45 kcal mol^{-1}.

Li et al.[277] used a (9-5-1) NN to reduce the RMS error in the calculated absorption energies for 150 organic molecules obtained with TDDFT/B3LYP methods and a 6-31G(d) basis set from 10.8 kcal mol^{-1} to 5.1 kcal mol^{-1}.

Table 10-5 NN Predictive results for 31 DFT/B3LYP calculations with 6-311G(3df,3pd) basis sets using CD molecular descriptors and two DFT results with smaller basis sets. All energies are in kcal mol^{-1}. Data are taken from Table VII in Reference 79.

Molecule	DFT/B3LYP/LBS	NN Prediction
eclipsed Hexafluoroethane	3.45	2.65
Methylisocyanate (linear barrier)	1.70	2.20
cis-1,1-difluoroethane	2.72	3.26
gauche-1,2-difluoroethane	−1.04	−0.21
cis-acetaldehyde	1.23	1.19
cis-trans-acetic acid	5.34	5.91
trans-gauche-acetic acid	0.46	0.41
cis-methyl formate	−5.07	−5.43
cis-trans-2-fluoroethanol	2.36	2.30
cis-cis-2-fluoroethanol	3.73	4.07
eclipsed ethane	2.64	2.74
cis-dimethyldiazene	6.92	7.25
cis-trans-ethanol	1.37	1.31
trans-gauche-ethanol	3.05	3.47
dimethyl ether	32.89	33.00
cis-cis-ethylene gycol	3.63	3.68
cis-ethyl hydroperoxide	5.27	5.23
cis-acrolein	2.06	1.74
gauche(+)-*gauche*(−)–1,3-difluoropropane	−2.37	−2.35
cis-allyl alcohol	1.84	1.92
cis-propanal	−1.86	−2.21
gauche-butane	0.85	0.83
cis-formic acid	−4.03	−5.87
methane flat	136.94	133.75
cyanic acid	24.26	26.21
water	30.83	32.21
ammonia flat	5.32	6.00
N_2O_2	10.44	11.27
HCOOH→H_2O + CO	10.01	8.98
CO + 0.5 O_2 → CO_2	−91.20	−90.74
1-butene → *trans*-2-butene	−3.36	−3.33
Average absolute deviation = 0.57		

Other analogous applications of NNs to reduce the computational errors in various electronic structure calculations by training a set of results to experimental data include the approximation of C-H bond dissociation enthalpies,[278] Gibbs free energies of formation,[279] absorption wavelengths,[280,281] ionization energies, and electron affinities.[281]

10.2.5. Discussion and Evaluation of the Method

To properly evaluate the method proposed in Sections 10.2.2 and 10.2.3, the essential question is how accurately can a large NN trained to the entire set of 68,308 configuration energies in the database fit a testing set of vinyl bromide MP4(SDQ) energies using the same basis set. This result has been previous reported by Malshe et al.[126] Using a single, six-body (15-140-1) NN, they were able to fit the vinyl bromide database with an MAE of 0.065 eV. To date, this represents the best result anyone has been able to obtain for this system using any fitting method.[50,71] Consequently, if the median predictive NN is able to predict the MP4(SDQ) energies with an MAE equal to or less than 0.065 eV (6.27 kJ mol^{-1}), the best fitting result yet obtained, the method can be utilized with confidence to predict MP4(SDQ) energies using the HF results and thereby avoid the computational expense of actually executing the higher-level electronic structure calculations.

The results shown in Table 10-2 indicate that this level of accuracy is achieved when about 5% of the 68,308 points in the database are utilized for training the predictive NN. An additional 3% of the points must be used to provide a validation set to avoid overfitting. This means instead of having to compute the MP4(SDQ) energies at all 68,308 points, essentially the same accuracy could be achieved by computing the MP4(SDQ) energies for 5,465 configurations and just the HF energies for the remaining 62,843 configurations. On the machines employed by Malshe et al.[78] in the computations, each MP4(SDQ) and HF calculation using the basis set previously described required 82 and 12 s, respectively. Therefore, for this system, the NN predictive method converts 1,556 hr of computational time to 334 hr, a reduction of about 78%. While the actual computational times will vary with CPU speed and computer configuration, the percentage of time saved will be nearly independent of those factors.

It should be noted that the sampling error is small compared to the predictive error and, therefore, plays very little role in the evaluation of the utility of the method.

The reason the method works so well may be seen by evaluation of the sensitivity of the RMSE and MAE results to each of the 16 input elements of the NN. The first 15 specify the configuration of the NN. The last element, number 16, is the HF energy for the given configuration. The sensitivity of the MAE for the NN with respect to input element i, S_i, is defined to be the relative rate of change of the computed mean squared error (MSE) for the NN with respect to p_i normalized so that the largest sensitivity, $[S_i]_{max}$, is equal to unity. That is

$$S_i = \frac{\partial}{\partial p_i} N^{-1} \sum \left[V_{NN}^{MP4} - V_{Ab\,Initio}^{MP4} \right]^2 \Big/ [S_i]_{max}. \qquad (10\text{-}21)$$

The sensitivities for two cases, $f = 0.15$ and $f = 0.80$, are given in Table 10-6. The definitions of the input coordinates follow the atom numbering shown in Figure 10-6. As can be seen, when $f = 0.15$, the sensitivity of the MSE of the NN

Table 10-6 Sensitivity of the computed MSE of the NN to each of the 16 input elements for the median NNs trained with $f = 0.15$ and 10% of the database used for training and with $f = 80\%$ and 60% of the database used for training. The results are all normalized with the largest sensitivity being set equal to unity. The notation for the input elements follows the atom numbering given in Figure 10-6.[78] (Reprinted with permission from American Institute of Physics.)

	Sensitivity	
Input	10% Data Training	60% Data Training
r_{1-2}	0.196	0.368
r_{1-3}	0.428	0.118
r_{1-4}	0.522	0.685
r_{1-5}	0.265	1
r_{1-6}	0.042	0.156
r_{2-3}	0.170	0.036
r_{2-4}	0.570	0.248
r_{2-5}	0.084	0.693
r_{2-6}	0.187	0.173
r_{3-4}	0.775	0.133
r_{3-5}	0.533	0.168
r_{3-6}	0.510	0.275
r_{4-5}	0.274	0.504
r_{4-6}	0.441	0.118
r_{5-6}	0.365	0.067
HF (eV)	1	0.812

to the HF energy is by far the largest of the S_i for all input elements. Similar results were found by Balabin and Lomakina.[79] Physically, this means the HF energy is the most strongly correlated input element to the MP4(SDQ) energy. When it is present, it is far easier for the NN to accurately predict the MP4(SDQ) energies for a testing set than is the case when it is absent and only the configuration variables are present as input elements. As previously discussed, this factor can be used to our

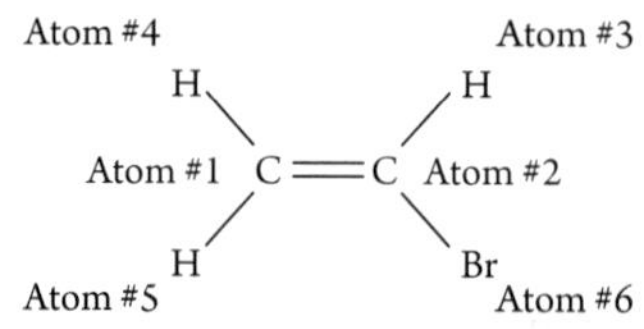

Figure 10-6 Atom notation used to specify the NN input elements for vinyl bromide[78] (reprinted with permission from American Institute of Physics).

advantage to greatly reduce the computational requirements for execution of the electronic structure calculations.

The sensitivities shown in Table 10-6 for the case with $f = 0.80$ indicate that as the size of the training set approaches the total size of the database, the sensitivity of the predicted MP4(SDQ) energy to the HF energy decreases. When $f = 0.80$, the HF energy is no longer the most strongly correlated input element to the computed MAE. This is not unexpected in view of the fact that if $f = 1$, there is no need to use the HF energy at all since all MP4(SDQ) energies will have been computed.

The importance of having the HF energy of the configuration in the input vector to the NN can be clearly seen by the results shown in Table 10-7. This table gives the median RMSE and MAE fitting errors when $f = 0.15, 0.25, 0.40$, or 0.70 when the predictive NN contains the 15 configuration coordinates but not the HF energy of the configuration. A comparison with the data for $f = 0.15$ given in Table 10-2, where the input vector contains the HF energy, shows that the RMSE and MAE median errors rise by factors of 3.03 and 2.76, respectively. With $f = 0.15$ and only 10% of the database in the training set, the 6,830 points in the training set are insufficient to adequately sample the configuration space of the vinyl bromide system. Hence, the fitting errors increase dramatically unless the HF energy is included as the 16th element of the input vector.

In the present evaluation of the method, Malshe et al.[78] have computed all 68,308 MP4(SDQ) energies to provide a benchmark for evaluation of the proposed method. In an actual application, the investigator will initially not have any sampling of configuration space executed nor will any high-level electronic structure calculations have been executed. If it is desired to utilize the present predictive NN method to obtain high-level energies from HF results and thereby reduce the computational requirements of the investigation, a reasonable 8-step

Table 10-7 Median error values, RMSE and MAE, using nine different random samplings for four different values of f for the 68,308 point database for vinyl bromide with a NN that does not contain the HF energy of the configuration as one of the input elements. Therefore, the (15-80-1) NN has only the 15 configuration coordinates in the input vector. All errors are given in units of eV. In each case, $f = (\% \text{ train} + \% \text{ val})/100$.[78] (Reprinted with permission from American Institute of Physics.)

% Train	% Val	% Test	RMSE Train	RMSE Val	RMSE Test	MAE Train	MAE Val	MAE Test
10	5	85	0.1436	0.3236	0.3010	0.0996	0.1521	0.1717
20	5	75	0.1108	0.1993	0.2133	0.0762	0.1005	0.1018
30	10	60	0.1223	0.1505	0.1559	0.0792	0.0879	0.0902
50	20	30	0.0945	0.1497	0.1686	0.0623	0.0758	0.0808

procedure (see Figure 10.7 for the flow chart) for the overall investigation might be as follows:

1. Adequately sample the configuration space of the system that is important in the reaction dynamics to obtain a database with M total configurations present. This can be done using several procedures. The two that follow lend themselves well to the present method:

   (A) Employ a semi-empirical PES with a novelty-sampling (NS) selection procedure[136] until the novelty sampling (NS) criteria for convergence of the PES are satisfied.

   (B) Employ direct dynamics[147] with a novelty-sampling selection procedure[136] until the novelty sampling criteria for convergence of the PES are satisfied. If this is done, full use of all the sampling methods described in Reference 147 will need to be employed.

2. Choose the value of f to be employed in the study.
3. Randomly select $f \times M$ points from the database.
4. Execute higher-level electronic structure calculations at the desired level of accuracy for the configurations in the $f \times M$ data set. Fit an NN of appropriate size to the database containing $f \times M$ configurations using the configuration coordinates and the corresponding HF energies as input vectors and the computed $f \times M$ high-level *ab initio* energies as targets. In this application, it will be necessary to divide the $f \times M$ points into training, validation, and testing sets.

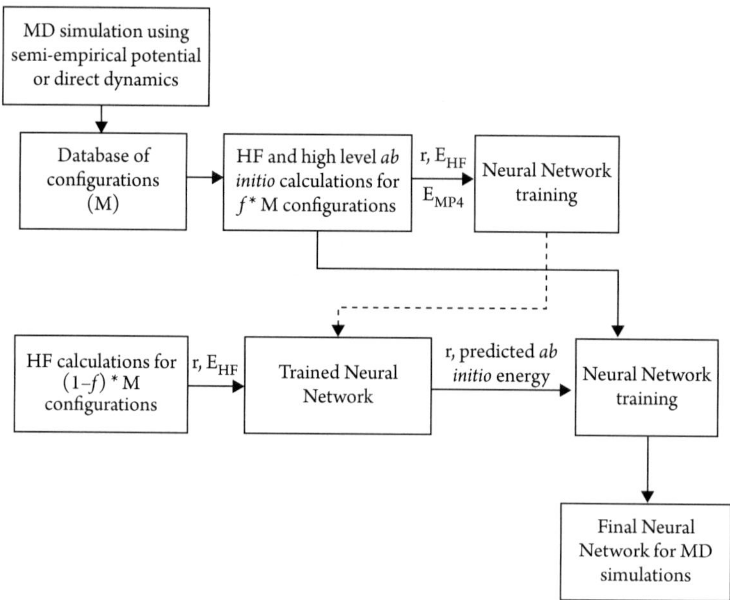

Figure 10-7 Flow chart for obtaining final neural network for MD simulations[78] (reprinted with permission from American Institute of Physics).

5. Compute the HF energies using the same basis set for all $(1\text{-}f) \times M$ configurations not selected in Step 3.
6. Use the NN fitted in Step 4 with an input vector comprising the configuration variables and the HF energies of the $(1\text{-}f) \times M$ configurations to predict the high-level energies for these configurations.
7. Fit a new NN to the database that now contains high-level electronic structure energies for M configurations. In this fitting, the input vector contains only the configuration variables.
8. Use the NN fitted in Step 7 to execute the dynamics studies.

The NN generated in Step 7 of this procedure can be expected to exhibit a fitting accuracy comparable to that which would have been obtained had the higher-level electronic structure calculations been executed for the entire M configurations in the database. As an illustrative example for vinyl bromide, Malshe et al.[78] have executed Step 7 using the MP4(SDQ) data predicted with $f = 0.15$. That is, a database was formed comprising the 10,246 points used for training and validation of the predictive NN with $f = 0.15$ plus the additional 58,062 MP4(SDQ) energies obtained from the predictive NN. This database was divided into training (80%), validation (10%), and testing (10%) sets, and a (15-140-1) NN, which was labeled NN($f = 0.15$), was fitted. A second (15-140-1) NN, which was labeled NN($f = 1$), was fitted to the MP4(SDQ) energies obtained from electronic structure calculations for all M points. The fitting accuracies of these two networks are given in Table 10-8.

As can be seen in Table 10-8, the RMSE and MAE errors for NN($f = 1$) and NN($f = 0.15$) are almost the same. The differences in the RMSE and MAE for the two networks are 0.0073 eV and 0.0013 eV, respectively, with the fitting error for NN($f = 1$) being the smaller of the two. These differences are about 7% and 2%, respectively, of the corresponding RMSE and MAE fitting errors.

It may be possible to make this predictive NN method even more effective. Since the HF energies computed with a smaller basis set should still be highly correlated to energies computed with higher-level electronic structure methods and large

Table 10-8 Testing set errors for the neural networks NN($f = 1$) and NN($f = 0.15$) as described in the text. All errors are given in eV. In each case, the architecture of the networks is (15-140-1).[78] (Reprinted with permission from American Institute of Physics.)

NN	RMSE	MAE
NN(f=1)	0.1062	0.0638
NN(f=0.15)	0.1187	0.0651

basis sets, the HF energies computed using a small basis set could be substituted for HF computations with the larger set. This procedure would further reduce the computational requirements. While the method has been tested only for the case of MP4(SDQ) energies for vinyl bromide, the fundamental basis for the method would seem to be valid for even higher-level *ab initio* methods such as QCISDtq, CCSDtq, or other methods. At the time of this writing, these possibilities were currently under investigation.

10.2.6. Summary

Malshe et al.[78] have shown that if the input vector to an NN includes both the configuration coordinates and the HF energies of a small subset of the configurations needed to adequately characterize the PES over all regions important in the reaction dynamics of a given system, MP4(SDQ) energies with the same basis set can be accurately predicted for the entire database using only the HF and MP4(SDQ) energies for the small subset and the HF energies for the remainder of the database. The predictive error is shown to be less than or equal to the NN fitting error if an NN is fitted to the entire database of higher-level electronic structure energies. The underlying basis for the method is the high degree of correlation existing between the HF and MP4(SDQ) energies.

Application to the 68,308 configurations comprising the database for the simultaneous unimolecular decomposition of vinyl bromide into six different reaction channels indicates that the subset can be as small as 8% of the total number of configurations in the overall database without loss of accuracy beyond that expected if an NN is employed to fit the higher-level energies for the entire database. The utilization of this procedure is shown to save about 78% of the computational time required for the execution of the MP4(SDQ) calculations for all 68,308 configurations in the total database. In this example, the actual CPU time saved would have been about 1,222 hrs for the computers employed in the calculations. It is also shown that a (15-140-1) NN fitted to a database formed by combining the energies obtained from the higher-level electronic structure calculations used to train the predictive NN with those obtained from the predictive NN has nearly the same fitting accuracy as a similar NN fitted to a database obtained by executing higher-level electronic structure calculations at all points.

The sampling error involved with selection of the subset is shown to be about 10% of the predictive error for the higher-level energies.

It is reasonable to expect that because the HF energies will always exhibit a very high degree of correlation with energies computed using higher-level electronic structure methods, this NN predictive procedure should work not only for MP4(SDQ) energies but also for more sophisticated and more computationally expensive methods such as QCISDtq or CCSDtq calculations.

The results obtained by Balabin and Lomakina[79] have demonstrated that equilibrium energies for a wide variety of different molecules computed using DFT methods with large basis sets can be accurately predicted using NN methods with four to six molecular descriptors combined with one or two DFT calculations using much smaller basis sets. The prediction accuracy averaged over 208 molecules is 2.5 ± 0.83 kJ mol^{-1}. When one considers the relatively small database and the wide variety of different chemical structures whose equilibrium energies are being predicted, such accuracy is excellent.

The methods and results presented in this chapter all point in the same direction: Neural networks are powerful tools that substantially reduce the computational effort needed to execute *ab initio* electronic structure calculations.

11

SUMMARY, CONCLUSIONS, AND FUTURE TRENDS

11.1 INTRODUCTION

Since the introduction of classical and semiclassical molecular dynamics (MD) methods in the 1960s and Gaussian procedures to conduct electronic structure calculations in the 1970s, a principal objective of theoretical chemistry has been to combine the two methods so that MD and quantum mechanical studies can be conducted on *ab initio* potential surfaces. Although numerous procedures have been attempted, the goal of first principles, *ab initio* dynamics calculations has proven to be elusive when the system contains five or more atoms moving in unrestricted three-dimensional space.

For many years, the conventional wisdom has been that *ab initio* MD calculations for complex systems containing five or more atoms with several open reaction channels are presently beyond our computational capabilities. The rationale for this view are (a) the inherent difficulty of high level *ab initio* quantum calculations on complex systems that may take numerous, large-scale computations impossible, (b) the large dimensionality of the configuration space for such systems that makes it necessary to examine prohibitively large numbers of nuclear configurations, and (c) the extreme difficulty associated with obtaining sufficiently converged results to permit accurate interpolation of numerical data obtained from electronic structure calculations when the dimensionality of the system is nine or greater.

Neural networks (NN) derive their name from the fact that their interlocking structure superficially resembles the neural network of a human brain and from the fact that NNs can sense the underlying correlations that exist in a database and properly map them in a manner analogous to the way a human brain can execute pattern recognition. Artificial neurons were first proposed in 1943 by Warren McCulloch, a neurophysiologist, and Walter Pitts, an MIT logician.[282] NNs have been employed by engineers for decades to assist in the solution of a multitude of problems. Nevertheless, the power of NNs to assist in the solution of numerous problems that occur in chemical reaction dynamics is just now being realized by the chemistry community. In fact, the first paper utilizing an NN in a problem involving chemical reaction dynamics did not appear in the literature until 1993,[40] a full 50 years after their initial conception in 1943. Seven state-of-the-art review articles in the field of reaction dynamics published in the 70s and 80s do not even mention the term "neural networks."[1-7]

A breakthrough occurred during the past decade, when efforts by a small number of research groups demonstrated that feedforward NNs can play an important role in facilitating the development of potential-energy surfaces (PES) of sufficient accuracy to execute accurate MD, Monte Carlo (MC), or quantum scattering calculations. Specifically, it has been found that NNs can greatly facilitate the use of potential energies and gradients obtained from first-principles electronic structure calculations to develop what are generally termed *ab initio* PESs for use in atomistic simulations. Likewise, NNs have been shown to be highly useful in making other methods, such as many-body expansions and genetic algorithms (GA), far more robust and powerful.

Neural network methods have been developed that greatly assist in the adjustment of parameters of empirical PESs to fit a database of some type. They have also been found to be highly useful in reducing the statistical error that is always present in MD studies. This has been accomplished by employing NNs to predict the results of trajectories without actually executing numerical integrations. When this is done, the results of a huge number of molecular trajectories can be obtained with far less computational effort. The result is a significant reduction in the statistical errors of the calculations and a corresponding increase of predictive accuracy. Most recently, the authors of this monograph have shown that NNs can be used to predict the results of higher-level (e.g., MP4) electronic structure calculations from lower-lever HF calculations. This allows the computational bottleneck of executing the extensive *ab initio* electronic structural calculations to be substantially reduced.

While all these methods have been reported by several research groups in the form of technical publications in leading international journals of the field, there is no succinct compilation of all of these results and methods that clearly describes both the methods and the techniques involved in their application. The discussion in this monograph provides both of these. In addition to the formal descriptions of the methods, example applications are provided to illustrate the power and limitations of NN methods. With the interest in this emerging field growing rapidly, it is clear that this approach will play an increasingly important role in MD, MC, and quantum mechanical studies of chemical reaction dynamics in the years to come.

The main objectives of this monograph are to provide the reader with a clear understanding of NNs along with descriptions of how they can be used to move MD investigations into the realm of *ab initio* calculations. The monograph also provides in detail a description of other methods for obtaining PESs from *ab initio* data and the methods by which NNs can be effectively employed to study non-adiabatic chemical reactions on *ab initio* potentials, to enhance GA calculations, to assist the execution of intramolecular energy transfer studies, to render many-body expansions far more robust and powerful, to predict the results of trajectories without actually integrating the equations of motions, and finally, to show how NNs can play a key role in solving the Schrodinger equation and in determining the electronic structure energies and gradients from high-level computations using primarily calculations conducted at the HF level of theory.

While there are hundreds of technical papers and tens of textbooks published on neural networks and on chemical reaction dynamics per se, this is not the case with the methods and applications of the neural networks in addressing problems in chemical reaction dynamics. Similarly, it would be hard to find even a handful of researchers, if any, with a chemical reaction dynamics background at an NN conference, or, NN researchers at a chemical reaction dynamics conference. This is because the chemical reaction dynamics researchers are, to some extent, unaware of the power of NNs, and NN researchers are unaware of the potential applications of NN to chemical reaction dynamics. As a result, each group in the midst of the other is not comfortable at a conference that is outside their field of activity. It is anticipated that this monograph, *Neural Networks in Chemical Reaction Dynamics,* will bridge the gap between them. In the future, it is anticipated that these two distinctly different research communities will come together, of necessity, to address the most challenging problems in chemical reaction dynamics using NNs.

Almost all of the powerful NN methods applied to chemical reaction dynamics that have been reported to date are from 1995 to the present. A complete literature review of NN methods and applications to the field of chemical reaction dynamics comprises, perhaps, no more than 50 papers. At present, there are hardly any review articles, monographs or books that summarize and detail all of the presently existing NN methods and procedures as they relate to chemical reaction dynamics that clearly describes their power and limitations and the techniques needed to implement the methods. This monograph seeks to fill that void.

The following summary highlights some of the topics covered in this monograph. These include (a) methods other than NNs for obtaining PESs from *ab initio* databases; (b) feedforward NN fitting of databases obtained from *ab initio* electronic structure calculations; (c) combined genetic algorithm and NN methods; (d) NN-assisted intramolecular energy transfer calculations; (e) configuration-space sampling methods; (f) the development of generalized PESs using a combination of many-body expansions, NNs, and moiety energy approximations; (g) the combined function and derivative approximation (CFDA) NN method; (h) data analysis and statistical error reduction; and (i) NN methods in quantum mechanical studies. In the next sections, the highlights of each topic covered in this monograph are summarized followed by some examples of future trends.

11.2. OTHER METHODS FOR OBTAINING PESs FROM *AB INITIO* DATA

Several research groups have developed powerful methods and techniques for obtaining PESs from databases resulting from *ab initio* electronic structure calculations. The first such general method was developed by Collins and co-workers.[23-25] This method employs Shepard techniques combined with a moving procedure that computes appropriate local coordinates at each point in an *ab initio* database and Bayesian weighting methods to accurately interpolate between the points in the database that are obtained by trajectory and other sampling methods. This modified or moving Shepard interpolation (MSI) method has been described in some detail in Chapter 2, which also reports many of the results that have been obtained.

Interpolative moving least squares (IMLS) methods[28-32] have also been discussed in detail in Chapter 2. These methods have the advantage of not requiring first and second derivatives to execute the fitting as does the MSI method. Numerous improvements of the methods have been described in Chapter 2 along with a comprehensive review of the results obtained with their use.

An IP method employing fitting polynomials that are invariant to permutation of identical atoms has been developed by Braams, Bowman, and co-workers.[36,37] Many of the details of the method have been discussed in Chapter 2 and a reasonably complete listing of the results obtained using the IP method have been given.

The results obtained using a reproducing kernel Hilbert space (RKHS) method developed by Rabitz, Ho, Schatz, Hollebeek, and co-workers[26,27,99,100,283-290] were described in Chapter 2. Using this method, the authors have developed perhaps the most accurate analytical PES for H_2O to date.

Finally, Chapter 2 describes a hybrid method that combines IMLS techniques with MSI Shepard interpolation. This IMLS/MSI method was developed by Ishida and Schatz.[33a, 33b] Its application to the $H + H_2$ exchange reaction and to the $O(^1D) + H_2$ exchange to yield $H + OH$ are discussed in the chapter.

11.3. CONFIGURATION-SPACE SAMPLING METHODS

The success or failure of any *ab initio* method to study chemical reaction dynamics depends critically on the sampling method employed to obtain the database of *ab initio* energies and gradients. Chapter 4 describes three robust methods—namely, trajectory and NS, sampling using direct dynamics (DD), and sampling using a gradient fitting method—for the execution of this sampling along with methods for testing for convergence of the database. Some investigators have also employed random configuration space sampling with appropriate constraints.[33a,33b]

The first of these methods is trajectory and NS. The trajectory portion of this method was developed by Collins.[23,25] It utilizes classical trajectories to explore and sample configuration space. The modified NS procedure was developed by Raff et al.[136] using a method originally proposed by Pukrittayakamee and Hagan.[138] Basically, the trajectory portion finds the configurations important in the reaction dynamics, and the NS portion provides a means for accepting or rejecting these configurations that provides a more uniform density of sampling points and incorporates considerations of the surface gradient in making the decision. It also provides a convenient means of testing for convergence.

The second method is sampling using direct dynamics (DD).[147] Trajectory methods previously proposed for obtaining *ab initio* databases are not self-starting. They generally require that the trajectory sampling of configuration space be initiated by using chemical intuition or a previously developed semi-empirical PES. This problem is avoided by using DD to initiate the sampling. The sampling procedure, therefore, becomes self starting. The method is described in Chapter 4, and illustrative examples are given in Chapter 5.

The third method—namely, configuration sampling using a grid fitting method —does not employ trajectory sampling or NS methods. Instead, it begins with a small database obtained by grid sampling methods and proceeds by adding more configuration points based on criteria requiring that the gradients predicted by an NN fitted to the current database accurately match those predicted by the *ab initio* electronic structure calculations.[124] The method, therefore, is essentially a gradient fitting procedure. Since the gradient is the important quantity in an MD simulation, the method is particularly well suited to MD studies. Chapter 4 gives details of the sampling method. Chapter 5 provides an illustrative example of its use.

11.4. FEEDFORWARD NN FITTING OF *AB INITIO* PESs

Since 1993, many investigators have employed NN methods to obtain PESs from *ab initio* databases.[40-48,122,124-126,136,147, 156,162,167,171,173,247] In Chapter 5 we have reviewed a selected sampling of these investigations.

In this chapter, we have reviewed and discussed NN studies involving near equilibrium structures that lead to an elucidation of the vibrational states of the system under study. Applications of this type to the H_3^+ molecular ion, H_2O, H_2O_2, and H_2CO are described.

Following this, we have discussed in detail the application of the CFDA method to the three-body exchange and abstraction reactions H + H'Br $\rightarrow$ HBr + H' and H + H'Br $\rightarrow$ H$_2$ + Br. This study is unique in that it is the first and only example of a fitting procedure that is capable of producing point-by-point matching of all trajectories computed on both the analytic surface and the NN fit to that surface. Therefore, in this sense, the fitting is exact.

Chapter 5 then addresses NN studies of four-body reactions involving *cis-trans* isomerizations and dissociation reactions. Two examples are discussed: the reactions of HONO and H_2O_2.

The most complex, gas-phase system that has been addressed to date is the unimolecular dissociation of vinyl bromide. This process involves six simultaneous, distinct reactions that include four distinct two-body dissociations and two distinct three-body dissociation reactions. Chapter 5 describes the complete details of the NN methods employed to execute these studies.

The investigation of non-adiabatic reactions involving multiple PESs is one of the more difficult topics in chemical reaction dynamics. The methods by which NNs can assist such investigations have been described in Chapter 5 using the dissociation of SiO_2 as an example.

Finally, we have reviewed and described in detail a new NN method developed by Behler and Parrinello[173] that incorporates complete permutation symmetry and permits the investigation of large molecular systems using NNs.

11.5. EXPANSION METHODS AND NNs

A potential may be defined as a sum of two-body, three-body, and many-body interaction terms as shown below:

$$V = \sum_{\substack{i>j}}^{N} f_2(i,j) + \sum_{\substack{i>j \\ k>i,j}}^{N} f_3(i,j,k) + \sum_{\substack{i>j \\ k>i,j \\ l>i,j,k}}^{N} f_4(i,j,k,l) + \cdots + \sum_{\substack{i>j \\ k>i,j \\ l>i,j,k, \\ \cdots \\ m>i,j,k,\ldots,l}}^{N} f_n(i,j,k,l,\ldots,m)$$

where V is the PES for the system and the f_i are expansion functions. Such expansions have been used by numerous research groups since the late 1970s.[52,53] The central reason for the use of such an expansion is an effort to represent an N-body system requiring the specification of 3N-6 coordinates with functions whose dimensionality is smaller. Unfortunately, the success of such expansions has been severely limited because of the unknown character of the expansion functions.

Chapter 6 of the monograph describes two NN methods both of which significantly enhance the power of such expansions. The central concept of both methods is the replacement of the expansion functions with NNs that are universal approximators for analytic functions.

The first method introduced by Manzhos and Carrington[49,50] uses the high-dimensional model representation (HDMR) that has been discussed and described in detail by Rabitz and co-workers[54-70] as the expansion. In this case, the summations of the expansion run over all combinations of the coordinates of the full N-body system. The expansion functions are termed "component functions" and each is characterized by its dimensionality. In previous applications, the parameters of the arbitrary component functions were determined by difficult Monte Carlo integrations over the database. Manzhos and Carrington[49] replaced these functions with NNs fitted by standard NN methods to the database. To simplify the problem further and address the combinatorial problem that plagues all expansion methods, they introduced "lumping," "parameter grouping," and "data blocking" methods.

Chapter 6 describes all of these in detail. The power of their approach is clearly shown by the excellent fit the method produces to the 71,969 point database characterizing the most difficult system yet fitted by *ab initio* methods, namely, vinyl bromide undergoing simultaneous reaction into six different channels.

The second approach, developed independently by Malshe et al.,[71] is similar to that proposed by Manzhos and Carrington[49] in that NNs replace the expansion functions. The primary difference is that instead of using an HDMR in which the summations run over the coordinates, Malshe et al.[71] use a many-body method in which the summations run over the atoms present in the system. This substantially reduces the combinatorial problem. A further reduction in the complexity of this problem is obtained using a "moiety-energy" (ME) approximation.[71] An algorithm is presented that efficiently adjusts all the coupled NN parameters to the database for the surface. This method is applied to four different systems of increasing complexity and the fitting accuracy was found to be good to excellent. Chapter 6 describes both expansion methods in detail.

11.6. GENETIC ALGORITHM AND IVR CALCULATIONS USING NN METHODS

In Chapter 7, it was shown that genetic algorithms (GA), in general, can be used to fit highly non-linear functional forms, such as empirical interatomic potentials from a large ensemble of data. The performance of a GA for fitting such functional forms is enhanced through the use of an NN to accelerate the computation of the objective function for the GA. Application of the new approach for fitting an ensemble of potentials computed from *ab initio* calculations to a specified functional form (here, the Tersoff potential functional form is used as an example) has shown that the computational efficiency achieved through the use of an NN can reduce computational time by over two orders of magnitude.[72] The potentials estimated from functions thus fitted were within 0.1% of the actual potential values. Furthermore, since the potential was fitted to a physically meaningful Tersoff functional form, the resulting potential function appears to have the ability to extrapolate over a reasonable range of the parameter space and may have a better accuracy in estimating the forces than that obtained from neural networks, which are often highly inaccurate when extrapolated. Hence, the method can be useful for rendering various MD simulations more tractable.

A GA-based parameterization process requires a repetitive computation of the objective function, which in this case is expressed as the mean squared error (MSE) taken over all configurations of interest between the potentials determined from the *ab initio* calculations and those estimated using the GA-selected parameter vector values. An NN is used to learn the relationship connecting the parameter vector and the objective function. The NN thus trained provides surrogate estimates of the objective function, thereby obviating the need to compute this function for every set of parameter values selected during the GA fitting process. In fact, the use of an

NN is the key to reducing the overall computational time for parameterization from a few weeks to a few hours. The GA has been found to quickly establish the correct "range" of the parameter values to optimize a non-linear and perhaps combinatorial objective function. Since the objective function has significant slope changes and long plateaus, a simplex (Nelder-Mead) search is used during the final stage of the parameterization process to ensure faster convergence to the optimal values. It has been shown that the approach can correctly identify the parameters used to fit Tersoff potential to within ±0.0023 eV (corresponding to a 95% confidence limit) accuracy. The results also show that that it is possible to fit *ab initio* potentials of 5-atom Si clusters to a Tersoff functional form within ±0.0507 eV accuracy. More important, the GA-parameterized fit to Tersoff functional form has excellent extrapolation capabilities compared to the use of NNs.

Chapter 7 has also reviewed and discussed a novel method developed by Sumpter and Noid[246] that has demonstrated that NNs can be used effectively to enhance and simplify IVR calculations. In an IVR study, it is generally necessary to convert the Cartesian coordinates and momenta in which the trajectories are integrated to the corresponding curvilinear coordinates and velocities needed to monitor the internal energy transfer processes. This transformation, which must be done repeatedly through out the integration of the trajectory, involves computationally intense matrix calculations. Sumpter and Noid show that much of this computational effort can be eliminated by training an NN to execute the calculations. Chapter 7 has reviewed the details of the manner by which this is accomplished using the H_2O_2 system as an example.[247,248]

11.7. NN METHODS FOR PARAMETER DETERMINATION OF EMPIRICAL PESs

In many studies, it is useful to employ parameterized empirical PESs. This is often the case when the system under investigation is extremely complex. In addition, the values of the fitted parameters of the empirical PES often have physical significance not easily found in more general methods involving NNs or moving interpolation or least-squares methods. If the empirical PES is extremely complex and highly parameterized, it is often extremely difficult to find the optimum values of the parameters of the PES.[16,250h] Malshe et al.[22] have shown that NNs can be used effectively to achieve this fitting to an *ab initio* database. Sumpter and Noid[74] have demonstrated how NNs can achieve the same thing using vibrational spectra rather than an *ab initio* database.

The method developed by Malshe et al.[22] provides the means by which the parameters of an empirical potential can be fitted to an *ab initio* database even in cases where a subset of the parameters are made functions of the configuration of the system. The method, which provides procedures to avoid local minima, will generally yield the best possible set of parameters for the chosen functional form of the PES. Furthermore, the entire procedure is automated so that a minimum of human time is involved in the fitting process.

Chapter 8 has described this NN method in detail and provided an illustrative example of the fitting of a Tersoff potential to a database comprising *ab initio* energies for 10,000 Si_5 clusters observed in the machining of a silicon workpiece.

For many decades, scientists have realized that a PES completely determines the form of the IR and Raman spectra for a molecule. Consequently, it has always been a goal of spectroscopists to invert this process, that is, to derive the PES from the measured IR and Raman. However, except for very small molecules, this goal has proven to be elusive. The NN method developed by Sumpter and Noid[74] provides a clever way to execute this inversion. Chapter 8 has described their method in detail.

11.8. COMBINED FUNCTION DERIVATIVE APPROXIMATION (CFDA) NN METHOD

Almost all MD studies reported to date have concentrated only on the fitting of the potential energy, although it is the force field that is the key component of the MD simulation. This is due to the following two factors. First, the *ab initio* computation of surface gradients requires far more computational effort than just the potential itself. Second, MSI[23-25] and IMLS[28-32] methods are not easily adaptable to fitting surface gradients, and adapting NN methods to the simultaneous fitting of both the potential and its gradients requires significant modification[121] to the usual NN fitting methods.[105,110] Since forces can be computed by differentiating the potential surface, it has been assumed in the past that an accurate potential surface will yield accurate forces, without specifically fitting the forces. While not unreasonable, this assumption is not always correct.

To address this problem, a new type of NN training algorithm was introduced in Chapter 3 that fits both the function and its gradient simultaneously.[121,122] This method is termed a combined function and derivative approximation (CFDA). When the CFDA is employed, the output layer contains only a single element, which is the function value corresponding to the input variables. The gradient at the input point is obtained by differentiation of the network. This feature greatly simplifies the network architecture by reducing the number of neurons and parameters. The output layer contains only a single neuron, which provides the network approximation to the potential. The fitted gradients are obtained directly by differentiation of the network.

An advantage of the CFDA method[121,122] is that a validation set is not needed because the fitting performance of the method has been found to be superior to that which can be obtained using either early stopping or Bayesian regularization. This positive result occurs because in the normal type of overfitting, the network fits each point in the database with near perfection by producing numerous maxima and minima between the points to force a fit at the training points. With such maxima and minima present, the fit of the gradient at the training points will be very poor. However, when the network is forced to fit both the function and its gradient at the training points, as it is in the CFDA method, this type of overfitting cannot occur and a validation set is not needed. Since it is the gradient of the potential energy that is of importance in MD simulations, the CFDA promises to be a major player in future applications of NNs to chemical reaction dynamics.

Applications of the CFDA to the reactions of H with HBr have shown that the fitting accuracy is extremely high (RMSE = 1.2 cm^{-1}). As a result, trajectories integrated on the NN fit and analytic PES show point-by-point agreement. To date, this is the only method that has achieved such a result. The method and results obtained using it have been reviewed and discussed in Chapter 3.

11.9. NN METHODS FOR DATA ANALYSIS AND STATISTICAL ERROR REDUCTION

The use of NNs to predict an outcome (or the output results) as a function of a set of input parameters has been gaining wider acceptance with the advances in computer technology as well as with an increased awareness of the potential of NNs. A neural network is first trained to learn the underlying functional relationship between the output and the input parameters by providing it with a large number of data points, where each data point corresponds to a set of output and input parameters.

In view of the success achieved in obtaining interpolated values of the PESs for multi-atomic systems using an NN trained by the *ab initio* energy values for a large number of configurations, it is reasonable to ask if we can successfully compute the results of an MD trajectory for a chemical reaction using an NN trained by the data obtained by previous MD simulations. If this can be done successfully, it becomes possible to execute a small number of trajectories, M, and then utilize the results of these trajectories as a database to train an NN to predict the final results of a very large number of trajectories N, where $N \gg M$, that can be used to increase the statistical accuracy of the MD calculations and to further explore the dependence of the trajectory results without actually having to perform any further numerical integrations. In effect, the NN replaces the computationally laborious numerical integrations.

Chapter 9 describes the results of an investigation of this intriguing possibility using the interaction of a carbon C_2 dimer with a diamond (100) surface as a test system.[75] MD simulations of C_2 collisions on a diamond surface at a given surface temperature were executed to compute the probabilities of C_2 chemisorption, scattering, and desorption probabilities as a function of the direction of the initial C_2 velocity vector (θ, ϕ), impact parameter (b), translational energy (E_{trans}), and rotational energy (E_{rot}). These data were then used as a database to train an NN to predict these probabilities for various input conditions. The reaction probabilities predicted by a trained NN committee are found to be in very good accord with those provided by MD calculations. After the NN committee is trained, the determination of the interpolated reaction probabilities as a function of the input parameters involves only the evaluation of simple analytical expressions. Thus, interpolation by using an NN committee can significantly reduce the burden of performing a large number of expensive MD calculations.

It was also found that NN fitting employing early stopping will fit the trend of the input data and not the statistical fluctuations associated with the input data. Thus, NNs can provide a computationally convenient method of averaging out statistical variations inherent in the MD calculations.

All these observations demonstrate that an ensemble of NNs can be an extremely powerful tool for interpolating reaction probabilities as a function of different variables in chemical reactions thereby avoiding the computational expense of integrating coupled sets of differential equations to compute MD trajectories.

11.10. OTHER APPLICATIONS OF NNS TO QUANTUM MECHANICAL PROBLEMS

Chapter 10 reviews several applications of NNs to the solution of computationally difficult quantum mechanical problems. The first topic discussed in the chapter involves the solution of the molecular vibrational Schrodinger equation. The usual expansion methods using fixed basis functions often results in expansions so large that they become intractable for larger molecules. Manzhos and Carrington[76,77] have developed an NN method in which the usual basis set expansion is replaced with a radial basis function NN in a framework that permits multiple vibrational wavefunctions and energies to be obtained in a single calculation. It is these last two features that distinguish this work from previous efforts to employ NNs in this capacity.[174-176] When an RBFNN is employed in this manner, the neurons of the network essentially play the role of the expansion functions, but since they have flexible form, convergence is obtained much more rapidly.

The complete details of the method have been described in Chapter 10, and the first application[77] of the method to a real molecule (H_2O) is described.

The second topic involves the use of trained NNs to predict higher-level electronic structure results from lower-level calculations[78] and to predict results using large basis sets from *ab initio* calculations employing much smaller basis sets.[79,273-281]

As the number and size of the atoms present in a molecular system increase and as the number of reactions and their complexity also increase, the computational requirements for the electronic structure calculations required to obtain a sufficiently converged database to obtain a PES become very large. At present, this is the factor that limits the complexity of reactions whose dynamics can be investigated using *ab initio* methods.

In Chapter 10, a new method for obtaining high-level electronic structure energies from low-energy HF energies using NNs is presented (Malshe et al.).[78] This work demonstrates that it is possible to obtain configuration energies equivalent to those computed at a high level with extended basis sets by employing NNs trained with lower-level HF calculations and some higher-level calculations. In test cases, the accuracy of the predicted higher-level *ab initio* energies is shown to be greater than the NN fitting accuracy for the database of higher-level energies.

Malshe et al.[78] have shown that if the input vector to an NN includes both the configuration coordinates and the HF energies of a small subset of the configurations needed to adequately characterize the PES over all regions important in the reaction dynamics of a given system, MP4(SDQ) energies with the same basis set can be accurately predicted for the entire database using only the HF and MP4(SDQ) energies for the small subset and the HF energies for the remainder of the database. The predictive error is shown to be less than or equal to the NN fitting error if an NN is fitted to the entire database of higher-level electronic structure energies. The underlying basis for the method is the high degree of correlation existing between the HF and MP4(SDQ) energies.

Using the same concept, Balabin and Lomakina[79] have demonstrated that DFT calculations at equilibrium using large basis sets can be obtained using NNs trained with DFT calculations with small basis sets and some large basis set calculations. Numerous investigators have shown that the accuracy of *ab initio* electronic structure results for equilibrium properties can be greatly improved by employing NNs trained using some experimental data.[273-281]

Although these investigations are still in their infancy, it appears likely that this NN method may provide a powerful tool to substantially circumvent the computational bottleneck to obtaining databases of high-level electronic structure results for many-atom systems undergoing simultaneous, complex reactions. Specifically, it appears likely that the procedure will work not only for MP4(SDQ) energies but also for more sophisticated and more computationally expensive methods, such as QCISDtq or CCSDtq calculations.

11.11. FUTURE TRENDS

With only about 50 papers presently in the literature exploring the use of NNs in the study of chemical reaction dynamics, it is clear that much work remains to be done. Specifically, the material presented in this monograph shows that future avenues of investigation will likely involve expansion of *ab initio* reaction dynamics to systems involving more than six atoms with multiple reaction channels simultaneously open.

Since it is the gradient of the potential rather than the potential itself that is of paramount importance in MD simulations, it is likely that future investigations will involve more emphasis on using sampling methods that place added weight on gradients rather than just upon energies. Likewise, NN methods that simultaneously fit the gradients as well as the potentials, such as CFDA, are likely to become the methods of choice. To date, the largest system to which CFDA has been applied is Si_5. In this application, no chemical reactions were occurring. Future studies will undoubtedly expand this range of application.

Expansion methods seem to be particularly powerful in that they offer the possibility of treating large systems with expansions of functions of much lower dimensionality. When the expansion functions are NNs, accurate representation of the PES can be expected. Therefore, it seems likely that future trends will involve increasing use of such expansions for larger and more difficult systems. Solids with a high degree of symmetry seem well-suited to such investigations.

The intriguing possibility of predicting the results of MD simulations using NNs and some sample of trajectories has been explored only for the very simple C_2-diamond (100) system. Expansion of this application to more challenging systems seems likely.

The ability to predict the results of high-level electronic structure calculations using NNs trained with a small subset of the total required database and HF energies for the remainder seems to offer the window through which much more complex systems can be investigated. It would also seem to provide the means by which systems of moderate complexity can be studied for the first time using the highest possible level of electronic structure calculations. Research in this direction will likely occur at an accelerated rate in the near future.

REFERENCES

1. Porter, R. N. 1974. Classical trajectory methods in molecular collisions. *Ann. Rev. Phys. Chem.* 25: 317.

2. Bunker, D. L. 1971. Classical trajectory methods. *Comput. Phys.* 10: 287.

3. Porter, R. N., and L. M. Raff. 1976. Classical trajectory methods in molecular collisions. In *Modern theoretical chemistry*, ed. W. H. Miller, 1. New York: Plenum Press.

4. Truhlar, D. G., and J. T. Muckerman. 1979. *Atom-molecule collision theory: A guide for the experimentalist*, ed. R. B. Bernstein, 505. New York: Plenum Press.

5. Hase, W. L. 1976. Dynamics of unimolecular reactions in modern theoretical chemistry. In *Dynamics of molecular collisions, Vol. 2, Part B*, ed. W. H. Miller, 121. New York: Plenum Press.

6. Kuntz, P. J. 1976. Features of potential energy surfaces and their effect on collisions. In *Dynamics of molecular collisions, Part B*, ed. W. H. Miller, 53. New York: Plenum Press.

7. Raff, L. M., and D. L. Thompson. 1985. The classical trajectory approach to reactive scattering. In *Theory of chemical reaction dynamics, Vol. III*, ed. M. Baer, 1. Boca Raton, FL: CRC Press.

8. Morse, P. M. 1929. Diatomic molecules according to the wave mechanics II vibrational levels. *Phys. Rev.* 34: 57.

9. Girifalco, L. A., and V. G. Weizer. 1959. Application of the Morse potential function to cubic materials. *Phys. Rev.* 114: 687.

10. Lennard-Jones, J. E. 1925, 1926. Forces between atoms and ions. *Proc of the Roy. Soc. (Lon)* A 109: 584 and A 110: 230.

11. Daw, M. S., and M. I. Baskes. 1984. Embedded-atom method: Derivation and application to impurities, surfaces, and other defects in metals. *Phys. Rev. B* 29: 6443.

12. Baskes, M. I. 1992. Modified embedded-atom potentials for cubic materials and impurities. *Phy. Rev. B* 46: 2727.

13. Tersoff, J. 1998. New empirical approach for the structure and energy of covalent systems. *Phys. Rev. B* 37: 6991;
    ibid. 1989. Modeling solid-state chemistry: Interatomic potentials for multicomponent systems. *Phys. Rev. B* 39(8): 5566;
    ibid. 1988. Empirical interatomic potential for silicon with improved elastic properties. *Phys. Rev. B* 38(14): 9902;
    ibid. 1986. New empirical model for the structural properties of silicon. *Phys. Rev. Lett.* 56(6): 632.

14. Bolding, B. C., and H. C. Anderson. 1990. Interatomic potential for silicon clusters, crystals, and surfaces. *Phy. Rev. B* 41: 10568.

15. Brenner, D. W., and B. J. Garrison. 1986. Dissociative valence force field potential for silicon. *Phy. Rev. B* 34(2): 1304.

16. (a) Rahaman, A., and L. M. Raff. 2001. Trajectory investigations of the dissociation dynamics of vinyl bromide on an *ab initio* potential-energy surface. *J. Phys. Chem. A* 105(11): 2156;

(b) Rahaman, A., and L. M. Raff. 2001. Theoretical investigations of intramolecular energy transfer rates and pathways for vinyl bromide on an *ab initio* potential-energy surface. *J. Phys. Chem. A* 105(11): 2147.

17. Car, R., and M. Parrinello. 1985. Unified approach for molecular dynamics and density-functional theory. *Phys. Rev. Lett.* 55(22): 2471.

18. Izvekov, S., Parrinello, M., Burnham, C. J., and G. A. Voth. 2004. Effective force fields for condensed phase systems from *ab initio* molecular dynamics simulation: A new method for force-matching. *J. Chem. Phys.* 120: 10896.

19. Carré, A., Horbach, J., Ispas, S., and W. Kob. 2008. New effective potential for silica. *Europhysics Lett.* 82: 17001.

20. Tangney, P., and S. J. Scandolo. 2002. An *ab initio* parametrized interatomic force field for silic. *J. Chem. Phys.* 117: 8898.

21. Ercolessi, F., and J. B. Adams. 1994. Interatomic potentials from first-principles calculations: The force-matching method. *Europhysics Lett.* 26: 583.

22. Malshe, M., Narulkar, R., Raff, L. M., Hagan, M., Bukkapatnam, S., and R. Komanduri. 2008. Parameterization of analytic interatomic potential functions using neural networks (NN). *J. Chem. Phys.* 129: 044111-1.

23. Ischtwan, N., and M. A. Collins. 1994. Molecular potential energy surfaces by interpolation. *J. Chem. Phys.* 100: 8080.

24. Farwig, R. 1987. In *Algorithms for approximations*, ed. J. C. Mason and M. G. Cox, p. 194. Oxford: Clarendon.

25. Collins, M. A. 2002. Molecular potential-energy surfaces for chemical reactions dynamics. *Theor. Chem. Acc.* 108: 313.

26. Ho, T., Hollebeek, T., Rabitz, H., Harding, L. B., and G. C. Schatz. 1996. A global H_2O potential energy surface for the reaction O (^1D) + $H_2 \rightarrow OH+H$. *J. Chem. Phys.* 105(10): 10472.

27. Pederson, L. A., Schatz, G. C., Hollebeek, T., Ho, T., Rabitz, H., and L. B. Harding. 2000. Potential energy surface of the $\bar{A}$ state of NH_2 and the role of excited states in the $N(^2D)$ + H_2 reaction. *J. Phys. Chem.* 104: 2301.

28. Maisuradze, G. C., Thompson, D. L., Wagner, A. F., and M. Minkoff. 2003. Interpolating moving least-squares methods for fitting potential energy surfaces: Detailed analysis of one-dimensional applications. *J. Chem. Phys.* 119: 10002.

29. Guo, Y., Kawano, A., Thompson, D. L., Wagner, A. F., and M. Minkoff. 2004. Interpolating moving least-squares methods for fitting potential energy surfaces: Applications to classical dynamics calculations. *J. Chem. Phys.* 121: 5091.

30. Kawano, A., Guo, Y., Thompson, D. L., Wagner, A., and M. Minkoff. 2003. Interpolating moving least-squares methods for fitting potential-energy surfaces: Further improvement of efficiency via cutoff strategies. *J. Chem. Phys.* 119: 10002.

31. Guo, Y., Harding, L. B., Wagner, A. F., Minkoff, M., and D. L. Thompson. 2007. Improving the accuracy of interpolated potential energy surfaces by using an analytical zeroth-order potential function. *J. Chem. Phys.* 126: 104105.

32. (a) Dawes, R., Thompson, D. L., Guo, Y., Wagner, A. F., and M. Minkoff. 2007. Interpolating moving least-squares methods for fitting potential energy

surfaces: Computing high-density potential energy surface data from low-density *ab initio* data points. *J. Chem. Phys.* 126: 184108.

(b) Guo, Y., Tokmakov, I., Thompson, D. L., Wagner, A. F., and M. Minkoff. 2007. Interpolating moving least-squares methods for fitting potential energy surfaces: Improving efficiency via local approximants. *J. Chem. Phys.* 127: 214106.

33. (a) Ishida, T., and G. C. Schatz. 1999. A local interpolation scheme using no derivatives in quantum-chemical calculations. *Chem. Phys. Lett.* 314: 369.

(b) *ibid.* 2003. A local interpolation scheme using no derivatives in potential sampling: Application to $O(^1D) + H_2$ system. *J. Comp. Chem.* 24: 1077.

34. Huang, X., Braams, B. J., Carter, S., and J. M. Bowman. 2004. Quantum calculations of vibrational energies of $H_3O_2^-$ on an *ab initio* potential. *J. Am. Chem. Soc.* 126(16): 5042.

35. Chen, C., Shepler, B. C., Braams, B. J., and J. M. Bowman. 2007. Quasiclassical trajectory calculations of the $OH + NO^2$ association reaction on a global potential energy surface. *J. Chem. Phys.* 127: 104310.

36. Huang, X., Braams, B. J., and J. M. Bowman. 2005. *Ab initio* potential energy and dipole moment surfaces for $H_5O_2^+$. *J. Chem. Phys.* 122: 044308.

37. Xie, Z., Braams, B. J., and J. M. Bowman. 2005. *Ab initio* global potential energy for $H_5^+ \rightarrow H_3^+ + H_2$. *J. Chem. Phys.* 122: 244307.

38. Brown, A., Braams, B. J., Christoffel, K., Zhong, J., and J. M. Bowman. 2003. Classical and quasiclassical spectral analysis of CH_5^+ using an *ab initio* potential energy surface. *J. Chem. Phys.* 119: 8790.

39. Zhang, X., Braams, B. J., and J. M. Bowman. 2006. An *ab initio* potential surface describing abstraction and exchange for $H + CH_4$. *J. Chem. Phys.* 124: 021104.

40. Blank, T. B., and S. D. Brown. 1993. Data processing using neural networks. *Anal. Chim. Acta.* 277: 273.

41. Blank, T. B., Brown, S. D., Calhoun, A. W., and D. J. Doren. 1995. Neural network models of potential energy surfaces. *J. Chem. Phys.* 103: 4129.

42. Hobday, S., Smith, R., and J. BelBruno. 1999. Applications of genetic algorithms and neural networks to interatomic potentials. *Nuclear Inst. Meth. Phys. Res. B* 153: 247.

43. Tafeit, E., Estelberger, W., Horejsi, R., Moeller, R., Oettl, K., Vrecko, K., and G. J. Reibnegger. 1996. Neural networks as a tool for compact representation of *ab initio* molecular potential energy surfaces. *J. Mol. Graphics* 14: 12.

44. Gassner, H., Probst, M., Lauenstein, A., and K. Hermansson. 1998. Representation of intermolecular potential functions by neural networks. *J. Phys. Chem. A* 102(24): 4596.

45. Brown, D. R., Gibbs, M. N., and D. C. Clary. 1996. Combining *ab initio* computations, neural networks, and diffusion Monte Carlo: An efficient method to treat weakly bound molecules. *J. Chem. Phys.* 105: 7597.

46. Lorenz, S., Groß, A., and M. Scheffler. 2004. Representing high-dimensional potential-energy surfaces for reactions at surfaces by neural networks. *Chem. Phys. Lett.* 395: 210.

47. Manzhos, S., Wang, X., Dawes, R., and T. Carrington Jr. 2006. A nested molecule independent neural network approach for high-quality potential fits. *J. Phys. Chem. A* 110: 5295.

48. Manzhos, S., and T. Carrington Jr. 2006. Using neural networks to represent potential surfaces as sums of products. *J. Chem. Phys.* 125: 194105.

49. Manzhos, S., and T. Carrington Jr. 2006. A random-sampling high dimensional model representation neural network for building potential energy surfaces. *J. Chem. Phys.* 125: 084109.

50. Manzhos, S., and T. Carrington Jr. 2008. Using neural networks, optimized coordinates and high dimensional model representations to obtain a vinyl bromide potential surface. *J. Chem. Phys.* 129: 224104.

51. Kuhn, B., Rizzo, T. R., Luckhaus, D., Quack, M., and M. A. Suhm. 1999. A new six-dimensional analytical potential up to chemically significant energies for the electronic ground state of hydrogen peroxide. *J. Chem. Phys.* 111: 2565.

52. Murrell, J. N., Carter, S., Rantos, S., Huxley, P., and A. J. C. Varandas. 1984. *Molecular potential energy functions.* Toronto, Canada: John Wiley.

53. Murrell, J. N., and S. Carter. 1984. Approximate single-valued representations of multi-valued potential energy surfaces. *J. Phys. Chem.* 88: 4887.

54. Ho, T.-S., and H. Rabitz. 2003. Reproducing Kernel Hilbert space interpolation methods as a paradigm of high dimensional model representations: Application to multidimensional potential energy surface construction. *J. Chem. Phys.* 119: 6433.

55. Geremia, J. M., Weiss, E., and H. Rabitz. 2001. Achieving the laboratory control of quantum dynamics phenomena using nonlinear functional maps. *Chem. Phys.* 267(1–3): 209.

56. Geremia, J. M., and H. Rabitz. 2001. The Ar–HCl potential energy surface from a global map-facilitated inversion of state-to-state rotationally resolved differential scattering cross sections and rovibrational spectral data. *J. Chem. Phys.* 115(19): 8899.

57. Geremia, J. M., Rabitz, H., and C. Rosenthal. 2001. Constructing global functional maps between molecular potentials and quantum observables. *J. Chem. Phys.* 114(21): 9325.

58. Shenvi, N., Geremia, J. M., and H. Rabitz. 2004. Efficient chemical kinetic modeling through neural network maps. *J. Chem. Phys.* 120(21): 9942.

59. Alis, O. F., and H. Rabitz. 2001. Efficient implementation of high dimensional model representations. *J. Math. Chem.* 29(2): 127.

60. Rabitz, H., Alis, O. F., Shorter, J., and K. Shim. 1999. Efficient input-output model representations. *Comp. Phys. Comm.* 117(1–2): 11.

61. Hayes, M. Y., Li, B., and H. Rabitz. 2006. Estimation of molecular properties by high-dimensional model representation. *J. Phys. Chem. A* 110(1): 264.

62. Rabitz, H., and O. F. Alis. 1999. General foundations of high dimensional model representations. *J. Math. Chem.* 25: 197.

63. Geremia, J. M., and H. Rabitz. 2001. Global, nonlinear algorithm for inverting quantum-mechanical observations. *Phys. Rev. A* 64(2): 022710.

64. Li, G., Rosenthal, C., and H. Rabitz. 2001. High dimensional model representations. *J. Phys. Chem. A* 105(33): 7765.

65. Li, G., Wang, S.-W., Rosenthal, C., and H. Rabitz. 2001. High dimensional model representations generated from low dimensional data samples. I. MP-cut-HDMR. *J. Math. Chem.* 30(1): 1.

66. Li, G., Artamonov, M., Rabitz, H., Wang, S.-W., Georgopoulos, P. G., and M. Demiralp. 2003. High-dimensional model representations generated from low order terms—lp-RS-HDMR. *J. Comput. Chem.* 24: 647.

67. Li, G., Schoendorf, J., Ho, T.-S., and H. Rabitz. 2004. Multicut-HDMR with an application to an ionospheric model. *J. Comput. Chem.* 25(9): 1149.

68. Li, G., Wang, S.-W., and H. Rabitz. 2002. Practical approaches to construct RS-HDMR component functions. *J. Phys. Chem. A* 106(37): 8721.

69. Li, G., Hu, J., Wang, S.-W., Georgopoulos, P. G., Schoendorf, J., and H. Rabitz. 2006. Random sampling-high dimensional model representation (RS-HDMR) and orthogonality of its different order component functions. *J. Phys. Chem. A* 110(7): 2474.

70. Wang, S.-W., Georgopoulos, P. G., Li, G., and H. Rabitz. 2003. Random sampling-high dimensional model representation (RS-HDMR) with nonuniformly distributed variables: Application to an integrated multimedia/multipathway exposure and dose model for trichloroethylene. *J. Phys. Chem. A* 107: 4707.

71. Malshe, M., Narulkar, R., Raff, L. M., Hagan, M., Bukkapatnam, S., Agrawal, P. M., and R. Komanduri. 2009. Development of generalized potential-energy surfaces (GPES) using many-body expansions, neural networks, and moiety-energy approximations. *J. Chem. Phys.* 130: 184102-1.

72. Bukkapatnam, S., Malshe, M., Agrawal, P. M., Raff, L. M., and R. Komanduri. 2006. Parameterization of interatomic potential functions using a genetic algorithm accelerated with a neural network. *Phys. Rev. B* 74: 224102.

73. Sumpter, B. G., Getino, C., and D. W. Noid. 1994. Theory and applications of neural computing in chemical science. *Annu. Rev. Phys. Chem.* 45(1): 439.

74. Sumpter B. G., and D. W. Noid. 1992. Potential energy surfaces for macromolecules. A neural network technique. *Chem. Phys. Lett.* 192: 455.

75. Agrawal, P. M., Samadh, A. N. A., Raff, L. M., Hagan, M. T., Bukkapatnam, S. T., and R. Komanduri. 2005. Prediction of MD simulation results using feed-forward neural networks: Reaction of C_2 dimer with an activated diamond (100) surface. *J. Chem. Phys.* 123: 224711-1.

76. Manzhos, S., and T. Carrington Jr. 2009. An improved neural network method for solving the Schrödinger equation. *Can. J. Chem.* 87: 864.

77. Manzhos, S., and T. Carrington Jr. 2009. Using a neural network based method to solve the vibrational Schrödinger equation for H_2O. *Chem. Phys. Lett.* 474: 217.

78. Malshe, M., Narulkar, R., Raff, L. M., Hagan, M., Bukkapatnam, S., and R. Komanduri. 2009. Accurate prediction of higher-level electronic structure energies for large databases using neural networks, HF energies, and small subsets of the database. *J Chem. Phys.* 131: 124123-1.

79. Balabin, R. M., and E. I. Lomakina. 2009. Neural network approach to quantum chemistry data: Accurate prediction of density functional energies. *J. Chem. Phys.* 131: 074104.

80. Thompson, K. C., Jordan, M. J. T., and M. A. Collins. 1998. Polyatomic molecular potential energy surfaces by interpolation in local internal coordinates. *J. Chem. Phys.* 108: 8302.

81. Lancaster, P., and K. Salkaukas. 1986. *Curve and surface fitting: An introduction.* Chapter 10. London: Academic.

82. Wilson, E. B. Jr., Decius, J. C., and P. C. Cross. 1955. *Molecular vibrations.* New York: Dover.

83. Ben-Israel, A., and T. N. Greville. 1974. *Generalized inverses: Theory and applications.* New York: Wiley-Interscience.

84. Press, W. H., Teukolsky, S. A., Vetterline, W. T., and B. P. Flannery. 1986. *Numerical recipes in fortran: The art of scientific computing*. 52. Cambridge: Cambridge University Press.

85. Jordon, M. J. T., Thompson, K. C., and M. A. Collins. 1995. Convergence of molecular potential energy surfaces by interpolation: Application to the $OH+H_2 \rightarrow H_2O+H$ reaction. *J. Chem. Phys.* 102: 9669.

86. Thompson, K. C., and M. A. Collins. 1997. Molecular potential-energy surfaces by interpolation: Further refinements. *J. Chem. Soc., Faraday Trans.* 93: 871.

87. Jordon, M. J. T., and R. G. Gilbert. 1995. Classical trajectory studies of the reaction $CH_4+H \rightarrow CH_3+H_2$. *J. Chem. Phys.* 102: 5669.

88. Schranz, H. W., Nordholm, S., and G. Nyman. 1991. An efficient microcanonical sampling procedure for molecular systems. *J. Chem. Phys.* 94: 1487.

89. Bettens, R. P. A., and M. A. Collins. 1998. Potential energy surfaces and dynamics for the reactions between $C(^3P)$ and $H_3^+(^1A_1')$. *J. Chem. Phys.* 108: 2424.

90. Bettens, R. P. A., Hansen, T. A., and M. A. Collins. 1999. Interpolated potential energy surface and reaction dynamics for $O(^3P)+ H_3^+ (^1A_1')$ and $OH^+(^3\Sigma^-)+H2(^3\Sigma_g^+)$. *J. Chem. Phys.* 111: 6322.

91. Collins, M. A., and R. P. A. Bettens. 1999. Potential energy surface for the reaction $BeH_2 + H \rightarrow HeH + H_2$. *Phy. Chem. Chem. Phys.* 1: 939.

92. Collins, M. A., and R. P. A. Bettens. 1998. Interpolated potential energy surface and dynamics for the reactions between $N(^4S)$ and $H_3^+(^1A_1')$. *J. Chem. Phys.* 109: 9728.

93. Bettens, R. P. A., and M. A. Collins. 1999. Learning to interpolate molecular potential energy surfaces with confidence: A Bayesian approach. *J. Chem. Phys.* 111: 816.

94. Kawano, A., Guo, Y., Thompson, D. L., Wagner, A. F., and M. Minkoff. 2004. Improving the accuracy of interpolated potential energy surfaces by using an analytical zeroth order potential function. *J. Chem. Phys.* 120: 6414.

95. Collins, M. A., and L. Radom. 2003. Proton-transport catalysis, proton abstraction, and proton exchange in $HF + HOC^+$ and $H_2O + HOC^+$ and analogous deuterated reactions. *J. Chem. Phys.* 118: 6222.

96. Moyano, G. E., and M. A. Collins. 2003. Interpolated potential energy surface and classical dynamics for $H_3^+ + HD$ and $H_3^+ + D_2$. *J. Chem. Phys.* 119: 5510.

97. Maisuradze, G. G, Kawano, A., Thompson, D. L., Wagner, A. F., and M. Minkoff. 2004. Interpolating moving least-squares methods for fitting potential energy surfaces: Analysis of an application to a six-dimensional system. *J. Chem. Phys.* 121: 10329.

98. Anderson, E., Bai, Z., Bischof, C., et al. 1999. *LAPACK users' guide*, 3rd ed. Philadelphia, PA: SIAM.

99. Hollebeek, T., Ho, T-S., and H. Rabitz. 1999. Constructing multidimensional molecular potential energy surfaces from *ab initio* data. *Annu. Rev. Phys. Chem.* 50: 537.

100. Ho, T-S., Hollebeek, T., Rabitz, H., Chao, S. D., Skodje, R. T., Zyubin, A. S., and A. M. Mebel. 2002. A globally smooth *ab initio* potential surface of the 1A' state for the reaction $S(^1D) + H_2$. *J. Chem. Phys.* 116: 4124.

101. Liu, B. 1973. Ab initio potential energy surface for linear H. *J. Chem. Phys.* 58: 1925.

102. Siegbahn, P., and B. Liu. 1978. An accurate three-dimensional potential energy surface for H_3. *J. Chem. Phys.* 68: 2457.

103. Truhlar, D. G., and C. J. Horowitz. 1978. Functional representation of Liu and Siegbahn's accurate *ab initio* potential energy calculations for $H+H_2$. *J. Chem. Phys.* 68: 2466.

104. Ishida, T., and G. C. Schatz. 1998. Monte Carlo sampling methods for determining potential energy surfaces using Shepard interpolation. The O (1D)+ H2 system. *Chem. Phys. Lett.* 298: 285.

105. Hagan, M. T., Demuth, H. B., and M. H. Beale. 1996. *Neural network design.* Boston, MA: PWS.

106. Bishop, C. M. 1992. Exact calculation of the Hessian matrix for the multilayer perceptron. *Neural Comp.* 4(4): 494.

107. Haykin, S. 1999. *Neural networks: A comprehensive foundation,* 2nd ed. Englewood Cliffs, NJ: Prentice-Hall.

108. Hornik, K. M., Stinchcombe, M., and H. White. 1989. Multilayer feedforward networks are universal approximators. *Neural Networks* 2(5): 359.

109. Stone, M. H. 1948. The generalized Weierstrass approximation theorem. *Math. Mag.* 21(4): 167; 21(5): 237.

110. Pinkus, A. 1999. Approximation theory of the MLP model in neural networks. *Acta Numerica* 8: 143.

111. Niyogi, P., and F. Girosi. 1999. Generalization bounds for function approximation from scattered noisy data. *Adv. Comp. Math.* 10: 51.

112. Werbos, P. J. 1974. Beyond regression: New tools for prediction and analysis in the behavioral sciences. Ph.D. diss., Harvard University, Cambridge, MA. 1994. Also, published as *The roots of backpropagation.* New York: John Wiley.

113. Rumelhart, D. E., Hinton, G. E., and R. J. Williams. 1986. Learning representations by back-propagating errors. *Nature* 323: 533.

114. Shanno, D. F. 1990. Recent advances in numerical techniques for large-scale optimization. In *Neural networks for control,* ed. W. Thomas Miller, Richard S. Sutton, and Paul J. Werbos. Cambridge, MA: MIT Press.

115. Scales, L. E. 1985. *Introduction to non-linear optimization.* New York: Springer-Verlag.

116. Charalambous, C. 1992. Conjugate gradient algorithm for efficient training of artificial neural networks. *IEEE Proceedings* 139(3): 301.

117. Hagan, M. T., and M. Menhaj. 1994. Training feedforward networks with the Marquardt algorithm. *IEEE Transactions on Neural Networks* 5(6): 989.

118. Sarle, W. S. 1995. Stopped training and other remedies for overfitting. *Proceedings of the 27th Symposium on the Interface.* Pittsburgh, PA.

119. MacKay, D. J. C. 1992. A practical framework for backpropagation networks. *Neural Comput.* 4: 448.

120. Foresee, F. D., and M. T. Hagan. 1997. Gauss Newton approximation to Bayesian regularization. *Proceedings of the 1997 International Conference on Neural Networks,* Houston, Texas.

121. Pukrittayakamee, A., Hagan, M., Raff, L. M., Bukkapatnam, S., and R. Komanduri. 2007. Fitting a function and its derivative. In *Intelligent Engineering Systems through Artificial Neural Networks (ANNIE2007)* 17.

121b. Pukrittakamee, A., Hagan, M., Raff, L., Bukkapatnam, S., and R. Komanduri. A network pruning algorithm for combined function and derivative approximation. International Joint Conference on Neural Networks (IJCNN), IEEE publication (2009) 2553-2560

122. Pukrittayakamee, A., Malshe, M., Hagan, M., Raff, L. M., Narulkar, R., Bukkapatnam, S., and R. Komanduri. 2009. Simultaneous fitting of a potential-energy surface and its

corresponding force fields using feedforward neural networks (NN). *J. Chem. Phys.* 130: 134101-1.

123. Malshe, M., Raff, L. M., Hagan, M., Bukkapatnam, S., and R. Komanduri. 2010. Input vector optimization of feed-forward neural networks for fitting *ab initio* potential-energy databases. *J. Chem. Phys.* 132: 204103.

124. Le, H. M., Huynh, S., and L. M. Raff. 2009. Molecular dissociation of hydrogen peroxide (HOOH) on a neural network *ab initio* potential surface with a new configuration sampling method involving gradient fitting. *J. Chem. Phys.* 131: 014107-1.

125. Le, H. M., and L. M. Raff. 2008. Cis→trans, trans→cis isomerizations and N-O bond dissociation of nitrous acid (HONO) on an *ab initio* potential surface obtained by novelty sampling and feed-forward neural network fitting. *J. Chem. Phys.* 128: 194310-1.

126. Malshe, M., Raff, L. M., Rockley, M. G., Hagan, M., Agrawal, P. M., and R. Komanduri. 2007. Theoretical investigation of the dissociation dynamics of vibrationally-excited vinyl bromide on an *ab initio* potential-energy surface obtained using modified novelty sampling and feedforward neural networks II: Numerical application of the method. *J. Chem Phys.* 127: 134105.

127. Press, W. H., Teukolsky, S. A., Vetterling, W. T., and B. P. Flannery. 1992. *Numerical recipes.* Cambridge: Cambridge University Press.

128. Ito, Y. 1993. Approximation of differentiable functions and their derivatives on compact sets by neural networks. *Math. Scient.* 18: 11.

129. Li, X. 1996. Simultaneous approximations of multivariate functions and their derivatives by neural networks with one hidden layer. *Neurocomputing* 12: 327.

130. Ferrari, S., and R. F. Stengel. 2005. Smooth function approximation using neural networks. *IEEE Transactions on Neural Networks* 16(1): 24.

131. Basson, E., and A. P. Engelbrecht. 1999. Approximation of a function and its derivatives in feedforward neural networks. *International Joint Conference on Neural Networks,* Washington 1: 419.

132. Hanselmann, T., Zaknich, A., and Y. Attikiouzel. 1999. Learning functions and their derivatives using Taylor series and neural networks. *International Joint Conference on Neural Networks,* Washington, 1: 409.

133. Witkoskie, J. B., and D. J. Doren. 2005. Neural network models of potential energy surfaces: Prototypical examples. *J. Chem. Theory Comput.* 1: 14.

134. Dennis, J. E., and R. B. Schnabel. 1983. *Numerical methods for unconstrained optimization and nonlinear equations.* Englewood Cliffs, NJ: Prentice-Hall.

135. Marquardt, D. 1963. An algorithm for least-squares estimation of nonlinear parameters. *J. Soc. Ind. Appl. Math.* 11(2): 431.

136. Raff, L. M., Malshe, M., Hagan, M., Doughan, D. I., Rockley, M. G., and R. Komanduri. 2005. *Ab initio* potential-energy surfaces for complex, multichannel systems using modified novelty sampling and feed-forward neural networks. *J. Chem. Phys.* 122 (084104): 1.

137. Sudhakaran, M. P., and L. M. Raff. 1985. Quasiclassical trajectory studies of H (D)+ HBr (DBr) abstraction and exchange reactions. *Chem. Phys.* 95(2): 165.

138. Pukrittayakamee, A., and M. T. Hagan. Submitted 2009. Multilayer network confirmation using novelty detection.

139. Jolliffe, I. T. 1986. *Principal component analysis.* New York: Springer-Verlag.

140. Komanduri, R., Chandrasekaran, N., and L. M. Raff. 2001. Molecular dynamics simulation of the nanometric cutting of silicon. *Philo. Mag. B* 81: 1989.

141. (a) *Gaussian 98 (Revision A.1)*. 2001. Pittsburgh, PA: Gaussian.

    (b) Frisch, M. J., Trucks, G. W., Schlegel, H. B., Scuseria, G. E., Robb, M. A., Cheeseman, J. R., Montgomery, J. A. Jr., Vreven, T., Kudin, K. N., Burant, J. C., Millam, J. M., Iyengar, S. S., Tomasi, J., Barone, V., Mennucci, B., Cossi, M., Scalmani, G., Rega, N., Petersson, G. A., Nakatsuji, H., Hada, M., Ehara, M., Toyota, K., Fukuda, R., Hasegawa, J., Ishida, M., Nakajima, T., Honda, Y., Kitao, O., Nakai, H., Klene, M., Li, X., Knox, J. E., Hratchian, H. P., Cross, J. B., Bakken, V., Adamo, C., Jaramillo, J., Gomperts, R., Stratmann, R. E., Yazyev, O., Austin, A. J., Cammi, R., Pomelli, C., Ochterski, J. W., Ayala, P. Y., Morokuma, K., Voth, G. A., Salvador, P., Dannenberg, J. J., Zakrzewski, V. G., Dapprich, S., Daniels, A. D., Strain, M. C., Farkas, O., Malick, D. K., Rabuck, A. D., Raghavachari, K., Foresman, J. B., Ortiz, J. V., Cui, Q., Baboul, A. G., Clifford, S., Cioslowski, J., Stefanov, B. B., Liu, G., Liashenko, A., Piskorz, P., Komaromi, I., Martin, R. L., Fox, D. J., Keith, T., Al-Laham, M. A., Peng, C. Y., Nanayakkara, A., Challacombe, M., Gill, P. M. W., Johnson, B., Chen, W., Wong, M. W., Gonzalez, C., and J. A. Pople, 2004. *Gaussian 03 (Revision C.02)*. Wallingford, CT: Gaussian.

142. Ludwig, J., and D. G. Vlachos. 2007. *Ab initio* molecular dynamics of hydrogen dissociation on metal surfaces using neural networks and novelty sampling. *J. Chem. Phys.* 127: 154716.

143. Lorenz, S., Gross, A., and M. Scheffler. 2004. Representing high-dimensional potential-energy surfaces for reactions at surfaces by neural networks. *Chem. Phys. Lett.* 395(4-6): 210.

144. Ishida, T., and G. C. Schatz. 1997. Automatic potential energy surface generation directly from *ab initio* calculations using Shepard interpolation: A test calculation for the H+ H system. *J. Chem. Phys.* 107: 3558.

145. Collins, M. A., and D. H. Zhang. 1999. Application of interpolated potential energy surfaces to quantum reactive scattering. *J. Chem. Phys.* 111: 9924.

146. Nguyen, K. A., Rossi, I., and D. G. Truhlar. 1995. A dual-level Shepard interpolation method for generating potential energy surfaces for dynamics calculations. *J. Chem. Phys.* 103: 5522.

147. Agrawal, P. M., Malshe, M., Narulkar, R., Raff, L. M., Hagan, M., Bukkapatnum, S., and R. Komanduri. 2009. A self-starting method for obtaining analytic potential-energy surfaces from *ab initio* electronic structure calculations. *J. Phys. Chem. A* 113: 869.

148. Raff, L. M. 1988. Intramolecular energy transfer and mode specific effects in unimolecular reactions of 1, 2 difluoroethane. *J. Chem. Phys.* 89: 5680; Raff, L. M. 1989. Intramolecular energy transfer and mode-specific effects in unimolecular reactions of 1,2-difluoroethane. 90: 6313.

149. Guan, Y., and D. L. Thompson. 1989. Mode specificity and the influence of rotation in cis-trans isomerization and dissociation in HONO. *Chem. Phys.* 139(1): 147.

150. Head-Gordon, M., Pople, J. A., and M. J. Frisch. 1988. MP2 energy evaluation by direct methods. *Chem. Phys. Lett.* 153(6): 503.

151. Frisch, M. J., Head-Gordon, M., and J. A. Pople. 1990. A direct MP2 gradient method. *Chem. Phys. Lett.* 166(3): 275.

152. Frisch, M. J., Head-Gordon, M., and J. A. Pople. 1990. Semi-direct algorithms for the MP2 energy and gradient. *Chem. Phys. Lett.* 166(3): 281.

153. Head-Gordon, M., and T. Head-Gordon. 1994. Analytic MP 2 frequencies without fifth-order storage: Theory and application to bifurcated hydrogen bonds in the water hexamer. *Chem. Phys. Lett.* 220(1-2): 122.

154. Saebo, S., and J. Almlof. 1989. Avoiding the integral storage bottleneck in LCAO calculations of electron correlation. *Chem. Phys. Lett.* 154: 83.

155. Demuth, H. B., and M. Beale. 2000. *Users' guide for the neural network toolbox for MATLAB, ver. 4.0.* Natick, MA: Mathworks.

156. Hobday, S., Smith, R., and J. BelBruno. 1999. Application of neural networks to fitting interatomic potential functions. *Modell. Simul. Mater. Sci. Eng.* 7: 397.

157. Lorenz, S., Scheffler, M., and A. Gross. 2006. Descriptions of surface chemical reactions using a neural network representation of the potential-energy surface. *Phys. Rev. B* 73: 115431.

158. Gross, A., Wilke, S., and M. Scheffler. 1995. Six-dimensional quantum dynamics of adsorption and desorption of H_2 at Pd (100): Steering and steric effects. *Phys. Rev. Lett.* 75: 2718.

159. Wilke, S., and M. Scheffler. 1995. Poisoning of Pd (100) for the dissociation of H_2: A theoretical study of co-adsorption of hydrogen and sulphur. *Surf. Sci.* 329: L605.

160. Wilke, S., and M. Scheffler. 1996. Potential-energy surface for H_2 dissociation over Pd (100). *Phys. Rev. B* 53: 4926.

161. Gelb, A. 1974. *Applied optimal estimization.* Cambridge, MA: MIT Press.

162. Prudente, F. V., Acioli, P. H., and J. J. S. Neto. 1998. The fitting of potential energy surfaces using neural networks: Application to the study of the vibrational levels of H_3^+. *J. Chem. Phys.* 109: 8801.

163. Dykstra, C. E., and W. C. Swope. 1979. The H: Potential energy surface. *J. Chem. Phys.* 70: 1.

164. Meyer, W., Botschwina, P., and P. Burton. 1986. *Ab initio* calculation of near equilibrium potential and multipole moment surfaces and vibrational frequencies of H and its isotopomers. *J. Chem. Phys.* 84: 891.

165. Rocha Filho, T. M., Oliveira Jr., Z. T., Malbouisson, L. A. C., Gargano, R., and J. J. Soares Neto. 2003. The use of neural networks for fitting potential energy surfaces: A comparative case study for the H_3^+ molecule. *Int. J. Quant. Chem.* 95: 261.

166. Cencek, W., Rychlewski, J., Jaquet, R., and W. Kutzelnigg. 1998. Sub-microhartree accuracy potential energy surface for H_3^+ including adiabatic and relativistic effects. I. Calculation of the potential points. *J. Chem. Phys.* 108: 2831.

167. Prudente, F. V., and J. J. Soares Neto. 1998. The fitting of potential energy surfaces using neural networks. Application to the study of the photodissociation processes. *Chem. Phys. Lett.* 287: 585.

168. Pradhan, A. D., Kirby, K. P., and A. Dalgarno. 1991. Theoretical study of HCl^+: Potential curves, radiative lifetimes, and photodissociation cross sections. *J. Chem. Phys.* 95: 9009.

169. Prudente, F. V., Costa, L. S., and J. J. Soares Neto. 1997. Discrete variable representation and negative imaginary potential to study metastable states and photodissociation processes. Application to diatomic and triatomic molecules. *J. Mol. Struct.* 394: 169.

170. Sumpter, B. G., and D. W. Noid. 1992. Potential energy surfaces for macromolecules. A neural network technique. *Chem. Phys. Lett.* 192: 455.

171. Agrawal, P. M., Raff, L. M., Hagan, M. T., and R. Komanduri. 2006. Molecular dynamics investigations of the dissociation of SiO_2 on an *ab initio* potential energy surface obtained using neural network methods. *J. Chem. Phys.* 124: 134306.

172. Kuntz, P. J., Nemeth, E. M., Polanyi, J. C., Rosner, S. D., and C. E. Young. 1966. Energy distribution among products of exothermic reactions. II. Repulsive, mixed, and attractive energy release. *J. Chem. Phys.* 44: 1168.

173. Behler, J., and M. Parrinello. 2007. Generalized neural-network representation of high-dimensional potential-energy surfaces. *Phys. Rev. Lett.* 98: 146401.

174. Lagaris, I. E., Likas, A., and D. J. Fotiadis. 1997. Artificial neural network methods in quantum mechanics. *Comput. Phys. Commun.* 104: 1.

175. Sugawara, M. 2001. Numerical solution of the Schrödinger equation by neural network and genetic algorithm. *Comput. Phys. Commun.* 140: 366.

176. Nakanishi, H., and M. Sugawara. 2000. Numerical solution of the Schrödinger equation by a microgenetic algorithm. *Chem. Phys. Lett.* 327: 429.

177. Burns, J. A., and G. M. Whitesides. 1993. Feed-forward neural networks in chemistry: Mathematical systems for classification and pattern recognition. *Chem. Rev.* 93: 2583.

178. Handley, C. M., and P. L. A. Popelier. 2010. Potential energy surfaced fitted by artificial neural networks. *J. Phys. Chem. A* 114: 3371.

179. Zupan, J., and J. Gassteiger. 1999. *Neural networks in chemistry and drug design*, 2nd ed. Weinheim, NY: John Wiley-VCH.

180. Carney, G. D., and R. N. Porter. 1974. H: Geometry dependence of electronic properties. *J. Chem. Phys.* 60: 4251.

181. Ceperley, D. M., and B. Bernu. 1988. The calculation of excited state properties with quantum Monte Carlo. *J. Chem. Phys.* 89: 6316.

182. Cencek, W., Rychlewsk, J., Jaquet, R., and W. Kutzelnigg. 1998. Sub-microhartree accuracy potential energy surface for H_3^+ including adiabatic and relativistic effects. I. calculation of the potential points. *J. Chem. Phys.* 108(7): 2831.

183. Jaquet, R., Cencek, W., Kutzelnigg, W., and J. Rychlewski. 1998. Sub-microhartree accuracy potential energy surface for H_3^+ including adiabatic and relativistic effects. II. rovibrational analysis for H_3^+ and D_3^+. *J. Chem. Phys.* 108: 2837.

184. Jensen, P. 1989. The potential energy surface for the electronic ground state of the water molecule determined from experimental data using a variational approach. *J. Mol. Spectrosc.* 133: 438.

185. Carter, S., Handy, N. C., and J. Demaison. 1997. The rotational levels of the ground vibrational state of formaldehyde. *Mol. Phys.* 90(5): 729.

186. Manzhos, S., and T. Carrington Jr. 2007. Using redundant coordinates to represent potential energy surfaces with lower-dimensional functions. *J. Chem. Phys.* 127: 014103.

187. Sathyamurthy, N., and L. M. Raff. 1975. Quasiclassical trajectory studies using 3D spline interpolation of *ab initio* surfaces. *J. Chem. Phys.* 63: 464.

188. McDonald, P. A., and J. S. Shirk. 1982. The infrared laser photoisomerization of HONO in solid N and Ar. *J. Chem. Phys.* 77: 2355.

189. Shirk, A. E., and J. S. Shirk. 1983. Isomerization of HONO in solid nitrogen by selective vibrational excitation. *Chem. Phys. Lett.* 97: 549.

190. Khriachtchev, L., Lundell, J., Isoniemi, E., and M. Räsänen. 2000. HONO in solid Kr: Site-selective trans cis isomerization with narrow-band infrared radiation. *J. Chem. Phys.* 113: 4265.

191. Guan, Y., Lynch, G. C., and D. L. Thompson. 1987. Intramolecular energy transfer and cis-trans isomerization in HONO. *J. Chem. Phys.* 87: 6957.

192. Guo, Y., and D. L. Thompson. 2003. A theoretical study of cis-trans isomerization in HONO using an empirical valence bond potential. *J. Chem. Phys.* 118: 1673.

193. Møller, C., and M. S. Plesset. 1934. Note on the approximation treatment for many-electron systems. *Phys. Rev.* 46: 618.

194. Ditchfield, R., Hehre, W. J., and J. A. Pople. 1971. Self consistent molecular orbital methods. IX. An extended gaussian type basis for molecular orbital studies of organic molecules. *J. Chem. Phys.* 54: 724.

195. Hehre, W. J., Ditchfield, R., and J. A. Pople. 1972. Self-consistent molecular orbital methods. XII. Further extensions of Gaussian-type basis sets for use in molecular orbital studies of organic molecules. *J. Chem. Phys.* 56: 2257.

196. Hariharan, P. C., and J. A. Pople. 1974. Accuracy of AHn equilibrium geometries by single determinant molecular orbital theory. *Mol. Phys.* 27: 209.

197. Gordon, M. S. 1980. The isomers of silacyclopropane. *Chem. Phys. Lett.* 76: 163.

198. Petersson, G. A., Bennett, A., Tensfeldt, T. G., Al-Laham, M. A., Shirley, W. A., and J. Mantzaris. 1988. A complete basis set model chemistry. I. the total energies of closed-shell atoms and hydrides of the first-row elements. *J. Chem. Phys.* 89(4): 2193.

199. McLean, A. D., and G. S. Chandler. 1980. Contracted gaussian basis sets for molecular calculations. I. Second row atoms, Z = 11–18. *J. Chem. Phys.* 72: 5639.

200. Krishnan, R., Binkley, J. S., Seeger, R., and J. A. Pople. 1980. Self consistent molecular orbital methods. XX. A basis set for correlated wave functions. *J.Chem. Phys.* 72: 650.

201. Murto, J., Rasanen, M., Aspiala, A., and T. Lotta. 1985. *Ab initio* calculations of HONO: Energies, geometries and force fields on different levels of theory. *J. Mol. Struct.* 122(3-4): 213.

202. Richter, F., Hochlaf, M., Rosmus, P., Gatti, F., and H. D. Meyer. 2004. A study of the mode-selective trans-cis isomerization in HONO using *ab initio* methodology. *J. Chem. Phys.* 120: 1306.

203. Dunning Jr., T. H. 1989. Correlation consistent basis sets for use in molecular calculations. I. The atoms boron through neon and hydrogen. *J. Chem. Phys.* 90: 1007.

204. Becke, A. D. 1993. Density-functional thermochemistry. III. The role of exact exchange. *J. Chem. Phys.* 98(1): 5648.

205. Andersson, K., Malmqvist, P. A., Roos, B. O., Sadlej, A. J., and K. Wolinski. 1990. Second-order perturbation theory with a CASSCF reference function. *J. Phys. Chem.* 94(14): 5483.

206. Andersson, K., Malmqvist, P., and B. O. Roos. 1992. Second order perturbation theory with a complete active space self consistent field reference function. *J. Chem. Phys.* 96: 1218.

207. Harding, L. B. 1991. Theoretical studies of the hydrogen peroxide potential surface. 2. An *ab initio*, long-range, Hydroxyl(2.PI.) + Hydroxyl(2.PI.) potential. *J. Phys. Chem.* 95(22): 8653.

208. Liu, D. K., Letendre, L. T., and H. L. Dai. 2001. 193 nm photolysis of vinyl bromide: Nascent product distribution of the CHBr CH (vinylidene)+ HBr channel. *J. Chem. Phys.* 115: 1734.

209. Doughan, D. I., Raff, L. M., Rockley, M. G., Hagan, M., Agrawal, P. M., and R. Komanduri. 2006. Theoretical investigation of the dissociation dynamics of vibrationally excited vinyl bromide on an *ab initio* potential-energy surface obtained using modified novelty sampling and feedforward neural networks. *J. Chem. Phys.*, 124: 054321.

210. Doughan, D. I., Raff, L. M., Rockley, M. G., Hagan, M., Agrawal, P. M., and R. Komanduri. 2006. Retraction: Theoretical investigation of the dissociation dynamics of vibrationally excited vinyl bromide on an *ab initio* potential-energy surface obtained using modified novelty sampling and feedforward neural networks. *J. Chem. Phys.* 125: 079901.

211. Tully, J. C. 1990. Molecular dynamics with electronic transitions. *J. Chem. Phys.* 93: 1061.

212. Chu, T. S., Zhang, R. Q., and H. F. Cheung. 2001. Geometric and electronic structures of silicon oxide clusters. *J. Phys. Chem. B* 105(9): 1705.

213. Nayak, S. K., Rao, B. K., Khanna, S. N., and P. Jena. 1998. Atomic and electronic structure of neutral and charged SiO clusters. *J. Chem. Phys.* 109: 1245.

214. Lide, D. R., ed. 2004–2005. *CRC handbook of chemistry and physics.* Boca Raton, FL: CRC Press.

215. Lu, W. C., Wang, C. Z., Nguyen, V., Schmidt, M. W., Gordon, M. S., and K. M. Ho. 2003. Structures and fragmentations of small silicon oxide clusters by *ab initio* calculations. *J. Phys. Chem. A* 107(36): 6936.

216. Schmidt, M. W., Baldridge, K. K., Boatz, J. A., Elbert, S. T., Gordon, M. S., Jensen, J. H., Koseki, S., Matsunaga, N., Nguyen, K. A., Su, S. J., Windus, T. L., Dupuis, M., and J. A. Montgomery. 1993. General atomic and molecular electronic structure system. *J. Comput. Chem.* 14(11): 1347.

217. Chase, M. W. Jr., ed. 1998. *J. Phys. Chem. Ref. Data, Monograph No. 9, NIST-HANAF Thermochemical Tables,* 4th ed.

218. Puskorius, G. V., and L. A. Feldkamp. 1994. Neurocontrol of nonlinear dynamical systems with Kalman filter trained recurrent networks. *IEEE Trans. Neur. Net.* 5: 279.

219. M. F. Moller. 1993. A scaled conjugate gradient algorithm for fast supervised learning. *Neural Networks* 6: 525.

220. Vanderbilt, D. 1990. Soft self-consistent pseudopotentials in a generalized eigenvalue formalism. *Phys. Rev. B* 41: 7892.

221. Mujica, A., Rubio, A., Munoz, A., and R. J. Needs. 2003. High-pressure phases of group IV, III-V, and II-VI compounds. *Rev. Mod. Phys.* 75: 863.

222. Duane, S., Kennedy, A. D., Pendleton, B. J., and D. Roweth. 1987. Hybrid Monte Carlo. *Phys. Lett. B* 195: 216.

223. Clamp, M. E., Baker, P. G., and S. A. Brass. 1994. Hybrid Monte Carlo: An efficient algorithm for condensed matter simulation. *J. Comput. Chem.* 15: 838.

224. Laio, A., and M. Parrinello. 2002. Escaping free energy minima. *Proc. Natl. Acad. Sci. U. S.* 99(20): 12562.

225. Martonak, R., Laio, A., and M. Parrinello. 2003. Predicting crystal structures: The Parrinello-Rahman method revisted. *Phys. Rev. Lett.* 90: 075503.

226. Kuhne, T. D., Krack, M., Mohamed, F. R., and M. Parrinello. 2007. Efficient and accurate Car-Parrinello-like approach to Born-Oppenheimer molecular dynamics. *Phys. Rev. Lett.* 98: 066401.

227. Bazant, M. Z., Kaxiras, E., and J. F. Justo. 1997. Environment-dependent interatomic potential for bulk silicon. *Phys. Rev. B* 56: 8542.

228. Goedecker, S. 2002. Optimization and parallelization of a force field for silicon using OpenMP. *Comput. Phys. Commun.* 148: 124.

229. Lenosky, T. J., Sadigh, B., Alonso, E., Bulatov, V. V., Diaz de la Rubia, T., Kim, J., Voter, A. F., and J. D. Kress. 2000. Highly optimized empirical potential model of silicon. *Model. Simul. Mater. Sci. Eng.* 8: 825.

230. Carter, S., Culik, S. J., and J. M. Bowman. 1997. Vibrational self-consistent field method for many-mode systems: A new approach and application to the vibrations of CO adsorbed on Cu (100). *J. Chem. Phys.* 107: 10458.

231. Carter, S., and J. M. Bowman. 2000. Variational calculations of rotational–vibrational energies of CH_4 and isotopomers using an adjusted *ab initio* potential. *J. Phys. Chem. A* 104: 2355.

232. Carter, S., and N. C. Handy. 2002. On the representation of potential energy surfaces of polyatomic molecules in normal coordinates. *Chem. Phys. Lett.* 352(1–2): 1.

233. Manzhos, S., Yamashita, K., and T. Carrington Jr. In press 2009. Fitting sparse multidimensional data with low-dimensional terms. *Comp. Phys. Comm.*

234. Holland, J. H. 1975. *Adaptation in natural and artificial systems: An introductory analysis with applications to biology, control, and artificial intelligence.* Cambridge, MA: MIT Press.

235. Tang, K. S., Man, K. F., Kwong, S., and Q. He. 1996. Genetic algorithms and their applications. *IEEE Signal Process Mag.* 13(6): 22.

236. Bäck, T., Hammel, U., and H. P. Schwefel. 1997. Evolutionary computation: Comments on the history and current state. *IEEE Trans. Evol. Comput.* 1(1): 3.

237. Buche, D., Schraudolph, N. N., and P. Koumoutsakos. 2005. Accelerating evolutionary algorithms with Gaussian process fitness function models. *IEEE Trans. Syst. Man Cybern. Part C Appl. Rev.* 35(2): 183.

238. Yan, S., and B. Minsker. 2006. Optimizing groundwater remediation designs under uncertainty using dynamic surrogate models. In *World Environmental and Water Resources Congress.* Omaha, Nebraska: Environmental and Water Resources Institute (EWRI) of ASCE.

239. Kim, J. O., and P. K. Khosla. 1992. A multi-population genetic algorithm and its application to design of manipulators. In *IEEE/RSJ Int. Conf. on Intelligent Robots and Systems,* 279. Raleigh, NC: Robotics Institute.

240. Hacioglu, A. 2005. A novel usage of neural network in optimization and implementation to the internal flow systems. *Aircraft Engineering and Aerospace Technology: An International Journal* 77(5): 369.

241. Agrawal, P. M., Raff, L. M., and R. Komanduri. 2005. Monte Carlo simulations of void-nucleated melting of silicon via modification in the Tersoff potential parameters. *Phys. Rev. B* 72: 125206-1.

242. Bridges, C. L., and D. E. Goldberg. 1991. An analysis of multipoint crossover. *Proceedings of the Foundation of Genetic Algorithms,* 301.

243. Nelder, J. A., and R. Mead. 1965. A simplex method for function minimization. *Comput. J.* 7(4): 308.

244. Wilson, E. B., Decius, J. C., and P. C. Cross. 1955. *Molecular vibrations*. Reprint, New York: Dover.

245. McIntosh, D. F., and K. H. Michaellion. 1979. The Wilson GF matrix method of vibrational analysis. Part I, II, III, *Can. J. Spec.*, 24(1, 2, 3): 1.

246. Sumpter, B. G., Getino, C., and D. W. Noid. 1992. A neural network approach to the study of internal energy flow in molecular systems. *J. Chem. Phys.* 97: 293.

247. Getino, C., 1990. Ph.D. diss., Universidad Complutense de Madrid, Madrid, Spain.

248. Getino, C., Sumpter, B. G., Santamaria, J., and G. S. Ezra. 1990. *Ab initio* study of stretch-bend coupling in hydrogen peroxide. *J. Am. Chem. Soc.* 94: 3995.

249. Hertz, J., Krogh, A., and R. G. Palmer. 1996. *Introduction to the theory of neural computation*. Reading, MA: Addison-Wesley.

250. (a) Sewell, T. D., and D. L. Thompson. 1990. Classical trajectory studies of the unimolecular decomposition of the 2 chloroethyl radical. *J. Chem. Phys.* 93: 4077.

(b) Marks, A. J. 1994. A classical trajectory study of CD_2HNC isomerization. *J. Chem. Phys.* 100: 8096.

(c) Rice, B. M., and D. L. Thompson. 1990. Classical dynamics studies of the unimolecular decomposition of nitromethane. *J. Chem. Phys.* 93: 7986.

(d) Alimi, R., Apkarian, V. A., and R. B. Gerber. 1993. Effect of pressure on molecular photodissociation in matrices: Molecular dynamics simulations of Cl~ 2 in Xe. *J. Chem. Phys.* 98(1): 331.

(e) Lendvay, G., and G. C. Schatz. 1992. Trajectory studies of collisional relaxation of highly excited CS_2 by H_2, CO, HCl, CS_2, and CH_4. *J. Chem. Phys* 96: 4356.

(f) Lendvay, G., and G. C. Schatz. 1993. Collisional energy transfer from highly vibrationally excited SF. *J. Chem. Phys.* 98: 1034.

(g) Budenholzer, F. E., Chang, M. Y., and K. C. Huang. 1994. A classical trajectory study of IVR in dimethyl peroxide. 2. Dissociation lifetimes and comparison with statistical calculations. *J. Phys. Chem.* 98(48): 12501.

(h) Raff, L. M. 1987. A semiempirical potential-energy surface for the 1,2-difluoroethane system. *J. Phys. Chem.* 91: 3266; Raff, L. M. 1988. Computational studies of the bimolecular reaction dynamics of the ethene + fluoride system. *J. Phys. Chem.* 92: 141; Raff, L. M. 1989. *J. Chem. Phys.* 90: 6313; Raff, L. M. 1990. Energy transfer and reaction dynamics of matrix isolated 1, 2 difluoroethane D. *J. Chem. Phys.* 93: 3160; Raff, L. M. 1991. Theoretical studies of the reaction dynamics of the matrix isolated F+ Cis D ethylene system. *J. Chem. Phys.* 95: 8901; Raff, L. M. 1992. Effects of lattice morphology upon reaction dynamics in matrix isolated systems. *J. Chem. Phys.* 97: 7459.

251. Tesauro, G., Touretzky, D. S., and T. K. Leen, eds. 1995. *Advances in neural information processing systems 7*. Cambridge, MA: MIT.

252. Maren, A., Harsten, C., and R. Pap. 1990. *Handbook of neural computing applications*. New York: Academic Press.

253. Wasserman, P. D. 1993. *Advanced methods in neural computing*. New York: Van Nostrand Reinhold.

254. Nami, Z., Misman, O., Erbil, A., and G. S. May. 1997. Semi-empirical neural network modeling of metal-organic chemicalvapor deposition. *IEEE Trans. Semicond. Manuf.* 10(2): 288.

255. Kim, S. K., White, J. M., Agrawal, P. M., and D. L. Thompson. 2001. Photodissociation of methyl nitrite on Ag (111): Simulation. *J. Chem. Phys.* 115: 7657.

256. Perry, M. D., and L. M. Raff. 1994. Theoretical studies of elementary chemisorption reactions on an activated diamond ledge surface. *J. Phys. Chem.* 98(16): 4375.

257. Agrawal, P. M., Thompson, D. L., and L. M. Raff. 1988. Computational studies of heterogeneous reactions of SiH_2 on Si (111) surfaces. *Surf. Sci.* 195(1–2): 283.

258. Agrawal, P. M., Raff, L. M., and D. L. Thompson. 1987. Effect of the lattice model on the dynamics of dissociative chemisorption of H_2 on a Si(111) surface. *Surf. Sci.* 188: 402.

259. Rice, B. M., NoorBatcha, I., Thompson, D. L., and L. M. Raff. 1987. The dynamics of dissociative chemisorption of H on a Si (111) surface. *J. Chem. Phys.* 86: 1608.

260. Raff, L. M., NoorBatcha, I., and D. L. Thompson. 1986. Monte Carlo variational transition state theory study of recombination and desorption of hydrogen on Si (111). *J. Chem. Phys.* 85: 3081.

261. NoorBatcha, I., Raff, L. M., and D. L. Thompson. 1985. Monte Carlo random walk study of recombination and desorption of hydrogen on Si (111). *J. Chem. Phys.* 83: 1382.

262. Brenner, D. W., Shenderova, O. A., Harrison, J. A., Stuart, S. J., Ni, B., and S. B. Sinnott. 2002. A second-generation REBO potential energy expression for hydrocarbons. *J. Phys. Condens. Matter* 14: 783.

263. Allen, M. P., and D. J. Tildesley. 2002. *Computer simulation of liquids.* Oxford: Clarendon Press.

264. Berendsen, H. J. C., Postma, J. P. M., Van Gunsteren, W. F., DiNola, A., and J. R. Haak. 1984. Molecular dynamics with coupling to an external bath. *J. Chem. Phys.* 81: 3684.

265. Amari, S., Murata, N., Muller, K. R., Finke, M., and H. H. Yang. 1997. Asymptotic statistical theory of overtraining and cross-validation. *IEEE Trans. Neural Networks* 8(5): 985.

266. Krogh, A., and J. Vedelsby. 1995. *Advances in neural information processing system,* vol. 7, ed. G. Tesauro, D. Touretzky, and T. Teen. Boston, MA: MIT Press.

267. Arbib, Michael A., ed. 1995. *The handbook of brain theory and neural networks.* Boston, MA: MIT Press, p. 666.

268. Liao, Y., Fang, S-C., and H. L. W. Nuttle. 2003. Relaxed conditions for radial-basis function networks to be universal approximators. *Neural Networks* 16: 1019.

269. Wu., W., Nan, D., Long, J., and Y. Ma. 2008. Letter to the editor: A comment on relaxed conditions for radial-basis function networks to be universal approximators. *Neural Networks* 21: 1464.

270. Boutry, C., Elad, M., Golub, G. H., and P. Milanfar. 2005. The generalized eigenvalue problem for nonsquare pencils using a minimal perturbation approach. *SIAM J. Matrix Anal. Appl.* 27: 582.

271. Jensen, P. 1980. The potential energy surface for the electronic ground state of the water molecule determined from experimental data using a variational approach. *J. Mol. Spectrosc* 133: 438.

272. Choi, S. E., and J. C. Light. 1992. Highly excited vibrational eigenstates of nonlinear triatomic molecules. Application to H_2O. *J. Chem. Phys.* 97(10): 7031.

273. Hu, L. H., Wang, X. J., Wong, L. H., and G. H. Chen. 2003. Combined first-principles calculation and neural-network correction approach for heat of formation. *J. Chem. Phys.* 119: 11501.

274. Duan, X. M., Li, Z. H., Song, G. L., Wang, W. N., Chen, G. H., and K. N. Fan. 2005. Neural network correction for heats of formation with a larger experimental training set and new descriptors. *Chem. Phys. Lett.* 410: 125.

275. Wu, J., and X. Xu. 2007. The X1 method for accurate and efficient prediction of heats of formation. *J. Chem. Phys.* 127: 214105.

276. Wu, J., and X. Xu. 2008. Improving the B3LYP bond energies by using the X1 method. *J. Chem. Phys.* 129: 164103.

277. Li, H., Shi, L. L., Zhang, M., Su, Z., Wang, X. J., Hu, L. H., and G. H. Chen. 2007. Improving the accuracy of density-functional theory calculation: The genetic algorithm and neural network approach. *J. Chem. Phys.* 126: 144101.

278. Urata, S., Takada, A., Uchimaru, T., Chandra, A. K., and A. Sckiya. 2002. Artificial neural network study for the estimation of the C–H bond dissociation enthalpies. *J. Fluorine Chem.* 116: 163.

279. Wang, X. J., Hu, L. H., Wong, L. H., and G. H. Chen. 2004. A combined first-principles calculation and neural networks correction approach for evaluating Gibbs energy of formation. *Mol. Simul.* 30: 9.

280. Li, G. Z., Yang, J., Song, H. F., Yang, S. S., Lu, W. C., and N. Y. Chen. 2004. Semiempirical quantum chemical method and artificial neural networks applied for λ_{max} computation of some azo dyes. *J. Chem. Inf. Comput. Sci.* 44: 2047.

281. Wang, X. J., Wong, L. H., Hu, L. H., Chan, C. Y., Su, Z., and G. H. Chen. 2004. Improving the accuracy of density-functional theory calculation: The statistical correction approach. *J. Phys. Chem.* A 108: 8514.

282. McCulloch, W., and W. Pitts. 1943. A logical calculus of the ideas imminent in nervous activity. *Bulletin of Mathematical Biophysics* 5: 115.

283. Ho, T-S., and H. Rabitz. 1996. A general method for constructing multidimensional molecular potential energy surfaces from ab initio calculations. *J. Chem. Phys.* 104: 2584.

284. Hollebeek, T., Ho, T-S., Rabitz, H., and L. B. Harding. 2001. Construction of reproducing kernel Hilbert space potential energy surfaces for the 1 A" and 1 A' states of the reaction N $(^2 D)$+ H$_2$. *J. Chem. Phys.* 114: 3945.

285. Schatz, G. C., Papaioannou, A., Pederson, L. A., Harding, L. B., Hollebeek, T., Ho, T.-S., and H. Rabitz. 1997. A global A-state potential surface for H$_2$O: Influence of excited states on the O (^1D) + H$_2$ reaction. *J. Chem., Phys.* 107: 2340.

286. Pederson, L. A., Schatz, G. C., Ho, T.-S., Hollebeek, T., Rabitz, H., Harding, L. B., and G. Lendvay. 1999. Potential energy surface and quasiclassical trajectory studies of the N (D)+ H reaction. *J. Chem. Phys.* 110: 9091.

287. Alagia, M., Balucani, N., Cartechini, L., et al. 1999. Exploring the reaction dynamics of nitrogen atoms: A combined crossed beam and theoretical study of N (D)+ D ND+ D. *J. Chem. Phys.* 110: 8857.

288. Ho, T.-S., and H. Rabitz. 2000. Proper construction of *ab initio* global potential surfaces with accurate long-range interactions. *J. Chem. Phys.* 113: 3960.

289. Ho, T.-S., Rabitz, H., and G. Scoles. 2000. Reproducing kernel technique for extracting accurate potentials from spectral data: Potential curves of the two lowest states $X^1\Sigma_g^+$ and $a^3\Sigma_u^+$ of the sodium dimer. *J. Chem. Phys.* 112: 6218.

290. Higgins, J., Hollebeek, T., Reho, J., Ho, T.-S., Lehmann, K. K., Rabitz, H., Scoles, G., and M. Gutowski. 2000. On the importance of exchange effects in three-body interactions: The lowest quartet state of Na. *J. Chem. Phys.* 112: 5751.

ABOUT THE AUTHORS

Dr. Lionel M. Raff is a Regents Professor at Oklahoma State University (OSU), a rank he has held for 33 years. He received his B.S and M.S. degrees from the University of Oklahoma graduating with Special Distinction and his Ph.D. from the University of Illinois while working with Professor Aron Kuppermann in the area of electron impact spectroscopy. Professor Raff was a National Science Foundation Postdoctoral Research Fellow at Columbia University, New York City, where he worked in the area of molecular dynamics with Professor Martin Karplus. In addition to holding a faculty position at Oklahoma State University, he was a Visiting Scientist at the Los Alamos Scientific Laboratories in 1973. Dr. Raff has authored or co-authored over 200 articles primarily in the *Journal of Chemical Physics*, the *Journal of Physical Chemistry*, and *Physical Review* in the areas of gas-phase molecular dynamics, Monte Carlo methods, gas-surface interactions, gas-phase quantum scattering, and *ab initio* molecular dynamics. He is the author or co-author of six book chapters and three books including the comprehensive textbook *Principles of Physical Chemistry* (Prentice Hall, Upper Saddle River, NJ, 2001). He has been elected to Outstanding Educators of American. In 1979, he received the Oklahoma Chemist Award and the following year the Oklahoma Scientist Award from the Oklahoma Academy of Sciences. In 1993, he received the Oklahoma Medallion for Excellence in College/University Teaching from the Oklahoma Foundation for Excellence. Forty-four students have received advanced degrees working under his direction.

Dr. Ranga Komanduri is a Regents Professor and holds the A. H. Nelson, Jr. Endowed Chair in Engineering in the School of Mechanical and Aerospace Engineering, Oklahoma State University, Stillwater. He received B.E. (Mechanical) and M.E. (Heat Power) degrees from Osmania University, Hyderabad, India, and Ph.D. (1972) and D. Eng. (1992) from Monash University, Melbourne, Australia. He came to the United States in 1972 as a research engineer and assistant professor in the Mechanical Engineering Department at Carnegie-Mellon University, Pittsburgh (1972–1977). He then moved to General Electric Corporate Research Development at Schenectady, New York (1977–1989). He was also an adjunct full professor at R. P. I., Troy, New York (1977–1986). He was on research sabbatical for about three years at the National Science Foundation serving as a program director for Materials Processing, and Manufacturing Processes Programs, Washington, DC (1986–1989). He joined OSU in 1989 as a professor and MOST Chair in Intelligent Manufacturing.

Dr. Komanduri's research interests are focused on various advanced manufacturing processes and materials. He developed a state-of-the-art research laboratory in this area located in the Advanced Technology Research Center (ATRC) at OSU. His research activities are broad and include finishing of advanced ceramics; hard, wear-resistant coatings on cutting tools, including low-pressure diamond coatings and multiple nanocoatings on cutting tools; molecular dynamics (MD) and Monte Carlo (MC) simulations of nanometric cutting and tribology; thermal aspects of various manufacturing processes, including polishing, machining, welding, heat treatment, and tribology; laser-assisted material processing; and EKG signal analysis for identification of cardiac disorders. He has published some 230 technical papers in archival journals, edited 22 conference proceedings, and holds 22 patents. He has delivered a number of keynote papers at various international conferences and has prepared several book chapters.

Dr. Komanduri has been very active in various professional societies. He was vice-president of the Manufacturing Group of ASME (1989–1993) and president of the North American Manufacturing Research Institution (NAMRI). He is a fellow of ASME, SME, and CIRP. His many honors include F. W. Taylor Medal of CIRP (the International Institution for Production Engineering Research) (1977); ASME Blackall Machine Tool & Gauge Award (1981); G.E. Managerial Award (1981); NSF Outstanding Performance Award (1989); ASME Pi Tau Sigma-Charles Russ Richards Memorial Award (1990); ASME William T. Ennor Manufacturing Technology Award (2002), and SME Albert M. Sargent Progress Award (2005). Awards at OSU include OSU President's Distinguished Service Award (2001); OSU Regents Distinguished Research Award (2003); and OSU Eminent Faculty Award (the highest honor at OSU, 2004), OSU Phoenix Award for outstanding graduate faculty (2009), and OSU-CEAT Halliburton Outstanding Faculty Award (2009).

Dr. Komanduri initiated an interdisciplinary MD/MC simulation group at OSU, which has been running successfully for nearly two decades. The group has published extensively (some 45 publications in this area) and the monograph titled *Neural Networks in Chemical Reaction Dynamics* is the latest addition.

On September 6, 2011, Professor Ranga Komanduri unexpectedly passed away. His coauthors have dedicated this monograph to his memory.

Dr. Martin Hagan is a professor in the School of Electrical and Computer Engineering at Oklahoma State University, Stillwater. He has taught and conducted research in the areas of statistical modeling and control systems for the last 30 years. His research has encompassed a variety of application areas: molecular dynamics, seismic signal processing, genetic pathway modeling, optimal portfolio management, electric load prediction, flight simulators, precision pointing systems, diesel engine control, adaptive flight control, and friction compensation. He has received grants from Boeing, Texas Instruments, Halliburton Energy Services, Cummins Engine Company, National Science Foundation, Air Force Office of Scientific Research, California Public Employees Retirement System (CalPERS), Amgen, and FlightSafety International. For the last 15 years his research has focused on the use of neural networks for non-linear filtering, prediction, and control. He is author, with Howard Demuth and Mark Beale, of a textbook, *Neural Network Design*, which has recently been translated into Chinese and Korean. He is also a co-author of the Neural Network Toolbox for MATLAB. He was a visiting scholar during 1994 at the University of Canterbury in Christchurch, New Zealand, and during 2005–2006 at the Laboratoire d'Analyse et d'Architecture des Systèms du Centre National de la Recherche Scientifique in Toulouse, France. He regularly teaches courses in stochastic processes, estimation theory, neural networks, system identification, and control systems. He was awarded the Oklahoma State University Regents Distinguished Teaching Award in 2000 and the Lockheed Martin Aeronautics Teaching Excellence Award in 2005.

Satish T. S. Bukkapatnam is a professor in the School of Industrial Engineering and Management, Oklahoma State University, Stillwater. He received the Ph.D. degree in Industrial and Manufacturing Engineering from Pennsylvania State University, University Park. His research addresses sensor-based modeling for improving quality and performance of manufacturing processes and machines, molecular dynamics simulation of machining and tribology, and chemical mechanical polishing of semiconductor materials for microelectronics applications. These systems include structural systems, human cardiovascular systems, and a variety of infrastructure systems of interest to information technology, transportation, medical, and defense enterprises. His research thus far has yielded 87 publications (49 in journals and 38 in refereed conference proceedings). It has attracted over $2.8M in contracts (as PI or Co-PI), and has been the basis for 5 Ph.D. and 8 M.S. theses. He is currently advising/co-advising five Ph.D. and three M.S. students, and mentoring three undergraduate students. He has been elected as the chair of the Quality, Statistics, and Reliability Section of the Institute of Operations Research and Management Science, invited as one of the 25 international experts to present at the National Science Foundation/Department of Defense–sponsored workshops on nanomanufacturing (2008) and dynamic systems (2007), appointed associate editor for two international journals, chaired or participated in 21 Ph.D. guidance committees, and organized multiple conferences (notably the Quality, Statistics and Reliability Cluster at the INFORMS conference 2008; Industrial Engineering Research Conference 2003). He received a student-elected best teacher award in 2002, best paper awards from IIE Manufacturing and Design Division (2009), INFORMS Transportation Society (TSL) in 2007, and Artificial Neural Network in Engineering in 1995, as well as SME's Dougherty Young Manufacturing Engineer Award in 2005.

SUBJECT INDEX